Advances in Intelligent Systems and Computing

Volume 528

Series Editor

Janusz Kacprzyk, Polish Academy of Sciences, Warsaw, Poland
e-mail: kacprzyk@ibspan.waw.pl

More information about this series at http://www.springer.com/series/11156

About this Series

The series "Advances in Intelligent Systems and Computing" contains publications on theory, applications, and design methods of Intelligent Systems and Intelligent Computing. Virtually all disciplines such as engineering, natural sciences, computer and information science, ICT, economics, business, e-commerce, environment, healthcare, life science are covered. The list of topics spans all the areas of modern intelligent systems and computing.

The publications within "Advances in Intelligent Systems and Computing" are primarily textbooks and proceedings of important conferences, symposia and congresses. They cover significant recent developments in the field, both of a foundational and applicable character. An important characteristic feature of the series is the short publication time and world-wide distribution. This permits a rapid and broad dissemination of research results.

Advisory Board

Wander Jager • Rineke Verbrugge • Andreas Flache
Gert de Roo • Lex Hoogduin • Charlotte Hemelrijk
Editors

Advances in Social Simulation 2015

 Springer

Editors
Wander Jager
University College Groningen
University of Groningen
Groningen, The Netherlands

Andreas Flache
Faculty for Social and Behavioural
   Sciences
Department of Sociology
University of Groningen
Groningen, The Netherlands

Lex Hoogduin
Faculty of Economics and Business
University of Groningen
Groningen, The Netherlands

Rineke Verbrugge
Faculty of Mathematics and Natural
   Sciences
Institute of Artificial Intelligence
University of Groningen
Groningen, The Netherlands

Gert de Roo
Faculty of Spatial Sciences
Department of Spatial Planning
   and Environment
University of Groningen
Groningen, The Netherlands

Charlotte Hemelrijk
Faculty of Mathematics and Natural
   Sciences
Department of Behavioural Ecology
   and Self-organisation
University of Groningen
Groningen, The Netherlands

ISSN 2194-5357   ISSN 2194-5365   (electronic)
Advances in Intelligent Systems and Computing
ISBN 978-3-319-83691-1   ISBN 978-3-319-47253-9   (eBook)
DOI 10.1007/978-3-319-47253-9

Printed on acid-free paper

This Springer imprint is published by Springer Nature
The registered company is Springer International Publishing AG
The registered company address is: Gewerbestrasse 11, 6330 Cham, Switzerland

Introduction

Social simulation is a rapidly evolving field. Social scientists are increasingly interested in social simulation as a tool to tackle the complex nonlinear dynamics of society. As such, it comes as no surprise that scientists employing social simulation techniques are targeting a wide variety of topics and disciplinary fields. The management of natural resources, financial-economical systems, traffic, biological systems, social conflict, and war—they are all examples of phenomena where nonlinear developments play an important role. Social simulation, often using the methodology of agent-based modeling, has proven to be a new and powerful methodology to address these processes, thus offering new insights in both the emergence and the management of nonlinear processes. Moreover, offering a formal and dynamical description of behavioral systems, social simulation also facilitates the interaction between behavioral sciences and other domain-related scientific disciplines such as ecology, history, agriculture, and traffic management, to just name a few examples. The increased capacity for simulating social systems in a valid manner contributes to the collaboration of different disciplines in understanding and managing various societal issues.

The European Social Simulation Association, founded in 2003, is a scientific society aimed at promoting the development of social simulation research, education of young scientists in the field, and application of social simulation. One of its activities is the organization of an annual conference. From September 14th to 18th in 2015, the 11th Social Simulation Conference was organized in Groningen, the Netherlands. The hosting organization was the Groningen Center for Social Complexity Studies.

This book highlights recent developments in the field of social simulation as presented at the conference. It covers advances in both applications and methods of social simulation. Because the field of social simulation is evolving rapidly, developments from a variety of perspectives have been brought together in this book, which has a multidisciplinary scope. Yet all the contributions in this book share a common interest: the understanding of how interactions between a multitude of individuals give rise to complex social phenomena, and how these phenomena in turn affect individual behavior. This multidisciplinarity is of vital importance, because it facilitates the communication between different disciplinary areas. The value of disciplinary collaboration and cross-fertilization in social simulation research is demonstrated by many contributions in this volume. To mention just one of the many areas for which this holds: insights from studying the socio-ecological dynamics of fisheries may prove to be relevant in understanding conflicts in human organizations as well.

Concerning the topics addressed in this book, the reader will find a wide variety of issues that are addressed using social simulation models. The topic of complexities of economic systems is addressed in a number of chapters, providing a perspective on our understanding of the nonlinear characteristics of economic systems on various levels. Opinion dynamics is another topic on which numerous contributions focus. Studying opinion dynamics is highly relevant to develop a deeper understanding of societal polarization, the emergence and resolution of conflict, and civil violence. A range of contributions addresses the interaction of humans with their environment, most notably the social dynamics of natural resource use and ecosystem management. Applied topics deal with fish stocks and land use. Closely related to this are contributions dealing with food production and consumption, a theme that in turn has important consequences for land use. Another field with important societal impact addressed by papers in this volume is transportation, where technology development and human behavior interact likewise. This is related to the rapid developments that we currently witness in systems for the production and consumption of energy. The energy transition can be seen as a typical example of a nonlinear process where social simulation contributes to a deeper understanding that may help to develop more effective managerial and societal strategies in the future. Besides looking at current societal and socio-ecological issues, social simulation is increasingly used to understand developments that happened in the past. In this book, the reader will find chapters demonstrating how social simulation, as a methodology, may be valuable in understanding historical developments.

Besides applications of social simulation models on topical domains, this book also covers relevant developments in the methodology of social simulation. An area that receives increasing attention in the literature is the empirical validation of simulation models. Various contributions address the question how empirical data can be used in further improving the reliability of social simulation models. Also attention is devoted to the use of behavioral theory in social simulation models, which requires a translation from more descriptive and correlational models to a formal dynamic model of behavior. Related to this is the topic of construction of artificial populations to be used in experimenting with models of societal processes. Finally, in making models more accessible for the general public, attention is given to running social simulation models in browsers, which would make them much more accessible.

This book is an important source for readers interested in cutting-edge developments exemplifying how simulation of social interaction contributes to understanding and managing complex social phenomena. The editors wish to thank all authors, the members of the scientific committee and the auxiliary reviewers who were responsible for reviewing all the papers submitted for the conference, as well as the organizers of the special sessions. For a list of all people involved in shaping the contents of the conference and reviewing the submissions, see the next pages. The papers published in this volume are a representative selection from a broader set of research papers presented at Social Simulation 2015.

Groningen, The Netherlands   Wander Jager
                                                                                  Rineke Verbrugge
                                                                                   Andreas Flache
                                                                                    Gert de Roo
                                                                                   Lex Hoogduin
                                                                               Charlotte Hemelrijk

PC Members

Shah-Jamal Alam
Floortje Alkemade
Frédéric Amblard
Tina Balke
Stefano Balietti
Riccardo Boero
Melania Borit
Giangiacomo Bravo
Edmund Chattoe-Brown
Emile Chappin
Guillaume Deffuant
Virginia Dignum
Frank Dignum
Bruce Edmonds
Corinna Elsenbroich
Andreas Ernst
Tatiana Filatova
Armando Geller
Rosanna Garcia
José-Ignacio García-Valdecasas
José-Manuel Galán
Nigel Gilbert
William Griffin
Rainer Hegselmann
Gertjan Hofstede
Luis Izquierdo
Marco Janssen
Bogumił Kamiński

Jean-Daniel Kant
Bill Kennedy
Andreas Koch
Friedrich Krebs
Setsuya Kurahashi
Jeroen Linssen
Iris Lorscheid
Michael Mäs
Ruth Meyer
Michael Möhring
Jean-Pierre Muller
Martin Neumann
Emma Norling
Mario Paolucci
Jakub Piskorski
Gary Polhill
Juliette Rouchier
Jordi Sabater-Mir
Frank Schweitzer
Roman Seidl
Jaime Sichman
Flaminio Squazzoni
Przemyslaw Szufel
Karoly Takacs
Shingo Takahashi
Richard Taylor
Pietro Terna
Klaus Troitzsch
Harko Verhagen
Nanda Wijermans

Auxiliary Reviewers

Adiya Abisheva
Floor Ambrosius
Apostolos Ampatzolglou
Priscilla Avegliano
Quang Bao Le
Gustavo Campos
Thomas Feliciani
Monica Gariup
Amineh Ghorbani
Bao Le

Robin Mills
Ivan Puga-Gonzalez
Mart van der Kam
Mark Kramer
Pavlin Mavrodiev
Keiko Mori
Vahan Nanumyan
Tomasz Olczak
Sjoukje Osinga
Francine Pacilly
Klara Pigmans
José Santos
Simon Schweighofer
Annalisa Stefanelli
Yoshida Takahashi
Keiichi Ueda
Harmen de Weerd
Nicolas Wider

Special Session Organizers

- ESSA@Work: Nanda Wijermans, Geeske Scholz, and Iljana Schubert
- Social Simulation and Serious Games: Jeroen Linssen and Melania Borit
- Simulation Model Analysis (SIGMA): Bogumił Kamiński and Laszlo Gulyas
- Social Conflict and Social Simulation: Armando Geller and Martin Neumann
- Applications in Policy Modelling: Petra Ahrweiler, Nigel Gilbert, Bruce Edmonds, and Ruth Meyer
- Cognitive Models in Social Simulation: Nanda Wijermans and Cara Kahl
- Social Simulations of Land, Water and Energy: Tatiana Filatova
- Simulating the Social Processes of Science: Bruce Edmonds
- Modelling Routines and Practices: Bruce Edmonds
- Qual2Rule—Using Qualitative Data to Inform Behavioural Rules: Melania Borit
- Modelling Social Science Aspects of Fisheries: Melania Borit
- Simulation of Economic Processes: Alexander Tarvid
- Affiliation, Status and Power in Society: Gert Jan Hofstede, Sjoukje Osinga, and Floor Ambrosius

Contents

From Field Data to Attitude Formation

Kei-Léo Brousmiche, Jean-Daniel Kant, Nicolas Sabouret, and
François Prenot-Guinard

Abstract This paper presents a multi-agent model for simulating attitude formation
and change based on perception and communication in the context of stabilization
operations. The originality of our model comes from (1) attitude computation that
evaluates information as part of a history relative to the individual and (2) a notion
of co-responsibility for attitude attribution. We present a military scenario of French
operations in Afghanistan along with polls results about the opinion of citizens
toward present Forces. Based on these field data, we calibrate the model and show
the resulting attitude dynamics. We study the sensibility of the model to the co-
responsibility factor.

Keywords Social simulation • Attitude formation and dynamics • Agent-based
modeling • Cognitive modeling • Calibration using field data

1 Introduction

The new conflicts that arouse during the two last decades have brought a deep shift
in military strategies [19]: most of the stabilization operations conducted by western
Forces involve opponents who blend themselves into the "human environment"
to turn the population in their favor. In order to counter them, one should not

K.-L. Brousmiche (✉)
LIP6 - CNRS UMR 7606, Université Pierre et Marie Curie, Paris, France

Airbus Defense & Space, 1 bd Jean-Moulin, Elancourt, France
e-mail: kei-leo.brousmiche@lip6.fr

J.-D. Kant
LIP6 - CNRS UMR 7606, Université Pierre et Marie Curie, Paris, France
e-mail: jean-daniel.kant@lip6.fr

N. Sabouret
LIMSI-CNRS, UPR 3251, Université Paris-Sud, Orsay, France
e-mail: nicolas.sabouret@limsi.fr

F. Prenot-Guinard
Airbus Defense & Space, 1 bd Jean-Moulin, Elancourt, France
e-mail: francois.prenot-guinard@airbus.com

© Springer International Publishing AG 2017
W. Jager et al. (eds.), *Advances in Social Simulation 2015*, Advances in Intelligent
Systems and Computing 528, DOI 10.1007/978-3-319-47253-9_1

only rely on tactical actions against them but also on non-kinetic actions such as reconstruction or specific communication actions that aim at altering the "hearts and minds" of the population. In this context, understanding the impact of actions performed by the intervention Force on the population's attitude is a major issue.

The concept of *attitude* derives from social psychology and could be defined as "a mental and neural state of readiness organized through experience" [2]. Multi-agent simulation of attitude dynamics seems a promising approach to study such complex social phenomenon since it is funded on individuals micro modeling and their interactions to analyze emergent macro trends [7]. While multiple agent-based models have been proposed to study attitude and opinion (i.e., expressed attitude) dynamics [3, 5, 20], the major difficulty relies in validation: can a given model correctly reflect the attitude dynamics of a population in a conflict zone? Opinion polls can provide target values. However, collecting field data to feed the simulation model and assessing the validity of its outcome (based on the expected values) can prove to be very difficult.

As part of our research, we have been given access to polls results about opinions of the population toward the different present Forces (foreign Force and Taliban) in an area of Afghanistan where French Forces conducted stabilization operations.[1] Along with these survey results, we have reconstituted the military actions sequences of each Force through a series of interviews with officers who were in charge of the situation. Our analysis of these data has brought light to attitude dynamics that go against classical approaches in social simulation. First, most models compute the attitude as the aggregation of the impact of each feature, seen as independent criteria or events [5, 20]. However, people do not evaluate each action (such as food provision, military patrol, and bombing attack) independently but toward what similar actions represents in general in terms of direct and indirect consequences for the population. It corresponds to Kahneman's memory-based, retrospective approach of evaluation [13]. Second, we noticed that populations attitude toward the UN army could decrease when it fails to accomplish its securing mission. For instance, in case of a bombing attack, victims will not only blame insurgents who are directly responsible but also the Security Force which have "failed" to prevent such an event. It corresponds in this case to the concept of "role-responsibility" in the sense of Hart [11]. In generally, people tend to attribute the responsibility of an action to other people, groups, or situational factors in addition to the direct responsible [12, 14].

In this paper, we propose a multi-agent simulation model based on field data that will help to better comprehend attitude dynamics in conflict zones where the population is confronted to antagonists Forces.

---

[1] These opinion surveys have been ordered by the French Ministry of Defense.

2 Related Works

Several researches have already proposed computational models of attitudes, from simple binary approaches [16] to more complex ones (e.g., [17]). However, as was pointed out by Castellano et al. [4], most of these models' focuses are limited to the interactions between individuals: they do not consider the construction mechanism of the attitude itself at a cognitive level. On this matter, Urbig and Malitz [20] proposed to represent attitudes as the sum of the evaluations of the object's features that can be seen as beliefs on the object, so as to take into account the cognitive aspect. While this model constitutes an interesting view on attitude formation, it has two limits with respect to our objectives.

First, the attitudes' values are not connected to the beliefs of each agent constituting their personal history. Indeed, their attitude revision mechanism is based on the bounded confidence model (e.g., [5]): when two individuals have attitude values close to each other, agents converge their attitudes. However, it could be possible to combine this with Fazio's model of attitude [9]. This model connects the attitude to its forming beliefs as a set of memory associations. Each of these evaluations is weighted by an accessibility value determining the evaluation's degree of reminiscence. By essence, this model maintains a balance between the cognitive representation of the object and its corresponding attitude.

Second, the attitude model of Urbig and Malitz does not embody an emotional component, while social psychologists have defined the attitude as embodying rational and affective components [9, 18]. This is the reason why Brousmiche et al. [3] attempted to combine Fazio's model with the Simplicity Theory proposed by Dessalles [6] which embodies an affective response mechanism. However, their model does not consider (1) the aggregation of beliefs into a personal history as proposed by Fazio and (2) the notion of role responsibility impacting attitudes toward other actors than perpetrators of an action.

In this paper, we will take the model presented in [3] as a starting point and extend it in order to take into account these two concepts by adding a mechanism of co-responsibility and modifying the aggregation method of the attitude computation.

3 Field Data

3.1 Scenario

In the course of the NATO intervention in Afghanistan to stabilize the country, the French Forces were tasked to maintain security in the regions of Kapisa and Surobi between 2008 and 2012. It is in this context that members of CIAE[2] were sent in the

---

[2]Joint Command for Human Terrain Actions: in charge of civil-military actions (e.g., reconstruction) and community outreach actions (i.e., attitude influence operations) that complement conventional security operations.

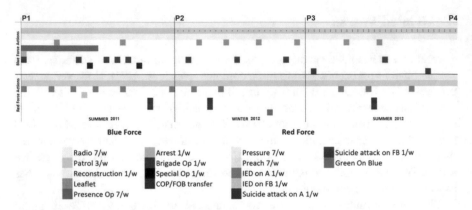

Fig. 1 Scenario of Blue and Red Forces actions. The time points P1, . . . ,P4 correspond to the dates of the opinion polls presented in the next section

area from October 2010 to September 2012. Through a set of six interviews with all the successive officers in charge (three colonels and three commanders) from the Joint Command for Human Terrain Actions, we managed to rebuild the sequence of the events that took place during their tenures, originating both from the NATO and from the Taliban insurgents. This sequence takes the shape of a scenario (see Fig. 1).

Each action is characterized by a reach, a frequency, and a payoff: how many people were directly affected by the action, how many times per week if it is frequent, and how each individual is impacted. These values were defined based on subjective assessments of this domain's subject matter experts (including members of CIAE). For instance, we defined the action "patrol" as being triggered by the Blue Force, affecting the population with a positive impact of 30 and a reach of 20 people, repeated three times per week in average. Similarly, "IED" (Improvised Explosive Devices) are done by Red Forces and affect one individual with a payoff of −100. The number of victims of each attack was defined according to open source records (e.g. [8]).

We can observe that both Forces have a constant background activity toward the population composed of non-kinetic actions. However, the Red Force activity is heavily decreased during winter which corresponds to the second period on the scenario. One reason is that local Taliban leaders leave the region to avoid the arid climate. Also, the surveillance systems, including drones, are more effective in the absence of foliage, making it more difficult for insurgents to place discretely the IEDs. On the Blue Force side, the activity decreases constantly due to the political decision taken after the big human losses on the first period. Indeed, the French government began to adopt a "retreat" strategy after the suicide attack of Joybar (July 13th 2011) which caused considerable human casualties among the Blue Forces.

Table 1 % of the population favorable with two questions at different dates

Questions	Polls dates			
	2/11(P1)	9/11(P2)	2/12(P3)	9/12(P4)
"The Blue Force contribute to security" (Q1)	40	32	24	19
"The Red Force is the principal vector of insecurity" (Q2)	27	60	27	37

3.2 Opinion Polls

In order to follow the progress of population's sentiments and to link them to foreign Forces activities, the French Ministry of Defense financed opinion polls in Afghan regions where the French forces were operating. Those surveys were conducted by Afghan contractors between February 2011 and September 2012 with an interval of approximately 6 months issuing into four measure points P1, P2, P3, and P4 of the opinion of the population of Kapisa toward the Blue Force and the Red Force on the period corresponding to our scenario (see Table 1 below).

Opinions toward Red Force in the context of security decrease in summer periods. This could be explained by their high activity level as exposed in the previous section. As for the Blue Force, the global opinion value keeps decreasing along with their decreasing activities. In overall, the opinion dynamics showed by these polls results are consistent with the scenario previously established.

4 Model

4.1 General Approach

Our model is based on the following principle: a simulation corresponds to the execution of actions (e.g. food distribution, construction of a bridge, bombing attack, etc.) by *actors* (e.g. UN Force, terrorists, or others) on a population. Individuals communicate about these actions with the others and form an attitude toward actors.

In our model, we consider a set of actors A and a set of individuals *Ind*. Actors represent Forces that act in the simulation and for which we want to analyze the attitudes evolution. Each of them corresponds to a computational automaton executing its actions list specified by the above scenario. Each individual $i \in Ind$ is represented by a computational agent and is characterized by its belief base that records facts he is aware of and his connections to other individuals. For each $i \in Ind$ and *actor* $\in A$, we denote $att(i, actor) \in \mathbb{R}$ the attitude of the individual i toward the actor *actor*.

Beliefs about actions will be the core element in our model: attitudes and communications will be based on these beliefs. We note $a(i)$ the belief of individual i about an action a.

Each $a(i)$ is a tuple: $\langle name(a), actor(a), coResp(a), bnf(a), payoff(a), date(a) \rangle$ with:

- *name* the unique name of the action (e.g., "patrol," "suicide attack")
- *actor* $\in A$ the actor who performed the action (Blue or Red Force)
- *coResp* $\in A$ the co-responsible actor of the action, if any (e.g., Blue Force will be co-responsible of "suicide attacks" performed by Red Force)
- *bnf* $\in Ind \cup A$ the beneficiary of the action, i.e., the individual or actor who undergoes the action
- *payoff* $\in \mathbb{R}$ the impact value of the action, negative when the action is harmful (e.g., attack) and positive when it is beneficial (e.g., food provision)
- *date* $\in \mathbb{N}$ the occurrence date of the action.

Attitudes are computed as the aggregation of evaluation of similar actions seized by the individual in his past. Two actions are *similar* if and only if they have the same name and actor. For instance, two distinct patrols done by the Blue Force in the same area are considered as similar. We call *general action* the meta action that includes similar actions. We denote $ga(i, a)$ the general action corresponding to the action a according to the individual i and $ga(i)$ the list of all general actions that i knows.

Actions can be perceived via direct perception (the agent is beneficiary of the action), actors communication toward the population (the agent receives a message from the actors), or intra-population communication (the agent receives a message from another individual).

4.2 Attitude Computation

When an agent receives a new information about an action a, it adds it to its belief base (if the action is not already present) and, possibly, communicates about it. Moreover, the agent revises its attitude toward the actor of the action. Our model of attitude construction is based on the model proposed by Fazio [9] (see Sect. 2). In short, Fazio proposes to compute the attitude as the average of *beliefs'* evaluations (i.e. how much this fact is beneficial) weighted by their accessibilities (i.e. how accessible is the information in the subject's mind).

First we compute the interest of an information to estimate its accessibility and its narrative interest (whether or not to communicate the action to other agents). Second, we evaluate the action based on its payoff. Third, we compute the impact of co-responsibility, if required. Finally, we aggregate these evaluations, weighted by their accessibilities, to compute the attitude.

Interest of an Action

In order to determine what to base their attitude on and what to communicate to other individuals, agents estimate a model of interest of the actions in their belief base. Following [3], our model of interest is based on the Simplicity Theory of Dessalles [6] which proposes to define the *narrative interest NI* of an information according to the emotion E and the surprise level S it causes to the individual using the following formula: $NI(a) = 2^{\alpha E(a) + (1-\alpha)S(a)}$ where E corresponds to the personal emotional response intensity of the individual when faced to an information and the surprise level S translates the sentiment of unexpectedness felt by the individual. The parameter $\alpha \in [0, 1]$ balances these two parts.

The emotional intensity E corresponds to the emotional amplitude experienced by the individual when exposed to an event and follows a logarithmic law in conformity with Weber–Fechner's law of stimuli. In our case, stimuli correspond to actions' payoff. A parameter of personal sensibility $\xi \in [0, 1]$ modulates the response intensity.

The surprise S experienced by an individual when exposed to an event derives from a level of raw unexpectedness (e.g. "It is surprising that a Taliban saves a citizen"). This level is reduced by a personal reference of unexpectedness based on a personal experience (e.g. "But I have once been saved by a Taliban before").

While $NI(a)$ corresponds to the narrative interest used as a communication heuristic, we also compute the information's interest corresponding to the accessibility of the information:

$$interest(a) = \log(NI(a)) = \alpha E(a) + (1 - \alpha)S(a)$$

Action Evaluation

Fishbein and Ajzen [1] advance that the evaluation of an action is weighted by the attitude toward its beneficiary. For instance, a beneficial action for my enemy is harmful to me. Therefore, we define the evaluation of an action belief as:

$$evaluation(a) = payoff(a) \times att(i, bnf(a))$$

Co-responsibility

In our case study, the Blue Force endorses the role of security guardian, thus they are co-responsible of all actions compromising the security of the population including Talibans' attacks from population's perspective. Thus, we introduce a co-responsibility mechanism that enables individual to attribute a fraction $\rho \in [0, 1]$, parameter of the simulation, of an action payoff to the co-responsible. This mechanism occurs when an individual faces an action a in which (1) there is a co-responsible actor, (2) its impact is negative (i.e. there is no co-responsibility for beneficial actions), and (3) its evaluation is negative. In that specific case, the individual adds a belief a' with $actor(a) = coResp(a)$ and $evaluation(a') = \rho \times evaluation(a)$.

Aggregation

Let $gaList(i, actor)$ be the list containing all the general actions performed by the actor in the belief base of agent i. The attitude $att(i, actor)$ of the individual i toward the *actor* is given at each time of the simulation by:

$$att(i, actor) = \sum_{ga(i) \in gaList(i,actor)} \left(\sum_{a(i) \in ga(i,a)} \left(\frac{evaluation(a) \times interest(a, i)}{|ga(i, a)|} \right) \right)$$

5   Experiments

In this section, we present the experimental results of our model. We aim to reproduce the results of opinion polls collected on the field using the established scenario of events that took place in Kapisa between September 2011 and September 2013. Since the polls did not asked directly the opinion toward the Red Force but "whether they represent a threat" (see Sect. 3.2), we decided to take the opposite of these results as the target attitude values.

5.1   Simulation Settings and Initialization

We input the action sequence presented in the scenario of both Red and Blue Forces into the simulation scheduler; one tick corresponds to 1 day: the simulation covers the period between the first and last opinion polls in 554 ticks. The two agents corresponding to each Force will then operate their actions according to the scenario. The artificial population representing the inhabitants of Kapisa is composed of 150 agents connected by an interaction network based on a small-world topology [15] with a degree of 4 (i.e., each individual has four neighbors in average).

Before running the actual simulation, we initialize the population with a personal history for each individual and an attitude corresponding to the value given by P1. Indeed, one of our model originality resides in the fact that the attitude depends on the agent's cognitive state characterized by its beliefs and accessibility values. Thus, we must give individuals an initialization belief with a certain reach and payoff for both attitudes toward Red and Blue Forces. These beliefs represent the interactions with Forces preceding to the simulation span. Another subtle point in our model is that individuals are surprised when they witness a totally new action, resulting in an overestimation of the action's impact. In order to habituate them to certain regular actions (such as patrols, preaches, and radio broadcasts) we need to run an initialization scenario before the actual one in which the population is confronted to these actions, until we reach a stable point (approximately 200 ticks).

5.2 Calibration Method

Once the simulation is properly initialized, we calibrate the model parameters using each opinion polls results as objectives. We have four points to calibrate per Force, thus totaling eight points of calibration. The model parameters are shared among all individuals of the population:

- α the weight of emotional sensibility toward the surprise factor
- ξ the level of sensibility to a stimuli (i.e., payoff)
- ρ the co-responsibility factor of Blue Forces for harmful Red actions.

We also have to determine the parameters of initialization actions to attain the first point P1: one positive and one negative action per Force. To do so, we fix their payoff values (negative for the harmful action and positive for the other) and calibrate their reaches.

We define our fitness function as the sum of differences' squares between each point of the opinion poll results and its corresponding percentage of favorable individuals in the simulation. We choose to minimize this fitness using the evolutionary algorithm CMA-ES that is one of the most powerful calibration method to solve this kind of problem [10]. Once the fitness stops progressing over 500 iterations, we interrupt the calibration process and save the parameters. Each calibration iteration is based on the average output on over 20 simulations replica since the model is stochastic.

5.3 Calibration Results

Figure 2 shows the results of our model once its parameters have been calibrated. Plain curves represent the objectives to reach that are based on the collected opinion polls results (see Sect. 3.2); dashed curves correspond to the simulation results, with $\alpha = 0.70$, $\xi = 0.08$, and $\rho = 0.15$ (as obtained by the calibration).

We can observe that the attitude dynamics tendencies are well reproduced. The average difference between results and objective points is 13.25 % with a maximum of 19 % for the last point. This gap between survey and simulation results could be explained by several factors in addition to the model itself: field data are generally inaccurate and capture only a limited part of reality.

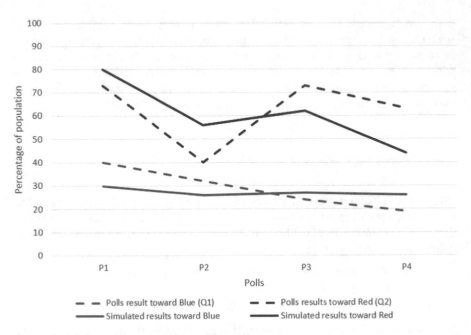

Fig. 2 Simulation results compared to opinion polls

First, the established scenario is based on subjective assessments of some Blue Force officers and do not capture all the military events that took place on the terrain. Adding to this, the parameters of action's models (i.e. payoffs, frequency, and reach) have been assessed based on qualitative appraisal of subject matter experts since there is no scientific method to assess them.

Second, the sampling of the opinion survey could not be maintained through the survey process, due to the dynamics of the conflict: certain villages could not be accessed constantly over time due to their dangerousness. Moreover, as it was pointed earlier, the questionnaire did not directly ask the opinion toward Red Force which might increase the gap between our model outputs and the polls results.

Finally, our field data is limited to the context of military events. Even if our study concerns attitudes toward Forces in the military/security context, other events might also have influenced these attitudes such as economic or daily activities.

In view of these limitations, the reproduction of the general tendencies of attitude dynamics between each polls seems encouraging. Besides, these results have been obtained by calibrating only three model parameters.

5.4 Attitude Dynamics

Agent-based simulation enables not only to reproduce aggregated data but also to analyze micro behaviors. Figures 3 and 4 below show the dynamics of population's attitudes means values between two polls along with their corresponding scenario.

The decreasing general tendency of the attitude toward the Red Force between P1 and P2 in Fig. 3 is due to the constant pressure activity that affects negatively the population. We can observe repercussions of each occurrence of action on the attitude dynamics. For instance, in Fig. 3 we can clearly see attitude decreasing toward Red Force at each IED (Improvised Explosive Device, gray blocs in the figure) and also that the first occurrence has the greatest impact since the population is surprised. Besides, we can notice that the impact of a suicide attack is much greater than other actions (mid-July 2011 and June 2013).

In Fig. 4, we can see that attitudes toward the Blue Force are also impacted by each of its actions. Moreover, we can notice that the curve greatly decreases when the suicide attack perpetuated by Reds occurs. This phenomenon is enabled by the mechanism of co-responsibility (see Sect. 4.2), the Blue Force is also responsible of these attacks, in a moderate way. Similarly, the attitude toward Blue Force decreases constantly since the background communication actions and patrols are not enough to counter their co-responsibility toward the pressure activity for Reds.

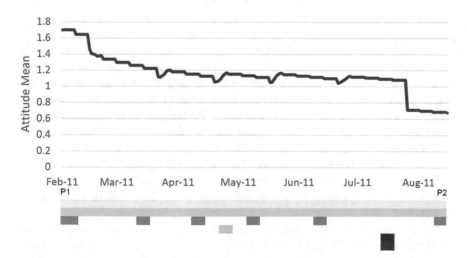

Fig. 3 Attitude mean toward Red Force between P1 and P2

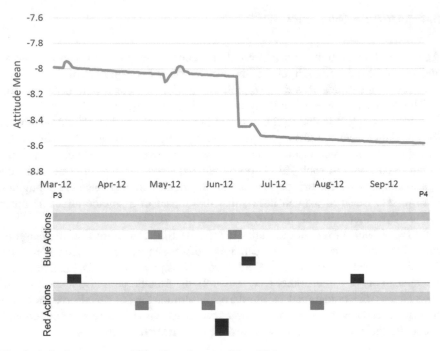

Fig. 4 Attitude mean toward Blue Force between P3 and P4

5.5 Role of Co-responsibility

To show the impact of the co-responsibility mechanism, we have performed a new simulation with $\rho = 0$ (i.e. no co-responsibility). Since the initial values of attitudes toward Blue Force depend on ρ during the initialization scenario, we had to re-calibrate the model parameters. Figure 5 shows the resulting attitude dynamics. As it was predictable, the attitude toward the Blue Force constantly increases as their only potentially negative actions (kinetic actions against the Red Force) are easily countered by communication toward the population or reconstruction actions. This is what was expected by the stakeholders when they decided to engage in Afghanistan.

The tendency of attitude toward Red Force remains the same since the co-responsibility only affects the Blue. We can notice that the simulated attitudes toward Reds is closer to opinion polls than in the first calibration. This is due to the fact that the red scenario is more simple (less action diversity).

6 Conclusion

We have collected information on the opinion dynamics and the events during the involvement of French army in Afghan war. These information were analyzed and processed with the support of subject matter experts. Based on these field

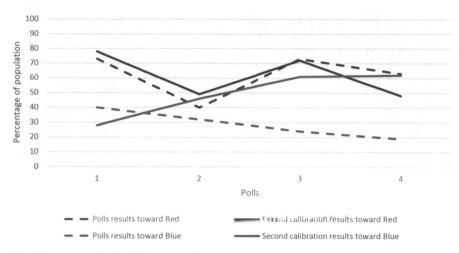

Fig. 5 Impact of co-responsibility

observations, we proposed a simulation model of attitude dynamics. This model embodies both cognitive and affective components in the formation of attitude and the diffusion of beliefs. While it was designed in the context of military operations, it can be applied to civilian use: the actors can represent any kind of active social object such as political parties, institutions, companies, or brands.

We also introduced a new concept of co-responsibility that reflects attitude behavior perceptible in conflict terrain. These components were aggregated through a new method that better understands the concept of attitude proposed by Fazio. We conducted a model calibration based on the collected data that showed encouraging results.

In future works, we intend to conduct deeper analysis of the field data to adapt the model and implement the simulation of different social groups that are present in this area. Moreover, we would like to implement a memory mechanism to let the agents "forget" some beliefs over time. This would be enabled, for instance, using the Peak-End mechanism of Kahneman [13]. Furthermore, we would like to add a behavioral component to enable population agents to express their attitudes through action's selection.

References

1. Ajzen, I., Fishbein, M.: The influence of attitudes on behavior. In: Albarrac, D., Johnson, B.T., Zanna, M.P. (eds.) The Handbook of Attitudes, pp. 173–221. Lawrence Erlbaum Associates Publishers, Mahwah, NJ (2005)
2. Allport, G.W.: Attitudes. In: Murchison, C. (ed.) Handbook of Social Psychology, pp. 798–844. Clark University Press, Worcester, MA (1935)

14   K.-L. Brousmiche et al.

3. Brousmiche, K.-L., Kant, J.-D., Sabouret, N., Prenot-Guinard, F., Fournier, S.: The role of emotions on communication and attitude dynamics: an agent-based approach. In: World Congress on Social Simulation (2014)
4. Castellano, C., Fortunato, S., Loreto, V.: Statistical physics of social dynamics. Rev. Mod. Phys. **81**(2), 591 (2009)
5. Deffuant, G., Neau, D., Amblard, F., Weisbuch, G.: Mixing beliefs among interacting agents. Adv. Complex Syst. **3**(1–4), 87–98 (2000)
6. Dimulescu, A., Dessalles, J.-L.: Understanding narrative interest: some evidence on the role of unexpectedness. In: Proceedings of the 31st Annual Conference of the Cognitive Science Society, pp. 1734–1739 (2009)
7. Drogoul, A., Ferber, J.: Multi-agent simulation as a tool for modeling societies: application to social differentiation in ant colonies. In: Artificial Social Systems, pp. 2–23. Springer, Berlin (1994)
8. EMA and ECPAD (Ministry of Defence). Ensemble vers l'autonomie. Cinq ans en Kapisa et Surobi. http://webdocs.ecpad.fr (2012)
9. Fazio, R.H.: Attitudes as object-evaluation associations of varying strength. Soc. Cogn. **25**(5), 603 (2007)
10. Hansen, N., Müller, S., Koumoutsakos, P.: Reducing the time complexity of the derandomized evolution strategy with covariance matrix adaptation (CMA-ES). Evol. Comput. **11**(1), 1–18 (2003)
11. Hart, H.L.A.: Punishment and Responsibility: Essays in the Philosophy of Law. Oxford University Press, Oxford (2008)
12. Jones, E.E., Harris, V.A.: The attribution of attitudes. J. Exp. Soc. Psychol. **3**(1), 1–24 (1967)
13. Kahneman, D., Kahneman, D., Tversky, A.: Experienced utility and objective happiness: a moment-based approach. Psychol. Econ. Decisions **1**, 187–208 (2003)
14. Kelley, H.: The processes of causal attribution. Am. Psychol. **28**(2), 107–128 (1973)
15. Milgram, S.: The small world problem. Psychol. Today **2**(1), 60–67 (1967)
16. Nowak, A., Szamrej, J., Latane, B.: From private attitude to public opinion: a dynamic theory of social impact. Psychol. Rev. **97**(3), 362 (1990)
17. Pahl-Wostl, C., Kottonau, J.: Simulating political attitudes and voting behavior. J. Artif. Soc. Soc. Simul. **7**(4), (2004). http://jasss.soc.surrey.ac.uk/7/4/6.html
18. Rosenberg, M.J., Hovland, C.I.: Cognitive, affective, and behavioral components of attitudes. Attitude Organ. Change: Anal. Consistency Among Attitude Components **3**, 1–14 (1960)
19. Smith, R.: The Utility of Force: The Art of War in the Modern World. Knopf, New York (2007)
20. Urbig, D., Malitz, R.: Drifting to More Extreme But Balanced Attitudes: Multidimensional Attitudes and Selective Exposure. ESSA, Toulouse (2007)

A Simple-to-Use BDI Architecture
for Agent-Based Modeling and Simulation

Philippe Caillou, Benoit Gaudou, Arnaud Grignard, Chi Quang Truong,
and Patrick Taillandier

Abstract With the increase of computing power and the development of
user-friendly multi-agent simulation frameworks, social simulations have become
increasingly realistic. However, most agent architectures in these simulations use
simple reactive models. Cognitive architectures face two main obstacles: their
complexity for the field-expert modeler, and their computational cost. In this paper,
we propose a new cognitive agent architecture based on the Belief-Desire-Intention
paradigm integrated into the GAMA modeling platform. Based on the GAML
modeling language, this architecture was designed to be simple-to-use for modelers,
flexible enough to manage complex behaviors, and with low computational cost.
This architecture is illustrated with a simulation of the evolution of land-use in the
Mekong Delta.

Keywords Cognitive model • BDI agent • Simulation Framework

P. Caillou (✉)
UMR 8623 LRI, University of Paris Sud, Paris, France
e-mail: caillou@lri.fr

B. Gaudou
UMR 5505 IRIT, University of Toulouse, Toulouse, France
e-mail: benoit.gaudou@utcapitole.fr

A. Grignard
UMI UMMISCO, University Pierre and Marie Curie/IRD, Paris, France

UMR 6266 IDEES, University of Rouen, Rouen, France
e-mail: agrignard@gmail.com

C.Q. Truong
UMI UMMISCO, University Pierre and Marie Curie/IRD Paris, France

CENRES, DREAM Team, Can Tho University Can Tho, Vietnam
e-mail: tcquang@ctu.edu.vn

P. Taillandier
UMR 6266 IDEES, University of Rouen Rouen, France
e-mail: patrick.taillandier@gmail.com

© Springer International Publishing AG 2017
W. Jager et al. (eds.), *Advances in Social Simulation 2015*, Advances in Intelligent
Systems and Computing 528, DOI 10.1007/978-3-319-47253-9_2

1 Introduction

Agent-based simulations are widely used to study complex systems. However, the problem of the agent design is still an open issue, especially for models tackling social issues, where some of the agents represent human beings. In fact, designing complex agents able to act in a believable way is a difficult task, in particular when their behavior is led by many conflicting needs and desires. A classic paradigm to formalize the internal architecture of such complex agents is the Belief-Desire-Intention (BDI) paradigm [3]. This paradigm allows to design expressive and realistic agents, yet it is barely used in social simulations. One explanation is that most agent architectures based on the BDI paradigm are too complex to be understood and used by non-computer scientists. Moreover, they are often very time-consuming in terms of computation and thus not adapted to simulations with thousands of agents.

In this paper, we propose a new architecture that is integrated into the GAMA platform. GAMA is an open-source modeling and simulation platform for building spatially explicit agent-based simulations [5, 6]. Its complete modeling language (GAML: GAma Modeling Language) and integrated development environment support the definition of large scale models (up to millions of agents) and make it usable even with low level programming skills. Our architecture was implemented as a new GAMA plug-in, and allows to directly and simply define BDI agents through the GAML language.

The paper is structured as follows: Scct. 2 proposes a state of the art of BDI architectures and their use in simulation context. Section 3 is dedicated to the presentation of our architecture. In Sect. 4, we present a simple case study using this architecture to study the land-use change in a village of the Mekong Delta (Vietnam). At last, Sect. 5 provides this paper with a conclusion and some perspectives.

2 State of the Art

The BDI approach has been proposed in Artificial Intelligence [3] to represent the way agents can do complex reasoning. It has first been formalized using Modal Logic [4] in order to disambiguate the various concepts (Belief, Desire, and Intention) and the logical relationships between them (concepts are detailed in Sect. 3.1).

2.1 BDI Frameworks

In parallel, BDI operational architectures have been developed in order to help the development of Multi-Agent Systems embedding BDI agents. Some of these BDI architectures are included in framework allowing to directly use them in different applications. A classic framework is the Procedural Reasoning System (PRS) [10]. This framework includes three main processes: the perception (in which agent acquires information from the environment), the central interpreter (which helps the agent to deliberate its goals and then to select the available actions), and the execution of intention (which represents agents reactions). This framework has been used as a base for many other frameworks. For instance, the JACK [7] commercial framework inherits many properties from PRS. JACK allows the user to define a multi-agent system with BDI agents using a dedicated language (a super-set of Java). It was used in many commercial applications (e.g., video-game, oil trading, etc.). Another classic framework for multi-agent system building is JADE [2]. This open-source Java framework integrates several add-ons dedicated to the definition of BDI agents. The most advanced framework is Jadex [12], that is an add-on of the JADE framework. In comparison to JACK, Jadex proposes an explicit representation of goals.

2.2 BDI Agents in Agent-Based Modeling and Simulation Platforms

BDI architecture's agents have been introduced in several agent-based modeling and simulation platforms. For example, Sakellariou et al. [14] have proposed an extension to Netlogo [21] to deal with BDI agents. The extension allows the model to add to agents a set of beliefs (information it gets by perception of communication) and intentions (what it wants to execute), and ways to manage these two sets. This very simple architecture is inspired by the PRS architecture (in particular using an intention stack) and is education-oriented. Its main aim was to allow modelers to manipulate BDI concepts in a simple language.

Singh and Padgham [15] went one step further in the integration between BDI architecture and agent-based modeling and simulation platforms. They propose a framework able to connect agents-based platforms and an existing BDI framework (such as JACK [7] or Jadex [12]). An application couples the Matsim platform [1] and the GORITE BDI framework [13]. Their framework aims at being generic and can be extended to couple any kind of simulation platforms and BDI frameworks. This approach is very powerful but remains computer-scientist-oriented, as it requires high programming skills to develop bridges between the framework and the platforms, and to write agents behaviors without a dedicated modeling language.

First attempts already exist to integrate BDI agents into the GAMA platform [6]. Taillandier et al. [16] proposed a BDI architecture where the choice of plans is formalized as a multi-criteria decision-making process: desires are represented by criteria that will be used to make a decision. Each plan is evaluated by each criterion according to the beliefs of the agent. However, this architecture was tightly linked to its application context (farmer decision making) and does not propose any formalism to model the agent beliefs and is rather limited concerning the way the plans are carried out: there is, for example, no possibility to have complex plans that require sub-objectives. Le et al. [8] proposed another architecture dedicated to simulation with a formalized description of beliefs and plans and their execution. However, the desires and plans have to be written in a very specific and complex way that can be difficult to achieve for some application contexts, in particular for non-computer scientists. In addition, this architecture has a scalability problem: it does not allow to simulate thousands of agents.

3 Presentation of the SimpleBDI Plug-In Architecture

3.1 Overview

Consider a simple Chopper-Fire model: A chopper agent patrols, looking for fires. When it finds one, it tries to extinguish it by dropping water, and when it has no more water, it goes to the nearest lake to refill its water tank. With a reactive agent model, defining an agent behavior means to define *What it does* (e.g., patrol, go to the fire, go to take water). This can be achieved both with reflexes or a finite state machine. Using a cognitive model means to define *what it wants* (e.g., to find fire, to extinguish a specific fire, and to get water) and *how to do it* (e.g., if I want to find a fire, I patrol (wandering randomly in my environment) and if I see a fire, I want it to be extinguished. If I want to put out a fire, go toward it and put water, and if I have no more water, get some). There are several advantages for using such cognitive approach: complex reasoning (planning), persistence (of the goals), easy to improve (both on what to do and how to do it), easy to use (the modeler can define goals instead of reactions), and easy to analyze (it is possible to know why—for what purpose—agents do what they do).

The architecture and the vocabulary can be summarized with this simple Fire-Chopper example: the Chopper agent has a general *desire* to patrol. As it is the only thing he wants at the beginning, it is its initial *intention* (what it is doing). To patrol, it wanders around (its *plan* to patrol). When it *perceives* a fire, it stores this information (it has a new *belief* about the existence of this fire), and it has a new *desire* (it wants the fire to be extinct). When it sees a fire, the patrol *intention* is put *on hold* and a new *intention* is selected (to put out the fire). To achieve this *intention*, the *plan* has two steps, i.e., two new *(sub)desires*: go to the fire and put water on the fire, and so on.

3.2 Vocabulary

The vocabulary introduced in the previous example can be summarized as follows.

Knowledge

Beliefs, Desires, and Intentions are described using **predicates**. A predicate has a name, and may also have a value (with no constraint on the type) and some parameters (each defined by a name and a value); For example, *Fire(true, (Position* :: (12, 16)))—a fire is present (value true) at position (12,16)—or *HaveWater(true)*—the Chopper has some water (value true).

- **Beliefs** (what it thinks). Beliefs is the internal knowledge the agent has about the world. The belief base is updated during the simulation. A belief is described by a predicate and is in general true or false. For example, the predicates *Fire(true, (Position* :: (12, 16))) is added when the agent perceives a fire at position (12,16).
- **Desires** (what it wants). Objectives that the agent would like to accomplish (for example, *Fire(false, Position* :: (12, 16)), the agent wants the previous fire to be put out). They are stored as a set of desires. A desire is fulfilled when it is present in the Belief base (or manually removed by the agent). Like the Belief base the Desire base is updated during the simulation. Desires can be related by hierarchical links (**sub/super-desires**) when a desire is created as an intermediary objective (for example, to the extinct a fire can have two sub-desires: go to the fire and put water on the fire). Desires have a **priority** value (that can be dynamic), used to select a new intention among the desires when necessary.
- **Intentions** (what it is doing). What the agent has chosen to do. The **current intention** will determine the selected plan. Intentions can be put **on hold** (for example, when they require a sub-desire to be achieved). For this reason, there is a stack of intention, the last one is the current intention, and the only one that is not on hold.

Behavior

- **Perception.** A perception is a function called at each iteration, where an agent can eventually update its belief or desire bases. It is technically identical to a **reflex** of a reactive architecture (a function called at each step).
- **Plan.** The agent has a set of **plans**, which are behaviors defined to reach specific desires. A plan can be instantaneous and/or persistent (*goToPosition*). Plans may have a **priority** value (that can be dynamic), used to select a plan when several possible plans are available.

3.3   *Thinking Process*

At each step, the agent applies the process described in Fig. 1. Roughly, the agent
will perceive the environment, then (1) **continue its current plan** if it is not finished,
or (2) if the plan is finished and its current intention is not fulfilled, it **selects a plan**,
or (3) if its current intention is fulfilled, it **selects a new desire** to add to its intention
stack. More precisely:

1.   **Perceive**: Reflexes/Perceptions are applied. This may update the Beliefs and
     add new Desires.
2.   **Is one of my intentions achieved?**: If one of my intentions is achieved,
     set current plan to nil and remove the intention and all its sub-desires from
     the desire and intention base (if I or someone else has extinguished *Fire*1,
     I remove not only the desire *Extinguish(Fire*1) from my desires, but also
     the sub-desires *MyPosition(Fire*1) and *WaterOn(Fire*1) if I have them). If the
     achieved intention super-desire is on hold, it is reactivated (its sub-desire just
     got completed).

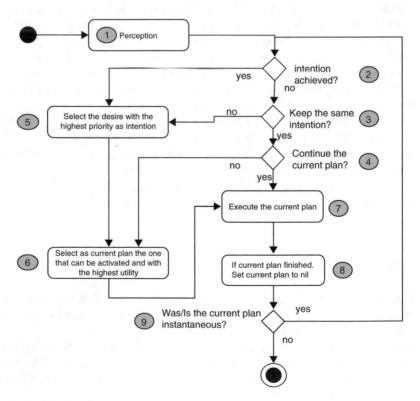

Fig. 1 Activity diagram

3. **Do I keep the current intention?**: To take into account the environment instability, an intention-persistence coefficient *ip* is applied: with a probability *ip*, the current intention is removed from the intention stack. More details about this coefficient are given in Sect. 3.4.

4. **Do I have a current plan?**: If I have a current plan, just execute it. As for the current intention stability, the goal is both persistence (I stick to the plan I have chosen) and efficiency (I don't choose at each step). For the same reason that the current intention is randomly removed, a plan-persistence coefficient *pp* is defined: with a probability *pp*, the current plan is just dropped.

5. **Choose a desire as new current intention**: If the current intention is on hold (or the intention base is empty), choose a desire as new current intention. The new intention is selected among the desires with the highest priority (and not already present in the intention base).

6. **Choose a plan as new current plan**: The new current plan is selected among the plans compatible with the current intention and with the highest priority.

7. **Execute the plan**: The current plan is executed.

8. **Is my plan finished?**: To allow persistent plans, a plan may have a termination condition. If it is not reached, the same plan will be kept for the next iteration.

9. **Was my plan instantaneous?**: Most multi-agent-based simulation frameworks (GAMA included) are synchronous frameworks using steps. One consequence is that it may be useful to apply several plans during one single steps. For example, if a step represents a day or a year, it would be unrealistic for an agent to spend one step to apply a plan like "To fight this fire, lets define two new sub-desires, go to the fire and put water on the fire, and put my objective on hold." This kind of plan (mostly reasoning) can be defined as **instantaneous**: in this case a new thinking loop is applied during the same agent step.

3.4 Properties

Persistence and Priority

Persistence coefficients and priority values are key properties of many BDI architectures. Agents with high persistence continue their current actions independently of the environment evolution (they are more stable and spend less time rethinking their plans and objectives). Less persistent agents are more adaptable and reactive but may lead to erratic and computationally costly behaviors. Priority values will determine both the selected desires and plans. The choice can be deterministic (highest priority selected) or probabilistic (highest priority has a higher probability to be selected). One advantage of the GAMA modeling framework for both persistence and priority coefficients is to allow to use dynamic or function-based variables. Plans and Desires priority values and agent persistence coefficients can be changed by the agent itself (for example, a plan could increase the persistence coefficient after evaluating the previous plan success). The modeler can also define functions to update a value. For example, the priority of a plan or desire could be defined as a function of the current step, which would make it more and more probable to be selected when the simulation advances.

Flexibility

One core objective when defining this architecture was to make it as simple-to-use and flexible as possible for the modeler. The modeler can use the architecture in its full potential (for example, dynamic coefficients as presented before), but he/she can also use only some parts of the architecture. It is, for example, possible to define only Desires and Plans, and no Beliefs (the effect would be that the intentions and desires achievement and removal will have to be done manually, i.e., defined by the modeler in the agent plans). Most parameters have default values and can be omitted. For example, the modeler doesn't have to know the existence of instantaneous plans (by default off), plan termination condition (by default true: always terminated at the end of its execution), or the possibility to define sub-desires or put intentions on hold.

GAMA Integration

The architecture is defined as a GAMA species architecture. The modeler only requires to define simpleBDI as agent architecture and define at least one plan to be operational. After that the modeler mostly defines plans to act and (usually one) reflexes to perceive. Many keywords are defined to help the user to update and manage both Belief, Desire, and Intentions bases and create/manage predicates. In the next section, we present an application of the architecture to a social simulation context.

4   Case Study: Land-Use Change in Coastal Area of the Mekong Delta

4.1   Context of the Case Study

The Mekong Delta region will be heavily influenced by the effects of global climate change [20]. Indeed, the sea level rise and salt water intrusion will strongly impact the life of people and the situation of agricultural production [18]. Nhan [11] pointed out that the environmental conditions significantly impact the agriculture and fisheries and that ordinary people tend to spontaneous change the land-use, which causes difficulties for land resource management and cultivation of farmers. Another difficulty comes from the behaviors of farmers that tend to adapt their food production to the market [17]. As showed in [9], the difference of planned and real production can be observed at the village level, where the land-use change has not evolved as expected. It is thus important to be able to understand the land-use planning at village level to be able to predict the evolution of land-use change at province level. In this context, we chose to study the evolution of land-use in the village of Binh Thanh. This coastal village of the Ben Tre province of the Mekong Delta is representative of regions with a mix of brackish and fresh water, where the land-use is strongly impacted by the irrigation planning.

4.2 Collected Data

We have collected data concerning the land-use of each parcel of this village in 2005 and in 2010 from the Department of Natural Resources and Environment of the Ben Tre province. In this area, six land-use types were defined: Rice, Rice–Vegetable, Rice–Shrimp, Annual crops, Industrial Perennial tree, and Aquaculture. We collected as well the soil map, the saltwater map, and the flood map of the regions and defined from them six land-unit types. From each of these land-unit types, we defined with the help of domain-experts a suitability value for each type of land-use (the lower the better). This suitability represents the adequacy between the land-unit type and the land-use type. For instance, producing industrial perennial tree on a salty soil is very difficult and the yield will be very low. Another data source that was built with domain-experts were the transition values for each type of land-use. This matrix allows to represent the technical difficulty to pass from one land-use type to another. This difficulty was evaluated using three values (1: easy, 2: medium, and 3: very difficult). Finally, we collected data concerning the evolution of benefit and cost of each land-use type per hectare from 2005 to 2010.

4.3 Implemented Model

The model was defined in order to simulate the evolution of the land-use of the Binh Thanh village. We make the assumption that each farmer has only one parcel and that he has to make a decision concerning the land-use of the parcel every year. A simulation step in this model represents then 1 year.

In this simple model, the main species of agents is the parcel species that represents the farmer and his/her parcel (5721 parcels for the study area). We use our SimpleBDI agent architecture for this species of agents.

A parcel agent has the following attributes:

- *shape*: geometry of the parcel (static)
- *profile*: the inclination of the farmer toward a change of production. It is used to set the value of the *intentionpersistence* (*ip*) variable. We defined

five possible values: innovator (2.5 %—*ip*: 0.0), early adopters (13.5 %—*ip*: 0.1), early majority (34 %—*ip*: 0.2), late majority (34 %—*ip*: 0.3), and laggard (16 %—*ip*: 0.5) (static)
- *land – unittype*: type of soil for the parcel (static)
- *neighbors*: list of parcels at a distance of 1 km (static)
- *land – use*: type of production (dynamic).

In addition to the parcel agents, we define a world agent that contains all the global variables:

- *profitmatrix*: for each year (from 2005 to 2010), for each land-use type, the benefit can be expected from 1 ha of production.
- *costmatrix*: for each year (from 2005 to 2010), for each land-use type, the cost of 1 ha of production.
- *suitabilitybylanduse*: for each land-unit type, the suitability to produce a given land-use type.
- *transitionmatrix*: difficulty to pass from a land-use to another one.

At each simulation step (i.e., every year), each parcel agent is activated in a random order.

In our model, each parcel agent has the following belief and desire base:

- *Beliefs*: pieces of knowledge concerning the expected profit for each land-use for each land-unit type. Each belief corresponds to a land-use type, a land-unit type, and a profit associated with it.
- *Desires*: for each type of production, the agent will have a desire to do it. In addition, the agent could have desires to give information (a belief) to the other farmers in its neighborhood concerning the expected price for a land-use type and a land-use unit (see below)

The priority of its "do a production" desires will be computed according to a multi-criteria analysis. This type of decision-making process is often used for land-use change models (see, for example, [16]). We defined 3 criteria for the decision: the profit, the cost, and the transition difficulty. Indeed, it is generally accepted that farmers tend to choose a production that maximizes their profits, that minimizes the cost—avoid risky productions—and that are easy to implement. More precisely, the criterion values are computed as follows for a given transition from *oldlu* to *lu* and a given *soil* type (i.e., land-unit type) and *year*:

$$Profit(lu, soil, year) = \frac{matrix\_profit(lu, year)}{(max\_profit(year) * matrix\_suitability(soil, lu))} \quad (1)$$

With:

$$max\_profit(year) = max(matrix\_profit(lu, year)) \quad (2)$$

$$Transition(old\_lu, lu) = \frac{(3 - transition\_matrix(old\_lu, lu))}{2} \quad (3)$$

```
plan do_production when: is_current_intention(predicate_do)  {
    predicate do_current <- get_current_intention();
    landuse <- string(do_current.value);
    string id_belief <- "profit,"+ landuse + "," + land_unit;
    predicate new_profit_belief <- new_predicate(id_belief, compute_profit(landuse));
    do replace_belief(get_belief_with_name(id_belief),new_profit_belief);
    do_predicates[landuse] <- do_predicates[landuse] with_priority updatePriority(landuse);
    do diffuse_information(new_profit_belief);
    do current_intention_on_hold();
}
```

Fig. 2 Production plan

```
action diffuse_information(predicate information) {
    loop people over: neighbours {
        predicate inform_people <- new_predicate("inform", people::information);
        do add_subintention(get current_intention(), inform_people, true);
    }
}
```

Fig. 3 Diffuse information action

To fulfill its desires, the agent can activate a dedicated plan: *do_production*. The GAML code of this plan is presented in Fig. 2. This plan is activated when the current intention is to produce something. First, the agent gets the current intention, and changes its land-use according to it. After that, it creates a new belief concerning the real profit that it got from this land-use and updates its old belief (concerning the profit of this land-use). Then, it diffuses its new belief to its neighborhood (call the *diffuse_information* action, see below) and puts its current intention on hold (wait to finish to diffuse the information before producing again).

Figure 3 presents the GAML code of the *diffuse_information* action. When this action is called, the agent does a loop on all the people in its neighborhood. For each of these people, the agent adds a new sub-intention to its current intention (produce a given land-use) to diffuse its new belief to this people.

In order to fulfill its information diffusion intention, the agent can activate a dedicated plan: *inform_people*. The GAML code of this plan is presented in Fig. 4. This plan is instantaneous, as we consider that the time taken to inform its neighborhood is insignificant in comparison to the simulation step (1 year). It is activated when the current intention is to inform someone. First the agent gets the people to inform and the information to diffuse from the current intention. After that, it asks the people to inform to receive the new information (call the *diffuse_information* action) and remove the current intention (and desire) from its intention base (desire base).

The complete source code of the model is available in the GAMA SVN [5].

```
plan inform_people when: is_current_intention(predicate_inform) instantaneous: true{
    predicate inform_current <- get_current_intention();
    land_parcel_bdi people_to_inform <- land_parcel_bdi(pair(inform_current.value).key);
    predicate information <- predicate(pair(inform_current.value).value);
    ask people_to_inform {
        do get_information(information);
    }
    do remove_intention(get_current_intention(),true);
}
```

Fig. 4 Information diffusion plan

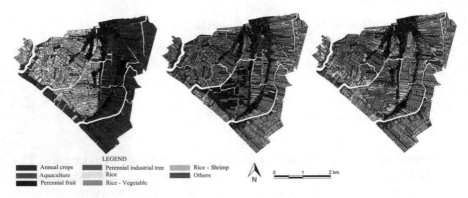

Fig. 5 Land-use for 2005 (*left*); land-use obtained for 2010 with the simulation (*middle*); observed land-use for 2010 (*right*)

4.4 Experiments

The different parameter values of the models were defined by using a genetic algorithm to find the parameter set that fits the best of the real data, i.e., minimization of the fuzzy kappa coefficient [19] computed by comparing the observed data in 2010 and the simulation result for the same date. The fuzzy kappa coefficient allows to compare two maps by taking into account the neighborhood of the parcels. This coefficient is between 0 (not similar at all) and 1 (totally similar).

Figure 5 shows simulation results obtained for the model and the observed data. As shown, the observed land-use is close to the real one.

To quantitatively evaluate the simulation results of the model, we used two indicators: the fuzzy kappa coefficient (local indicator) and the percent absolute deviation (global indicator). This second indicator that is often used to evaluate land-use change models is computed by the following formulae:

$$\text{PAD}(\%) = 100 \frac{\sum_{i=1}^{n} |\widehat{X}_i - X_i|}{\sum_{i=1}^{n} \widehat{X}_i} \tag{4}$$

with: $\widehat{X}_i$ the observed quantity of parcels with the land-use i and X_i the simulated quantity of parcels with the land-use i.

As our model is stochastic, we ran 100 times each model and computed the average fuzzy kappa coefficient (kappa) and percent absolute deviation (pad). We obtained for the pad a value of 35.98 % (the lower the better) and for the fuzzy kappa a value of 0.545 (the higher the better). These results are rather good and show that the model is able to reproduce in relevant way the real dynamic.

Concerning the computation time (on a Macbook pro computer from 2011), the mean duration of a simulation step was less than 0.6 s. This result is quite promising considering that we have more than 5700 BDI agents that can have many desires and that can activate many plans during the same simulation step with the information diffusion process.

5 Conclusion

In this paper, we have presented a new BDI architecture dedicated to simulation context. This architecture is integrated into the GAMA modeling and simulation platform and directly usable through the GAML language, making it easily usable even by non-computer scientists. We have presented a first simple application of this architecture concerning the land-use change in the Mekong Delta (Vietnam). This first application showed that our plug-in allows to built relevant models and to simultaneously simulate several thousand of agents.

If our architecture is already usable, some improvements are planned. First, we want to improve the inference capabilities of our architecture: when a new belief is added to the belief base, desire and intention bases should be updated in a efficient way as well. Second, we want to make it even more modular by adding more possibility concerning plans and desire choices and not just the plan/desire with the highest priority: let the possibility to make user-defined or with a multi-criteria decision process, etc. At last, we want to add the possibility to use high performance computing (distribute the computation on a grid or cluster) to decrease the computation time.

Acknowledgements This work is part of the ACTEUR ("Spatial Cognitive Agents for Urban Dynamics and Risk Studies") research project funded by the French National Research Agency.

References

1. Balmer, M., Rieser, M., Meister, K., Charypar, D., Lefebvre, N., Nagel, K., Axhausen, K.: Matsim–t: Architecture and simulation times. In: Multi-Agent Systems for Traffic and Transportation Engineering, pp. 57–78. IGI Global (2009). https://scholar.google.com/citations?view_op=view_citation&hl=en&user=6bkj2pkAAAAJ&citation_for_view=6bkj2pkAAAAJ:YsMSGLbcyi4C

2. Bellifemine, F., Poggi, A., Rimassa, G.: JADE–a FIPA-compliant agent framework. In: Proceedings of PAAM, London, vol. 99, p. 33 (1999)
3. Bratman, M.: Intentions, Plans, and Practical Reason. Harvard University Press, Cambridge (1987)
4. Cohen, P.R., Levesque, H.J.: Intention is choice with commitment. Artif. Intell. **42**, 213–261 (1990)
5. GAMA website (2015). http://gama-platform.org
6. Grignard, A., Taillandier, P., Gaudou, B., Vo, D., Huynh, N., Drogoul, A.: GAMA 1.6: advancing the art of complex agent-based modeling and simulation. In: PRIMA 2013: Principles and Practice of Multi-Agent Systems. Lecture Notes in Computer Science, vol. 8291, pp. 117–131. Springer, Berlin (2013)
7. Howden, N., Rönnquist, R., Hodgson, A., Lucas, A.: JACK intelligent agents-summary of an agent infrastructure. In: 5th International Conference on Autonomous Agents (2001)
8. Le, V.M., Gaudou, B., Taillandier, P., Vo, D.A.: A new BDI architecture to formalize cognitive agent behaviors into simulations. In: KES-AMSTA. Frontiers in Artificial Intelligence and Applications, vol. 252, pp. 395–403. IOS, Amsterdam (2013)
9. Ministry of Natural Resources and Environment. Detailing the establishment, regulation and evaluation planning, land-use planning (2009)
10. Myers, K.L.: User guide for the procedural reasoning system. SRI International AI Center Technical Report. SRI International, Menlo Park, CA (1997)
11. Nhan, D.K., Trung, N.H., Sanh, N.V.: The impact of weather variability on rice and aquaculture production in the Mekong delta. In: Stewart, M.A., Coclanis, P.A. (eds.) Environmental Change and Agricultural Sustainability in the Mekong Delta. Advances in Global Change Research, vol. 45, pp. 437–451. Springer, Netherlands (2011)
12. Pokahr, A., Braubach, L., Lamersdorf, W.: Jadex: a BDI reasoning engine. In: Multi-Agent Programming, pp. 149–174. Springer, Berlin (2005)
13. Rönnquist, R.: The goal oriented teams (gorite) framework. In: Programming Multi-Agent Systems, pp. 27–41. Springer, Berlin (2008)
14. Sakellariou, I., Kefalas, P., Stamatopoulou, I.: Enhancing NetLogo to simulate BDI commu-nicating agents. In: Artificial Intelligence: Theories, Models and Applications, pp. 263–275. Springer, Berlin (2008)
15. Singh, D., Padgham, L.: OpenSim: a framework for integrating agent-based models and simulation components. In: Frontiers in Artificial Intelligence and Applications. ECAI 2014, vol. 263, pp. 837–842. IOS, Amsterdam (2014)
16. Taillandier, P., Therond, O., Gaudou, B.: A New BDI Agent Architecture Based on the Belief Theory. Application to the Modelling of Cropping Plan Decision-Making. iEMSs, Manno (2012)
17. Tri, L.Q., Guong, V.T., Vu, P.T., Binh, N.T.S., Kiet, N.H., Chien, V.V.: Evaluating the changes of soil properties and landuse at three coastal districts in Soc Trang province. J. Sci. Cantho Univ. **9**, 59–68 (2008)
18. Tri, V.P.D., Trung, N.H., Thanh, V.Q.: Vulnerability to flood in the Vietnamese Mekong delta: mapping and uncertainty assessment. J. Environ. Sci. Eng. B **2**, 229–237 (2013)
19. Visser, H., de Nijs, T.: The map comparison kit. Environ. Model Softw. **21**(3), 346–358 (2006)
20. Wassmann, R., Hien, N.X., Hoanh, C.T., Tuong, T.P.: Sea level rise affecting the Vietnamese Mekong delta: water elevation in the flood season and implications for rice production. Clim. Change **66**(1–2), 89–107 (2004)
21. Wilensky, U., Evanston, I.: Netlogo. center for connected learning and computer based modeling. Technical Report, Northwestern University (1999)

Food Incident Interactive Training Tool: A Serious Game for Food Incident Management

Paolo Campo, Elizabeth York, Amy Woodward,
Paul Krause, and Angela Druckman

Abstract Food incidents, such as the horse meat scandal in 2013 and the *E. coli* outbreak in 2011, will always occur. Food business operators (FBOs) must be prepared when responding to such situations not only to protect their businesses, but more importantly to protect the consumers. The Food Standards Agency of United Kingdom has recommended that FBOs should not only have a response plan in place, but they should also have to consistently train and practice implementing these plans to effectively address and minimize the impact of food incidents. Traditional training exercises, such as tabletop or mock simulations, could entail considerable costs. While large companies may have the resources for these activities, smaller companies may not be in the same position. In this chapter we describe the development of a more accessible training tool for managing food incidents. The tool, which takes the form of a simple online serious game, called the Food Incident Interactive Training Tool, is co-designed with domain experts and stakeholders in the UK food industry. It engages players, specifically FBOs, in interactive simulations of food incidents, in which their decisions and actions within the game impact how the simulation unfolds. At the end of this study, we aim to determine the efficacy of the serious game in fulfilling its purpose, given the collaborative nature of the design process, as well as the simplicity of the end product.

Keywords Serious game • Food incident management • Simulation • Training tool • Participatory design

P. Campo (✉) • A. Woodward
Department of Sociology, University of Surrey, Guildford, UK
e-mail: p.campo@surrey.ac.uk; a.woodward@surrey.ac.uk

E. York • A. Druckman
Centre for Environmental Strategy, University of Surrey, Guildford, UK
e-mail: e.york@surrey.ac.uk; a.druckman@surrey.ac.uk

P. Krause
Department of Computer Science, University of Surrey, Guildford, UK
e-mail: p.krause@surrey.ac.uk

© Springer International Publishing AG 2017
W. Jager et al. (eds.), *Advances in Social Simulation 2015*, Advances in Intelligent Systems and Computing 528, DOI 10.1007/978-3-319-47253-9_3

1 Introduction

A food incident is defined by the Food Standards Agency (FSA) of UK as "any event where, based on the information available, there are concerns about actual or suspected threats to the safety or integrity of food and/or feed that could require intervention to protect consumers' interests" [1]. Incidents may arise, for example, from product mislabeling or adulteration [2]. The response to an incident is divided into four levels, namely, routine, serious, severe, and major [1]. The response level is escalated depending on several factors that include, among others, the complexity of the problem, the amount of resources required to address the situation and the potential or actual impact on the consumers, and the amount of media coverage [1]. For example, a routine incident may involve short term minor illnesses, little media coverage, and require local management and minimal resources, while a major incident may involve casualties spread across several countries, have a longer duration, involve cross-departmental and even transnational communication and coordination, and have considerable impact on resources, such as the case of the E. Coli contamination of fenugreek seeds in Germany in 2011 [1]. Some of the most noticeable incidents to have affected the UK in recent years are the bovine spongiform encephalopathy or mad cow disease, salmonella in eggs, Listeria in cheese, antibiotics and hormones in meat and pesticide and phthalates contamination, the *E. coli* outbreak and horse meat scandal.

While the risk of food incidents can be reduced, it cannot be avoided. This underscores the need of food business operators (FBOs) to have an incident management plan in place. While there is a distinction between a food incident and a crisis, with a crisis being a heightened incident, the approach in responding to them are the same [2]. Just as in other crisis management situations, food incident management is complex, involving many actors making decisions and actions amidst limited time and information in a dynamic and uncertain environment. Adding to this complexity is the ubiquity of digital communication. Information, regardless of its accuracy, may travel like wildfire and can significantly impact how a situation plays out [3].

Incident management plans or response protocols to food incidents are developed within companies, and therefore vary from one FBO to another. In general, however, the key steps remain the same. Regular training and practice of applying these plans is imperative to keep the incident management team up to date with the response procedures and be able to implement them efficiently [4]. Within the UK food industry at least, the usual approach to training is through bespoke table-top or mock simulations, either developed in-house or by a third party service provider, such as insurance and risk management enterprises [4]. These customized training exercises could entail considerable costs, to which smaller companies may not have enough resources for allocation.

Our initial interactions with industry stakeholders revealed a keen interest for simulation exercises on food incident management. Moreover, there is an apparent absence of a tool or serious game specifically designed for this important issue. This has inspired us to develop a web-based serious game, called the Food

Incident Interaction Training Tool or FIITT, that engages players in decision-making situations of simulated food incidents. Serious games are games that are designed specifically with a learning objective [5]. Various research projects have shown the value of serious games [6]. For example, not only do serious games attract the attention of users, but they also aim to continuously motivate participants [6]. Moreover, it can imbue the necessary attitudes and make the users feel more competent in performing tasks [6]. Being able to reuse the tool for practice is also notable [6]. These are important attributes to consider, especially when developing a training tool, so much so that serious games have been used in a myriad of domains and situations. For example, they have been used for military and crisis management, student education, and advertising [7]. Furthermore, serious games may be a cost-effective way of reaching a wide audience [8].

While we do not intend to replace the current training practices and established protocols with this game, we view it as a complementary tool to aid FBOs and, in particular, small FBOs, a solid foundation for effective management of food incidents.

2 Approach

At face-value, this online serious game may appear simply to be an electronic or computerized version of a table-top or mock simulation employed in training exercises. However, by adding gaming elements and complexity to the usual linear progression of mock simulations, we aim to make the tool more interesting and engaging to users and, at the same time, capture some of the intricacies of food incident management. A participatory design process is used to develop the game together with stakeholders.

2.1 Incident Simulation

The game is composed of several generic food incident scenarios designed using narrative scripts [9], in which we treat the scripts as models of food incident simulations. While useful, linear scripts are not enough to capture the complexities of management. As emphasized by the participants in our first design workshop, the game should not only be able to teach about making good decisions during an incident, but also be able to show the impact of decisions or actions that are not made, or not made on time. To address this issue, each of the scripts is expanded into a decision tree, resulting to multiple paths of simulation progression, having multiple outcomes. At each node of a tree, a player is tasked to make a decision or a set of decisions in response to the current situation of the scenario and in light of the available information provided to the player. While it may be impossible to map out all the possible situations of a food incident, and this may even be counter-productive [8], each scenario has been designed such that it highlights one or more of the key aspect of incident management, as identified by the participants of our preliminary stakeholder workshop, such as prioritizing consumers' welfare, effective communication, timeliness of decisions and actions, information mismatch or imbalance, and unusual or unforeseen situations.

2.2 Gamification

The FIITT serious game allows for an immersive user experience by recreating the tense setting of an incident, but doing so in a safe environment (no impact to the real world) [5]. As opposed to typical training tools, the gamification approach to the development of FIITT makes use of the advantages of serious gaming in engaging and motivating users by being entertaining and compelling. To this end, gaming elements, such as a scoring system, are added to keep the user motivated in using the tool. Having said this, we are aware that there should be a balance between the learning and gaming elements of the game. Gamification also allows for the players to take risk and explore the implications of different decisions, and thus failure becomes an integral part of the learning process [10].

2.3 Participatory Design

Participatory design is a process in which the stakeholders as much as possible are directly involved in the design of an artifact or system and ensure that the end result or product is fit-for-purpose [11]. The process involves three general stages, namely initial exploration of work, discovery process and prototyping [11]. With having stakeholders involved in the design process we perceive several advantages. Firstly, their knowledge and experience helps assess the realism of the dynamics and interactions being simulated. Secondly, they identify relevant and useful indicators for performance assessment, as well as how to (graphically) present these indicators. Thirdly, being involved in the design process could give them a sense of ownership of the game, which may help in its promotion and use within the food industry.

3 Preliminary Results

3.1 Participatory Workshop

A preliminary workshop was conducted in the 6th of March, 2015 in order to get food industry experts' opinion on the development of the tool and, at the same time, get their initial ideas on the tool's design. The participants of this workshop come from both the public and private sectors of the food industry, including a representative from the FSA, a testing laboratory, a food standards organization, and an incident management consultant. In this workshop, the objectives of the tool vis-à-vis the key aspects of food incident management, were identified as well as the potential users of the tool.

The workshop participants have identified food business operators (FBOs) as the potential main users of this tool, specifically, the FBOs' risk and crisis management team, if it exists. This also includes the organization's technical director and marketing and public relations department, which has a growing role in recent years in terms of managing social media communication and digital information. The tool can be used in various ways. For example, it can be used as means to assess the current strategy or decision-making skills or as a training tool towards making appropriate decisions during food incidents. In a team training setting, this tool can be used as a platform for discussion, negotiation and reflection on the dynamics and quality of decisions. In this setting, the training observer or evaluator is provided with guidelines on how to assess the team's performance. Other members of the organization not directly involved in incident management, as well as new employees, can use this tool to raise their awareness and learn more about the process of incident management. In an academic setting, this may be used as part of a teaching module for students taking up food-related courses.

To reflect the key aspects of incident management, in-game indicators or scoring system have also been identified by the participants to gauge the performance of the user. These indicators pertain to the internal and external communication of the organization, as well the perception of the media and the public to the incident and the actions taken by an FBO during this time. The number of victims that have been affected by the incident, i.e., those who have fallen seriously ill or died due to the incident, has also been identified as an important indicator. They have proposed that these indicators be implemented using a traffic lights system that would indicate the risk and impact of the decisions that are made in the game.

3.2   Game Design

FIITT (Fig. 1) is designed as a one-player game, with a selection of four game scenarios or stages that simulate four different food incidents. In each game scenario, the player plays the role of a food business operator, which is described at the beginning of each scenario. During a game scenario simulation, the player is provided information about the incident in form of messages, which may come from different actors involved in food business industry, such as a food standards body and trade organizations, internal communication messages, and a scenario status message. At certain points in the simulation, the player is asked to make a decision in response to the unfolding situation based on the available information. Each decision will have an impact on the game indicators or scores, the pathway of the simulation narrative, and the number of simulated days. The in-game scores

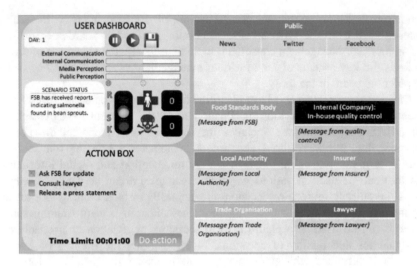

Fig. 1 Tentative graphical user interface of the FIITT

or indicators, as identified in the first stakeholder workshop, measure the player's performance. At the end of the scenario, a performance report is to be provided that compares a player's decisions against that of an expert's choice of an appropriate response, with accompanying explanations.

A set of features will later be added to the tool to extend its functionality. Firstly, a Save feature will be added to allow the player to pause and return to the gaming session at a later time. Secondly, a customization feature will allow the editing of existing simulations or to add new food incident scenarios into the game in order to fit the particular needs of an FBO. Lastly, a downloadable desktop application will also be added to allow for off-line use of the tool.

4 Challenges Ahead

With this simplified approach to developing a serious game for training on management of food incidents, it is anticipated that an engaging and effective training tool could be delivered while using minimal time and financial resources to both the developers and users. Nonetheless, as we are still in the initial stages of the research, we recognize several challenges ahead. First is on how the serious game is presented to the stakeholders, i.e., should we even call it a game rather than a tool? Involving stakeholders in a participatory exercise is a delicate process. Care must be taken in order to make the participants feel that the process they are engaged in is important to ensure their continued interest, support and participation. However, the term "game" has connotations that may affect the development process and how the end product is perceived. While some people in the food industry might view the "game" as innovative and therefore spark interest, others may view it as something that is frivolous and therefore not use it or not take it seriously (despite it being a "serious" game). Related to this is being able to strike a balance between the learning and the gaming elements. Without explicitly stating that this is a game, the gaming elements that should be designed for FIITT might be insignificant or overlooked by the stakeholders involved in the design process. At the moment, because we are not certain as to how the stakeholders will react to the term "game" or "serious game" we have presented FIITT as a tool. But because the FIITT's development is participatory, there is an opportunity to open these kinds of discussions with the potential users, which we intend to do. Finally, the level of realism of the food incident scenarios may be an issue with the users as the acceptance of these scenarios may differ from one user to another. These are some of the aspects of the serious game that will be discussed during succeeding stakeholder workshops.

5   Conclusion

The growth in the number of FBOs, especially small FBOs over the Internet, as well as the prohibitive cost of bespoke mock simulations, it is evident that there is a need for a more cost-effective way of delivering valuable training for managing food incidents that can reach a wider audience. A simple online serious game may be a cost effective way of reaching these FBOs and give them the necessary training that they need. Moreover, because it is a game, the entertainment value of the FIITT may increase the engagement of the users. By involving industry experts and stakeholders in the design process, this may ensure that the tool is fit-for-purpose.

References

1. Food Standards Agency: Food Incident Management Plan. London (2014), http://www.food.gov.uk/sites/default/files/FSA%20Incident%20Management%20Plan.pdf
2. Pourkomailian, M: Incident management: food safety requires competence. In: Food Safety Magazine (2013), http://www.foodsafetymagazine.com/magazine-archive1/december-2012january-2013/incident-management-food-safety-requires-competence/
3. Savanije, D.: Food safety summit: social media can destroy food brands, but you can stop it. Food Dive (2013), http://www.fooddive.com/news/food-safety-summit-social-media-can-destroy-food-brands-but-you-can-stop/126463/
4. Food Standards Agency: Principles for preventing and responding to food incidents. London. (2008), http://tna.europarchive.org/20110116113217/http://www.food.gov.uk/multimedia/pdfs/incidentsprinciples.pdf
5. Pallot, M., Le Marc, C., Richir, S., Schmidt, C., Mathieu, J.-P.: Innovation gaming: an immersive experience environment enabling co-creation. In: Cruz-Cunha, M.M. (ed.) Handbook of Research on Serious Games as Educational, Business and Research Tools. IGI Global, Hershey, PA (2012)

6. Correia, P., Carrasco, P.: Serious games for serious business: improving management process. In: Cruz-Cunha, M.M. (ed.) Handbook of Research on Serious Games as Educational, Business and Research Tools. IGI Global, Hershey, PA (2012)

7. Djaouti, D., Alvarez, J., Jessel, J.P., Rampnoux, O.: Origins of serious games. In: Minhua, M., Oikonomou, A., Jain, L.C. (eds.) Serious games and edutainment applications, pp. 25–43. Springer: Berlin (2011)

8. Di Loreto, I., Mora, S., Divitini, M.: Collaborative serious games for crisis management: an overview. In: IEEE 21st International Workshop on Enabling Technologies: Infrastructure for Collaborative Enterprises, pp. 352–357. Toulouse (2012)

9. Walker, W., Giddings, J., Armstrong, S.: Training and learning for crisis management using a virtual simulation/gaming environment. Cognit Tech Work **13**(3), 163–173 (2011)

10. Bhasin, K.: Gamification, game-based learning, serious games: any difference?. Learning Solutions Magazine. http://www.learningsolutionsmag.com/articles/1337/gamification-game-based-learning-serious-games-any-difference (2014)

11. Spinuzzi, C.: The methodology of participatory design. Techn Comm **52**(2), 163–174 (2005)

A Cybernetic Model of Macroeconomic Disequilibrium

Ernesto Carrella

Abstract In "On Keynesian Economics and the Economics of Keynes" (1968) Leijonhufvud described his cybernetic vision for macroeconomic microfoundations and the dynamics that move the economy from one equilibrium to another. Here I implement that vision in agent-based model.

I focus on the difference between price-led adjustments ("Marshallian" in the original text) and quantity-led adjustments ("Keynesian"). In a partial equilibrium model the two dynamics lead to the same equilibrium at the same speed through the same path.

In a general equilibrium model the Keynesian dynamics overshoots and undershoots with consequent over-employment and unemployment. The agents within the model act by trial and error using very limited knowledge which I simulate here through the use of PI controllers. When a demand shock is treated with increased labor flexibility or positive productivity shock the result is a deeper recession and longer disequilibrium.

Keywords Agent-based models • Economics • Flexibility • Macroeconomics • Market structure • Production • Cybernetics • PID control • Keynesian

1 Introduction

I extend here the Zero-Knowledge traders methodology [4] to macroeconomics. By modeling the economy as a process that agents try to control, I can study the effect that higher flexibility and adjustment speed have on the economy as a whole. More flexibility is not always beneficial as it can aggravate the disequilibrium and the social costs associated with a recession.

I use my model to study reforming the labor market during a recession. In Europe labor market reforms were touted both before [21] and after [3] the economic crisis as a way to boost economic growth. The point of reforming the labor market is to increase labor mobility and productivity. This paper challenges the notion that

E. Carrella (✉)
George Mason University, Fairfax, VA, USA
e-mail: ernesto.carrella@gmail.com

© Springer International Publishing AG 2017
W. Jager et al. (eds.), *Advances in Social Simulation 2015*, Advances in Intelligent Systems and Computing 528, DOI 10.1007/978-3-319-47253-9_4

increasing either is necessarily beneficial during a recession. I show how more mobility actually deepens the recession and output undershooting if it coincides with an exogenous drop in demand.

In Leijonhufvud's "Keynes and the Keynesians" the core difference between the Keynesian and Marshallian frameworks is how agents adapt to disequilibrium [14]. To Leijonhufvud, Marshallian agents react to mismatches in demand by first adjusting prices and only later changing production. Viceversa Keynesian agents react first by adjusting quantities and only later prices. I implement this idea and show how these differences have no effect in microeconomics but do so in macroeconomics.

This is not the first attempt to model the microfoundations of "Keynes and the Keynesians." Leijonhufvud himself cited the search models in "Information Costs, Pricing and Unemployment" [1] and Clower's false trades [6] as a way to achieve his vision. This chapter follows in the false trades tradition of allowing exchanges at wrong prices through a simple trial and error agent that corrects itself over time. It is the dynamics generated by this trial and error pricing that differentiate Keynesian disequilibrium from the Marshallian one.

2 Literature Review

I classify agents in macroeconomics on a spectrum that goes from complete feedback to complete feed-forward. Feedback agents are reactive, they manipulate control variables by inferring over time what their effect is on the other model variables. Feed-forward agents know perfectly the model and set all the control variables at the beginning of the simulation after having solved for the optimal path. These are sometimes referred as closed-loop (feedback) and open-loop (feed-forward) agents, respectively.

Modern economics focuses mostly on feed-forward agents. The Ramsey–Cass–Koopmans model [5, 12, 19] is an example of a pure feed-forward agent. In this model the agent is omniscient and the instant he's born he makes all the saving rates decisions for its infinite life by optimizing utility given the lifetime budget constraint. This omniscience is a fundamental driver of the model as it explains, for example, why permanent taxes do not crowd out investments while temporary taxes do [20].

In Prescott's Real Business Cycle growth model [18] the agent is an imperfect feed-forward control. In this model there are auto-regressive random technological shocks that cannot be predicted ahead of time. The agent then has a large feed-forward element that finds the optimal distribution of control strategies to implement and a small feedback process to choose the control strategies from this distribution as the random shocks occur.

Feed-forwarding optimization with feedback adaptation to uncertainty remains the standard macroeconomic approach to this day. The more uncertainty a model has, the larger the feedback element of the agent is but the focus is always on feed-forwarding. Learning models as in [8] are emblematic of this: agents manage model uncertainty by using statistics to learn the model so that they can then use the usual feed-forward control strategies on what they learned.

The only agent in economics that is still pure feedback is the central bank: the Taylor rule [23] is a simple feedback and adaptive rule to set interest rates. It is in fact a simplified PI controller [9]. Real economists allow simulated economists some slack in assuming not only that they face uncertainty but also that they are never able to reduce it by learning the full model.

The modern focus on feed-forwarding is surely a reaction to the feedback oriented methodology that preceded it. The Keynesian IS-LM [16] were almost pure feedback models. Consumption would be a fixed proportion of income, workers would be reacting the same fixed way to changes in prices and therefore could be fooled over and over again into generating a Philips curve [10]. Explicitly cybernetic models shared this top-down fixed feedback approach [7, 13, 17, 24]. Leijonhufvud called it the "Keynesian Revolution that didn't come off" [2] even though the approach survives in the field of system dynamics [22].

My agents are feedback only. The difference with the past is that my feedback mechanisms are there to allow a flexible agent to adapt to shocks rather than linking in a fixed way economic components as the IS-LM models wanted to do.

3 Microeconomics

3.1 Marshallian Agents

This is a brief summary of the Zero-Knowledge trader methodology. The unit of time is a "market day" as in Hicks [15]. Take a simple market for one type of good. Each market day, a firm produces y_t^s units of good, consumers buy y_t^d units at price p_t. The Marshallian firm is a price-maker that takes production as given and changes p_t every day in order to make production equal demand that is:

$$y^s = y^d \tag{1}$$

The agent has no knowledge of market demand and how his own price p_t affects it. It knows only that higher prices imply lower demand. It proceeds then by trial and error: it sets a price p_t and computes the error $e_t = y^s - y^d$ and uses it to set p_{t+1}. I simulate the trial and error process by a PI controller:

$$p_{t+1} = \alpha e_t + \beta \sum_{i=0}^{t} e_i \tag{2}$$

By manipulating p_t the Marshallian agent changes demand y^d until it equals supply y^s. Within each market day the Marshallian agent treats its own good supply as given but over time it can use the price it discovers to guide production. At the end of each day there is a small fixed probability (in this simulation $\frac{1}{20}$) to change supply y^s by adjusting labor hired. The decision is simple marginal optimization: increase production while MarginalBenefit > MarginalCosts and viceversa. The firm again adjusts by trial and error and with a separate PI controller whose error is $e_t = \frac{\text{MarginalBenefit}}{\text{MarginalCost}} - 1$.

In the original Zero-Knowledge paper I go through more complicated scenarios with multiple firms and markets, monopoly power, and learning. But this minimal setup is enough for this application. Here a firm has only two degrees of freedom: price set p and labor hired L. Each set by an independent PI controller.

3.2 Keynesian Agents

Keynesian firms function exactly as Marshallian ones except that they reverse the speed and the error of the two PI controllers. A Keynesian firm changes L every day trying to make $y^s = y^d$ and changes p with the small fixed probability trying to make Marginal Benefit = Marginal Cost.

Functionally the Keynesian firm tries to match supply and demand within a market day by changing y^s directly rather than changing p and therefore y^d as the Marshallian one.

3.3 In a Partial Equilibrium Scenario Keynesian and Marshallian Agents Perform Equally

There is one firm in the economy. It faces the exogenous daily linear demand

$$y_t^d = 100 - p_t \tag{3}$$

One person hired produces one unit of good a day:

$$y_t^s = L_t \tag{4}$$

There is infinite labor supply at $w = \$50$. The perfect competitive solution is $L = 50, y = 50$.

I run 1000 simulations for a Marshallian and a Keynesian firm each setting their PI parameters $\alpha, \beta \in [0.05, 0.2]$ and random initial price and labor $p_0, L_0 \in [1, 100]$. Firms in all simulations always find the equilibrium as shown in Fig. 1.

Define equilibrium day as the simulation day when the firm produces within 0.5 units of the equilibrium production. Figure 2 compares the equilibrium day distribution of Keynesian and Marshallian firms. A two-sided Kolmogorov–Smirnoff

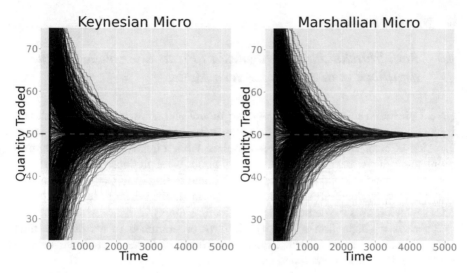

Fig. 1 The path of y traded for a 1000 Keynesian and Marshallian simulations. Regardless of initial conditions and PI parameters, the runs all reach equilibrium

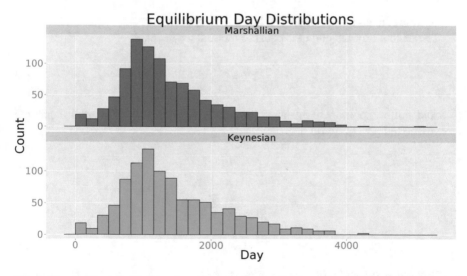

Fig. 2 The empirical distribution of equilibrium time for the Keynesian and Marshallian microeconomic simulations. There is no difference between them

test fails to reject that the two samples come from the same distribution (p-value is 0.263). In a partial equilibrium microeconomic scenario Keynesian and Marshallian firms perform equally well and at equal speed.

4 Macroeconomics

4.1 Both Marshallian and Keynesian Firms are Able to Reach Equilibrium in a Simple Macro Model

Here I present a minimal macroeconomic model and show how Marshallian and Keynesian dynamics diverge. The main reason they do is that Keynesian adjustment has side effects. Keynesian firms manipulate labor directly; in microeconomics hiring and firing workers affected how many goods were produced but in macroeconomics the good demand is equal to labor income so that hiring and firing workers affect how many goods are demanded as well. Marshallian firms instead manipulate prices which moves the demand without affecting supply.

There is a single firm in the world. It is programmed to act as in perfect competition and targets Marginal Benefits =*Marginal Costs*. It produces a single good with daily production function:

$$Y^S = a\sqrt{L} - b \tag{5}$$

It has access to an infinite supply of labor L at $w = 1$.

The demand for the output is equal to the real wages paid:

$$Y^D = \frac{L}{p} \tag{6}$$

Unsold output spoils, unused labor income is never saved. This market has the following unique equilibrium:

$$L = \frac{4b^2}{a^2} \tag{7}$$

$$p = \frac{2\sqrt{L}}{a} \tag{8}$$

$$y = b \tag{9}$$

When $a = 0.5$ and $b = 1$ the solution is

$$L = 16 \tag{10}$$

$$p = 16 \tag{11}$$

$$y = 1 \tag{12}$$

The computer simulation proceeds just like the previous microeconomic section except that demand here is endogenous and equal to wages paid. Two sample runs are shown in Fig. 3.

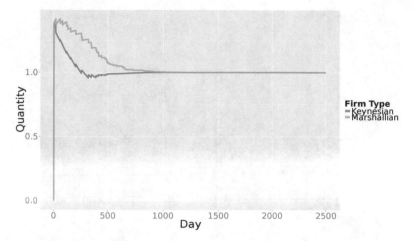

Fig. 3 Two sample runs of the economy Y with a Keynesian and a Marshallian firm

I run 100 simulations each for Keynesian and Marshallian firms, where the p and i parameters of the controllers are random $\sim U[0.05, 0.2]$. Both Keynesian and Marshallian firms are always able to achieve equilibrium.

4.2 Keynesian and Marshallian Firms Generate Very Different Dynamics When Reacting to a Demand Shock

As initial conditions matter, rather than studying the dynamics towards equilibrium ab ovo, I first let the model reach equilibrium then subject it to a demand shock and see how the firms differ in adapting to it. I run the same simulation as before, but after 10,000 days the output demand is shocked by $s\$$:

$$Y = \frac{L}{p} - s \tag{13}$$

When $s = 0.2$ the new equilibrium becomes

$$L = 10.24 \tag{14}$$

$$p = 12.8 \tag{15}$$

$$Y = 0.6 \tag{16}$$

Figure 4 shows the difference in adjustment dynamics between Keynesian and Marshallian firms. Marshallian firms react to the sudden drop in demand by lowering price so that quantity traded briefly recovers after the shock. Eventually though the

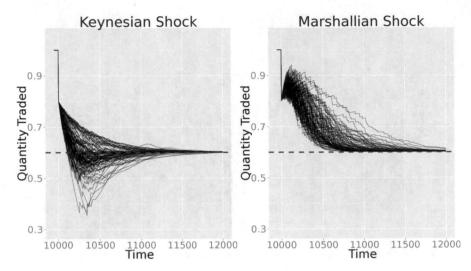

Fig. 4 A comparison between the adjustment dynamics after a demand shock of Keynesian and Marshallian firms. The Keynesian runs often undershoot and have larger output contractions than the same Marshallian firms in spite of the pre-shock and after-shock equilibria being the same

lower prices feed into the profit maximization PI which cuts production towards the new equilibrium. Keynesian firms instead react to the drop in demand by immediately firing workers. While firing workers lower supply it also decreases demand because unemployed workers don't consume. The Keynesian firm can't change supply without changing demand as well.

Keynesian firms reach the new equilibrium faster. Define equilibrium time as after how many days the output settles within 0.05 of equilibrium. Average equilibrium time is 570.2 days for a Keynesian firm and 808.37 days for a Marshallian one (which is a statistically significant difference). Moreover Keynesian firms tend to stay closer to equilibrium overall. To see this define deviation of output y from equilibrium y^* as:

$$\log(t) * (y_t - y^*)^2 \tag{17}$$

Then the average deviation for Keynesian economy is 4.076 while it is 20.971 in the Marshallian economy. Figure 5 shows the difference. On the other hand, output drops 10 % or more below the new equilibrium in 29 Keynesian runs out of 100. Marshallian firms never undershoot.

Keynesian adjustment is less efficient and creates larger social losses in spite of reaching equilibrium faster. In Fig. 6 I compare firm profits and labor income during disequilibrium versus what they would be if the adjustment was immediate. Labor income is higher in the Marshallian world (on average 2010.957$ per run compared to 1024.894$ in the Keynesian world). This is because the disequilibrium involves firing unnecessary workers and the longer it takes the more the workers benefit from

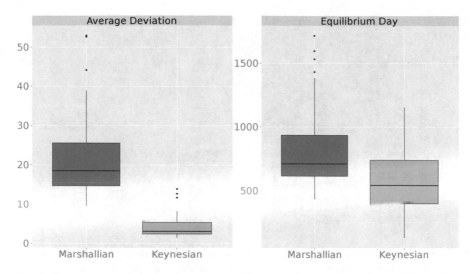

Fig. 5 Box-plot comparison of deviation and equilibrium day between Keynesian and Marshallian macro

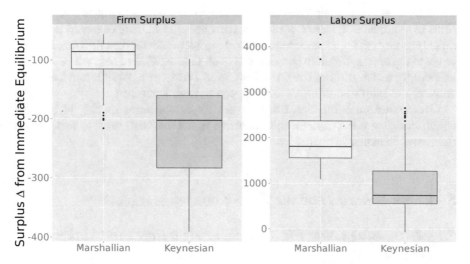

Fig. 6 The difference in surpluses between Marshallian and Keynesian firms. The surplus is measured as a difference in $ (or wage units) compared to what it would be if it moved immediately to the new equilibrium

the disequilibrium. What is less obvious is that the Marshallian firm is also better off than the Keynesian one as Fig. 6 shows.

The reason Marshallian firms can over-produce for longer and still make consistently less losses than Keynesian firms is that Marshallian disequilibrium dynamics are less wasteful. To see this focus on market day equilibria, each day the difference between what is produced and what is sold is wasted. Figure 7 shows

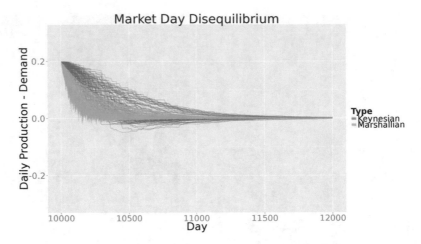

Fig. 7 The difference between what is produced and what is sold each day, regardless of what the profit-maximizing equilibrium is. The larger the deviation from 0 the more the waste

the daily waste and in particular how it is larger with Keynesian firms. Keynesian firms over-produce and waste because of their inability to match demand to supply quickly as any cut in production cuts demand as well. Marshallian firm takes longer to get to the new equilibrium but proceeds over a more efficient path where demand and supply match most of the time. Keynesian firms get to equilibrium faster but demand and supply never match until the equilibrium is reached.

Overproduction is the signal that pushes Keynesian firms quickly to the new equilibrium, but it is a wasteful and expensive signal that costs more to society than the slower Marshallian alternative.

5   Labor Reforms and the Zero-Knowledge Agents

5.1   *Increasing Labor Flexibility During a Recession Makes it Worse*

In this section I model the world as Keynesian. I do so because price rigidities are a well-established empirical fact [11]. It is also advantageous to model labor market reforms and speed in the Keynesian world since the PID controlling production targets (and therefore labor) are not sticky.

I model labor flexibility in two ways. First, increasing flexibility may mean faster hiring and firing. I can replicate this in the model by increasing the parameters of the PI controlling the workforce so that it adjusts more aggressively. Alternatively increasing flexibility may mean increasing the productivity of labor. I can replicate this in the model by increasing the a parameter of the production function.

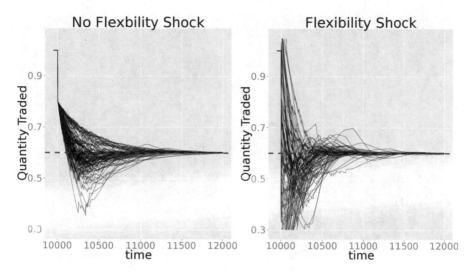

Fig. 8 Hundred Keynesian runs as in Fig. 4 and the same runs where concurrent to the demand shock we double the PI labor parameters. Overshooting becomes more likely and deeper. Ten runs fail to reach equilibrium when their flexibility is increased

Assume the world is Keynesian. Assume the same shock to demand as the previous section. Here I simulate what happens if concurrent to the demand shock there is a flexibility shock to the firm where its PI parameters double. I compare the same simulation with the same random seed with and without the flexibility shock. Notice that the economic equilibria has not changed, the difference can only be in dynamics.

Figure 8 shows the effect of increasing flexibility together with the demand shock. Higher flexibility results in higher chance of overshooting, 88 runs out of 100 have output dropping more than 10 % below equilibrium (compared to 29 without flexibility shock). Moreover in 10 runs the overshooting is so severe that the run ends on $Y = 0$ (which is a steady state) and never reaches the equilibrium. Figure 9 shows that the deviation from equilibrium is higher with higher flexibility (because of the severity of the overshooting) while there is no statistical significant difference in equilibrium time.

Figure 10 shows how labor surplus is lower when there is a flexibility shock. Overshooting is so severe that on average the labor surplus is negative. Note first that labor surplus was positive in the previous section: because the new equilibrium requires fewer workers and the agent takes time to get to the new equilibrium point some workers that should have been fired instantly profited from the disequilibrium. Higher flexibility fires workers faster, which reduces benefits from disequilibrium and when overshooting it fires too many so that labor overall is hurt by the disequilibrium rather than profiting from it. Firm surplus is higher with more flexibility; the difference in means is statistically significant.

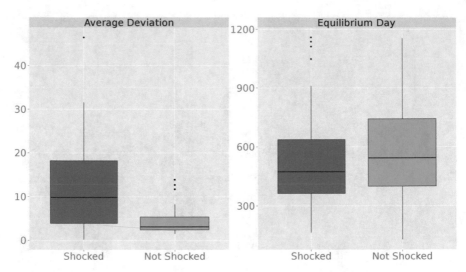

Fig. 9 Comparison between the Keynesian equilibrium metrics with and without flexibility shock

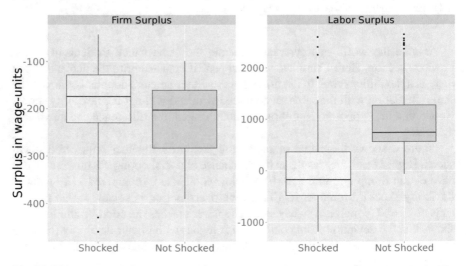

Fig. 10 Box-plot of surplus differences between runs with and without flexibility shock. The values are $ (or equivalently wage-units) differences between surpluses and what would the surplus be if the system immediately moved to the new equilibrium

More generally, what the right labor flexibility is in terms of speed is a tuning problem. What we want are the controller parameters that move the economy to the new equilibrium as fast as possible while minimizing overshooting. This is an empirical question and the answer depends on the kind of original equilibrium, production function, shock, and every other parameter. It is not the case that more flexibility and speed always make for a better economy.

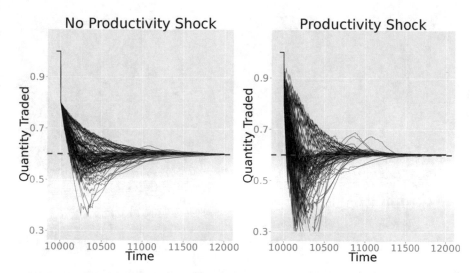

Fig. 11 The dynamics of 100 Keynesian simulations with paired random seeds and how they deal with demand shock with and without productivity shock

Turn to flexibility as an alias for productivity, assume again a Keynesian world. Concurrent with a demand shock the productivity a increases from 0.5 to 0.6. This changes the equilibrium L and p but not optimal output Y:

$$L = 7.11 \tag{18}$$

$$p = 8.88 \tag{19}$$

$$Y = 0.6 \tag{20}$$

Again I run 100 simulations with and without productivity shock, keeping fixed random seeds for comparison. In this case the only change is in the new equilibrium conditions, PI controllers are invariate. Figure 11 compares the two dynamics.

Increasing productivity makes the approach to equilibrium worse as shown in Fig. 12. More runs undershoot, 68 out of 100, and output deviation from equilibrium is higher with no improvement in equilibrium time. As shown in Fig. 13 there are no meaningful differences between in disequilibrium surpluses. Notice however that productivity shocks change the equilibrium the model coverges to which makes comparisons less clear.

While improving productivity is always a good long term policy there is no validation from this model that raising it makes disequilibrium any better.

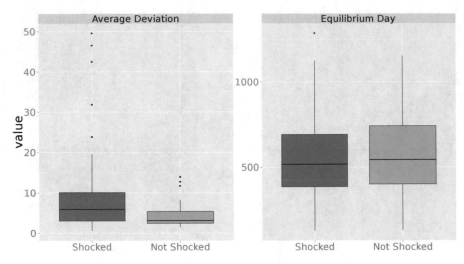

Fig. 12 Comparison between the Keynesian equilibrium metrics with and without productivity shock

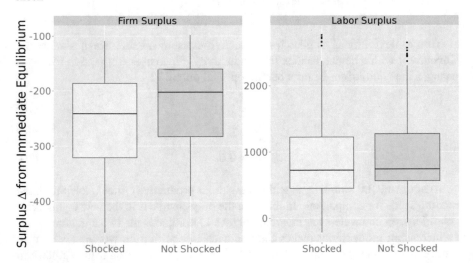

Fig. 13 Box-plot of surplus differences between runs with and without productivity shock. Notice that the productivity shock changes the equilibrium p and l so that the two classes of surpluses are compared to two different optimal points

6   Conclusion

To highlight disequilibrium dynamics this model was made very simple. Some of the assumptions present should be removed in future work. The first large assumption I made is infinite fixed wage labor supply. I did so here in order to simplify the decision process of the firm; in this model the firm only sets one price (output) and one production target. Had I added wages it would have made it impossible to compare Marshallian and Keynesian dynamics since there would be two prices to set concurrently. In that circumstance Marshallian firms would be quicker since they would set p and w quickly and target L slowly while Keynesian would have to set L quickly while p and w slowly.

One could argue that we can still use fixed wages around the equilibrium and salvage the shock comparisons in Sect. 4.2 by either assuming efficiency wages or some form of downward rigid wages as Modigliani's IS-LM [16]. But these would have to be microfounded rather than just assumed.

The second large assumption is the lack of utility microfoundations. If the consumer has a lexicographic utility where it prefers a world with no waste (that is demand equals supply) and splits ties according to the world that produces the most, then the simulation would be utility maximizing. But this is non-standard utility formulation and general results must not depend on these. I also gave no explanation for the demand shock.

The third limitation of the chapter is the lack of agents. Previous papers on the methodology had multiple firms competing with one another but here there is a single firm taking all the decisions. This was primarily to avoid any noise in the simulation except those caused by demand shocks. The same weakness is present regarding consumers and workers. A single force supplies labor and consumes wages; there are no distribution effects and no asymmetric cost to unemployment. All these assumptions are, I believe, minor. They are employed to remove noise from the model and further highlight the difference between Keynesian and Marshallian firms.

I believe this paper's result is timely. I show how increases in productivity and labor flexibility during a recession while improving the final economic equilibrium worsen the path the economy takes towards it. This kind of results can only be achieved through agent-based economics and simulation. This kind of results can only be achieved by focusing on disequilibrium. And these results are needed to chart a complete policy response to economic crises.

References

1. Alchian, A.A.: Information costs, pricing, and resource unemployment. Econ. Inq. **7**(2), 109–128 (1969)
2. Aoki, M., Leijonhufvud, A.: Cybernetics and macroeconomics: a comment. Econ. Inq. **14**(2), 251–258 (1976)
3. Bertola, G.: Labor market policies and European crises. IZA J. Labor Policy **3**(1), 1–11 (2014)
4. Carrella, E.: Zero-knowledge traders. J. Artif. Soc. Soc. Simul. **17**(3), 4 (2014)
5. Cass, D.: Optimum growth in an aggregative model of capital accumulation. Rev. Econ. Stud. **32**(3), 233–240 (1965)
6. Clower, R.W.: The keynesian counterrevolution: a theoretical appraisal. Theory Interest Rates **103**, 125 (1965)
7. Cochrane, J.L., Graham, J.A.: Cybernetics and macroeconomics. Econ. Inq. **14**(2), 241–250 (1976)

8. Evans, G.W., Honkapohja, S.: Learning and macroeconomics. Annu. Rev. Econ. **1**(1), 421–449 (2009)
9. Hawkins, R.J., Speakes, J.K., Hamilton, D.E.: Monetary policy and PID control. J. Econ. Interact. Coord. **10**, 1–15 (2014)
10. Heijdra, B.J., Van der Ploeg, F.: The Foundations of Modern Macroeconomics, 780 p, 1st edn. OUP, Oxford, New York (2002)
11. Klenow, P.J., Malin, B.A.: Microeconomic evidence on pricesetting. Working Paper No. 15826. National Bureau of Economic Research (2010)
12. Koopmans, T.: On the concept of optimal economic growth. Cowles Foundation Discussion Paper No. 163A. Cowles Foundation for Research in Economics, Yale University (1963)
13. Lange, O.: Introduction to Economic Cybernetics. Elsevier, New York (1970)
14. Leijonhufvud, A.: On Keynesian Economics and the Economics of Keynes: A Study in Monetary Theory. Oxford University Press, Oxford (1972)
15. Leijonhufvud, A.: Hicks on time and money. Oxf. Econ. Pap. **36**, 26–46 (1984). ArticleType: research-article/Issue Title: Supplement: Economic Theory and Hicksian Themes/Full publication date: November 1984/Copyright I' 1984 Oxford University Press (1984)
16. Modigliani, F.: Liquidity preference and the theory of interest and money. Econometrica **12**(1), 45–88 (1944)
17. Phillips, A.W.: Arnold Tustin's the mechanism of economic systems: a review. In: Phillips, A.W.H. (ed.) A Collected Works in Contemporary Perspective. Cambridge University Press, Cambridge (2000)
18. Prescott, E.C.: Theory ahead of business-cycle measurement. Carn.-Roch. Conf. Ser. Public Policy **25**, 11–44 (1986)
19. Ramsey, F.P.: A mathematical theory of saving. Econ. J. **38**(152), 543–559 (1928)
20. Romer, D.: Advanced Macroeconomics. McGraw-Hill/Irwin, Boston (2011)
21. Siebert, H.: Labor market rigidities: at the root of unemployment in Europe. J. Econ. Perspect. **11**(3), 37–54 (1997)
22. Sterman, J.: Business Dynamics: Systems Thinking and Modeling for a Complex World. McGraw-Hill/Irwin, Boston (2000)
23. Taylor, J.B.: Discretion versus policy rules in practice. Carn.-Roch. Conf. Ser. Public Policy **39**, 195–214 (1993)
24. Tustin, A.: The Mechanism of Economic Systems: An Approach to the Problem of Economic Stabilization from the Point of View of Control-System Engineering. Heinemann, Portsmouth (1957)

How Should Agent-Based Modelling Engage With Historical Processes?

Edmund Chattoe-Brown and Simone Gabbriellini

Abstract This chapter consists of two main parts. After an introduction, the first part briefly considers the way that historical processes have been represented in ABM to date. This makes it possible to draw more general conclusions about the limitations of ABM in dealing with distinctively historical (as opposed to merely dynamic) processes. The second part of the chapter presents a very simple ABM in which three such distinctively historical processes are analysed. These are the possible significance of unique individuals—the so-called Great Men, the invention and spread of social innovations from specific points in time and the creation of persistent social structures (also from specific points in time). The object of the chapter is to advance the potential applicability of ABM to historical events as understood by historians (rather than anthropologists or practitioners of ABM.)

Keywords Agent-based modelling • Social networks • Trade networks • Historical explanation • Book trade • Methodology • Social change • Innovation • Institutional evolution

1 Introduction

"The assumptions you don't realise you are making are the ones that will do you in."
(Trad. Anon.)

Like all disciplines, ABM has its particular perspectives and implicit assumptions. These may remain undiscovered while it keeps "safely" to certain known areas of research but may show up very clearly when it attempts to broaden its application. This chapter is a case study of that situation with respect to the study

E. Chattoe-Brown (✉)
Department of Sociology, University of Leicester, University Road, Leicester, Leicestershire, LE1 7RH, UK
e-mail: ecb18@le.ac.uk

S. Gabbriellini
Gemass, CNRS, Paris-Sorbonne, Paris, France
e-mail: simone.gabbriellini@cnrs.fr

© Springer International Publishing AG 2017
W. Jager et al. (eds.), *Advances in Social Simulation 2015*, Advances in Intelligent Systems and Computing 528, DOI 10.1007/978-3-319-47253-9_5

of history[1]. It arose from discussions with historians in the context of the evolution
of networks in the book trade but set us thinking about wider issues connected with
distinctively historical explanation. The chapter has a very simple structure. After
this introduction, which very briefly considers what we might mean by historical
analysis, the first part carries out a literature review of ABM applied to historical
topics or published in historical outlets. It uses this as a sample (small but nearly
complete) to induce general claims about the present "historical" uses of ABM.
These general claims can then be examined and challenged. The second part of
the chapter addresses these challenges and uses a simple ABM to show several
phenomena that can be seen as distinctively historical but have not been found in
historical ABM to date. As shown in the conclusion, this ABM is thus a jumping off
point for a possible dialogue between history and ABM about "historical models"
that may have mutual benefits[2].

Before analysing existing historical models in ABM and suggesting ways to
make them "more historical" it is necessary to say something about what history
might be. Understandably huge amounts of ink has been spilt on this topic by
historians and others so it is very important to avoid issues that ABM researchers
will probably never be qualified to comment on (like "how should history be done?")
Nonetheless, it is relatively straightforward to identify some assumptions without
which it is hard to see how a process could be defined as "historical". The first
is a changing population of agents. This is meat and drink to ABM but many
social science studies effectively work with fixed populations. Certainly over several
centuries, the kind of time scale we think of as most "typically" historical, the
entire population will inevitably have changed. The second assumption, related
to the first, is that history is the sort of time scale over which *ceteris paribus*
assumptions do *not* apply. We can see this by contrasting history with the sort of
cross sectional statistical analysis often carried out in sociology (for example). We
might find a particular association between class origins and educational success
in 1980. It might not be unreasonable to say, comparing that with the relationship
found in 1975, that the institutional framework (for example the structure of the
school system) was "more or less" unchanged and therefore that changes in the
parameters of the model reflect changes in society (more equality of opportunity
in the educational system for example). However, it is very clear that such a
comparison with *1780* would be more or less meaningless because the education
system would be so radically different (as would the logic of the entire class
structure). History is thus the time period over which it is *not* reasonable to take
social practices and institutions (or other social structures) as a fixed background

---

[1]Such challenges can be productive if they force ABM to reconsider its implicit assumptions
and broaden its scope. Conversely, a dogmatic reaction to such challenges is likely to harm the
likelihood of ABM being accepted more widely.

[2]Traditionally, history has been quite wary of "theory", let alone "models". Arguably history faces,
to an even greater degree, the sort of challenges that have made social scientists wary of modelling
based on statistics and equations. Some of these concerns will become clearer in the course of this
chapter.

to the wider social process. In fact, looking at existing non-historical research, it is not really clear how long that period might be and it is therefore important that ABM can cope with "historical time" if required. Clearly things do not remain constant over centuries but actually, for example, there was pretty radical change in the UK education system under Margaret Thatcher and while statisticians may choose to *treat* 5 years or so as a non historical period, whether that assumption is unreasonable (and more importantly whether their methods can *reveal* it to be unreasonable) is another matter. The final assumption that is worth noting (although it will not be relevant until *empirical* historical ABM are developed) is that the supply of data for historical analysis is fixed (even if some of it still remains to be discovered). Unlike much social science where, if data is missing it can be collected, the historical record is now fixed. These brief considerations set the scene for a consideration of existing "historical" ABM and how "more" historical variants of ABM might be developed.

2 ABM and "History": A Very Brief Analytical Review

Although examples of equation based theorising can be found on historical topics [1–3], these are relatively rare and will not be discussed further here. ABM approaches to history are a little more common (despite the shorter effective lifetime of these techniques) but their increased frequency seems to reflect a broader conception of history rather than a greater interest in the approach.[3] In particular, it seems that most apparently historical ABM actually involve what we might call an anthropological conception of history [6] in which agents behave according to fixed social rules that are simply played out over time. The clearest examples of these are research that is explicitly anthropological or archaeological [7–11]. However, other examples of ABM with this anthropological conception of history are those that (while dealing with clearly historical events) do so over sufficiently short time periods that behaviours and structures are also assumed to be constant [12, 13]. Formal theories involving genuine structuration (agents create structure which then acts on them) appear to be absent.[4]

---

[3]We will also exclude from this discussion research advocating ABM but based only on the work of others, for example [4, 5]. Instead we will discuss these examples directly. However, we will not discuss separately multiple publications on the same project. Also, for practical reasons, and without prejudice to its quality, we shall not discuss non-peer-reviewed work including research by Kerstin Kowarik and colleagues on prehistoric salt mining, Jim Doran on insurgency and Giuseppe Trautteur and Raniero Virgilio on the Battle of Trafalgar. More information on this work and actual copies of draft papers may be found on the Web but these sources are not academically robust. The authors would be glad to hear of any work (unpublished or not) that contradicts our hypotheses here so we can refine our analysis.

[4]The emphasis of this chapter is on publications by historians and/or those in historical outlets. However, the same appears to be broadly true of "historical" research (generously conceived) in ABM journals. For example, none of the models in Mithen's excellent chapter [14] appear to

Two apparent counter-examples to this claim prove not to be on closer examination. The first are what we might call possible worlds ABM. In this case, the ABM can be used to examine counterfactual cases ("suppose the flu had struck at a different time of year"—see [13] for example)[5]. This might give an impression of changing rules but the impression is mistaken because in each simulation run these are fixed and the results are compared across runs (a process that has no real historical equivalent[6]). What is not happening in these ABM is what appears to happen in history, namely that individual actions give rise to structures (for example institutions like "The Metropolitan Police") that subsequently have effects of their own. The other set of examples are referred to as "history-friendly" ABM [19]. In fact, this doesn't mean friendly to historical analysis but making use of stylised facts about history in designing ABM that are still arguably a-historical. While it is true that firms arrive and depart from markets and new drugs are created in a stylised form (for example), there is no sign of changes in regulatory or market structure. As before, in practice, the same rules are in operation at the end of the simulation run as at the beginning even though, at different times, they may give rise to more or less numerous firms or more or less novel drugs[7].

The most sophisticated case of a historical ABM we have been able to find is an article by Ewert et al. [21]. In investigating starvation, this represents both citizens and councillors as interacting agents to explore the effect of policy on hunger. However this article still involves an anthropological conception of history in that the rules of citizens and council seem to be fixed over the simulation run and merely played out over time. Even though this ABM does involve an explicit representation of a separate policy agency, it still doesn't allow for things like citizens banding together and overthrowing the council or the council reflecting on its past failings and innovating radically in its policy-making.

---

involve social innovation. [15] is a rare debatable counter example to our general claim in that, although cognitive complexity can increase (allowing for the evolution of hierarchy for example) the environment appears to remain fixed

[5]The status of counterfactuals in history is strongly contested to the point where it cannot plausibly be resolved here. Although there has been a lot of research based on this idea [16], it the approach has also been vigorously criticised not just in its details but also as a legitimately rigorous approach [17, 18].

[6]Again, we wish to avoid judgements about what history is. If one were asking "What caused WWI?" then the events that actually occurred would be all we could analyse. If, on the other hand, one were asking "What causes wars?" then one could construct a sample for study comparing different circumstances when war did and didn't occur.

[7]Another arguable case would be adaptive systems based on evolutionary algorithms [20]. On one hand, a firm might start with a "practice" (represented as a genetic programming tree perhaps) that involved calculating a price based on cost and "end up" simply following the price of a dominant firm (a very different kind of practice). On the other, the genetic programming grammar and the so-called "terminals" of the system (the variables on which firms calculate: for example own unit cost or price charged by another firm) remain fixed throughout a simulation run. It would seem to be an open question whether the "potential" novelty enshrined in a formal grammar is adequate to capture the novelty found in the real world. I would guess not.

These examples suggest that ABM appears to have blind spots about what makes history distinctive. Institutions and social practices both change over historical time. It is for this reason that the case study in the next section explicitly attempts to represent genuine social innovation and the development of institutions albeit in a stylised form.

3 A Case Study of "Historical" Network Dynamics

In this section we present a case study of an ABM representing the evolution of social networks and some variants of potential historical relevance. Before doing this, however, it may be helpful to provide a brief introduction to the idea of social networks.

Social networks are structures formally defined by two elements: Nodes and relations. The generality of the approach arises because both nodes and relations (often called ties) can be many different kinds of things. For example, nodes could be countries and relations the volume of trade that flows between them each year, nodes could be pupils in a class and relations could be how much they like to work (or play) together or nodes could be elements of the book trade (publishers, authors, dealers and customers) and the relations could be the buying and selling of books. The interest in this approach is the associations that can be found between network structure and other phenomena of social scientific interest, for example that less well-connected pupils in a classroom might also be the weakest academically[8]. Thus social networks enable us to understand how one conception of social structure (who is connected to who by various relations) may affect social processes and outcomes. To take a book-dealing example (the area of history that inspired this chapter), we can imagine a situation where every British county has always had (or at least ever since there was any market for books at all) a few people who were willing (and able) to buy books even if they have to come from a very distant city. These people can then have a network influence on others in the neighbourhood who they know or know them (perhaps by lending books, by a desire for social emulation or through education) to create a demand for books. This demand (though obviously coupled with non network factors like the general growth of literacy, the expansion of transport systems and so on) can reach the point where a local town can now support its own book dealer. This not only draws in more local customers who could not afford or would not countenance ordering books from London (thus increasing local demand further) but may also change the nature of the institutional distribution network. (Instead of London dealers selling direct to

---

[8]This can be contrasted with a "statistical" view often found in sociology that individual attributes are caused by (or at least associated with) other individual attributes. According to this view we should look for differences in educational attainment in terms of gender, ethnicity, class origin and so on rather than network position.

provincial customers, publishers may now start to deal directly with a growing circuit of provincial bookshops. This in turn may facilitate the sustainability of bookshops in other towns along their dealing routes.)

To gain a better understanding of what might be needed for a "more historical", ABM let us consider an evolving network where the nodes and relations are not supposed to represent any particular thing (though there are a number of historically interesting things—like book dealing networks—which they *might* represent). All we assume about the application of the ABM at this stage is that these nodes represent individuals (rather than organisations or nations) and that they have reasons to change over time.

Because ABM and the social process specifications that underlie them are intrinsically dynamic, we immediately face an interesting example of what history almost certainly isn't. Whatever processes we believe underlie the dynamics of social networks, it seems unlikely that history means nothing more than any event unfolding over time. Under that definition the network that forms between college students living in the same accommodation block over a month would be no more or less historical than the banking networks that grew up across Europe over centuries. Nor does it help to declare arbitrarily that things happening over months aren't historical while things happening over centuries are. Presumably what makes something historical is the kind of things that happen during the period of analysis (as suggested in the introduction). It is establishing what these kinds of things might be that underpins the argument of this chapter about a profitable debate between historians and ABM researchers.

To take an example, if nothing else, something that is certain to change over historical periods is the population. People will be born and die. History is obliged to deal with these demographic phenomena in a way that short term sociological studies often are not. Interestingly, existing (mathematical) SNA struggles with networks where the population of nodes changes and, perhaps surprisingly, often solves this problem by just eliminating such cases from analysis (see [22], p. 138). By contrast, the approach used in ABM is much more intuitive for this aspect of history. New agents appear in the world at birth (and can then form network ties) while existing agents die and disappear (and their network ties vanish.)

The baseline ABM for this case study is one in which (starting with an initial population of 500 agents) one new agent has a 0.3 chance to be born and a randomly chosen existing agent has a 0.3 chance to die in each time period. (This means that the population of agents who can form ties is constantly changing but on average the population remains reasonably stable. New agents start with no ties and when an agent dies all its ties are broken.) Each agent is assumed to have a fixed maximum capacity for close social relations (which can vary between 1 and 6)[9]. In each

---

[9]This assumption is empirically supported [23] and probably reflects the effort involved in maintaining close relations as well as the increasing pleasure to be derived from them. People can have millions of Facebook "friends" at the click of a mouse but usually have only a handful of really close non-family relationships that may have taken years to build.

time period one agent is selected at random and (with a small probability for each potential tie they have not yet made) tries to make that tie at random with another agent that also has capacity. If no such agent exists then these friendship attempts fail, at least in the current time period. (Because agents have capacity for few ties and the probability for making each potential tie is small, the upshot of this is that mostly, the chosen agent makes no new friendship tie in a time period, sometimes just one but almost never more than one. This reflects the fact that close friendships form rarely but are long lasting.)

As already suggested, the lack of realism in these assumptions hardly needs remarking. Nonetheless, even this arbitrary ABM does capture some key points about social networks. Firstly, most people have very few close ties. Secondly, there are reasons why networks change regardless of personal inclination (in this case just birth and death but in more complex ABM also migration, differential contact opportunities and so on[10].)

In fact, this baseline model is just a version of the Giant Component[11] model found in the NetLogo models library but without immortal agents[12]. Since the "result" of the standard GC model is the perhaps slightly counter-intuitive formation (fairly rapidly) of the GC, it is perhaps not surprising that what the "demographic" variant of this model produces is relatively stable component size distributions even though specific components are created and destroyed by the birth and death of crucial[13] agents[14].

In fact, this is just what we find. In our baseline ABM, the long-term behaviour that emerges from the system is a reasonably stable set of components (one fairly large and the others much smaller). We averaged the size of the giant component (expressed as a fraction of the population to control for demographic effects) for 4000 time periods (because the giant component size is quite noisy owing to births and deaths). This was done at two different points in the simulation run (time periods 12000–16000 and 20000–24000) for reasons that will be explained shortly. This exercise was repeated for ten different simulation runs so variation between

---

[10]Even this very simple ABM also allows for more subtle social process effects. For example, someone may want very much to make friends but be unable do so at one point in time. However, at another point, they may be in a position to make several friends rapidly. This mirrors the fact that although our number of friends may remain fairly constant, their actual identities often change when we undergo a major life event (like starting university or becoming a parent.) This phenomenon of "friendship churn" is empirically important but barely visible in formal SNA as far as we have discovered.

[11]The largest component in a network is called the Giant Component (hereafter GC).

[12]The simulations here were written in NetLogo [24]. The Giant Component Model code was also developed by Uri Wilensky [25] and has been further extended for research purposes with his permission.

[13]In this instance an agent is crucial if, by dying, it breaks a component into two smaller ones. An agent need not be crucial if there are multiple routes between other agents in the component (because components are defined just in terms of connectivity and not distance.)

[14]To use another example from book trading, dealers would bequeath their stock on death (and it might then be sold by legatees) so bookshops might persist with different owners/operators.

runs could be assessed. The result was that there was no significant difference between the average giant component sizes for each measurement period (0.65523 vs. 0.65502). The small difference that was observed was dwarfed by the variation in giant component sizes in different simulation runs (which was nonetheless quite small) suggesting that the simulation had reached a steady state with respect to the changing population when the measurements were taken. This case thus serves as a robust baseline for variants to the ABM presented shortly.

This process draws attention to another interesting advantage of ABM from the perspective of history. This is the ability to repeat simulation runs differing only in the instances of random numbers involved. This allows us to think about counterfactuals. What can we make of statements like "Had Hitler not risen to power, the disastrous state of Germany would have made it almost inevitable that some other demagogue would have done so." In repeated runs of a simulation we can say, for example, that although persistent completely connected networks are observed (in 2 out of 1000 runs), it is very much more likely that two or three stable components will be observed (in 990 out of 1000 runs[15].) Obviously, history is always what we have actually found happening but we cannot know how unlikely the observed case was relative to other conceivable outcomes. ABM at least allows us to talk about that state of affairs in a meaningful manner.

In the previous section, we sharpened the discussion about the role of ABM by suggesting why we might not consider just any dynamic process historical. Most historical theories seem to involve a situation where social practices and/or social structure change over the course of the analysis. We have already considered what happens when that change just involves the population (which already rules out a surprising amount of sociology and traditional SNA for example). This provides our baseline ABM. In the rest of this section, we consider three other things of potential historical relevance that might change.

The first, which is a slightly odd example (in that it is almost a part of the history of history), is a stylised attempt to incorporate Great Men (obviously, in fact, Great Persons but hereafter GM) into the ABM. The GM debate (between the views of Carlyle and Spencer) may seem on the face of it dated and perhaps even pointless [26]. Because of the problem with counterfactuals, can we really say meaningfully that without Napoleon or Emmaline Pankhurst a particular outcome (World War I, votes for women) would never have occurred?[16] Perhaps not, but we may nonetheless be able to make better sense of this debate using an ABM. In fact, of course (like all important debates), this one echoes very deep matters, in particular the role of individual agency as against "social forces" or the net effects of many individuals doing what they do habitually or without reflection. Given the ships, the

---

[15] And in all simulation runs for each variant ABM, we found no qualitative variations in behaviour (there was never an occasion when the GC size remained higher after the death of a unique individual for example) allowing us to be provisionally confident in the outcomes reported.

[16] And if not "never", then how long does it take before the event is no longer "the same?" Could the rise of dictator in Germany in 1990 still be "counted" as the inevitable working out of an alternate history where Hitler did *not* come to power?

institutions and the sailors, could anyone have won Trafalgar or did it take someone with the arguably unique calibre of Nelson? (Assuming that his calibre is not simply deduced from winning Trafalgar!) Are leaders simply figureheads for events which (for much larger reasons) would almost certainly have occurred at around that time anyway? (Hopefully nobody is going to appoint a complete idiot to be in charge of a fleet during a period of major naval engagements but maybe all victory took was actually any one of a number of recognisably competent admirals.) In a historical context it is quite hard even to make sense of such questions let alone answer them. In an ABM, by contrast, it is much easier. Recall that, so far, we have assumed that all agents have a fairly small maximum capacity for forming ties (1–6). Suppose at a particular moment in time we create an agent with a much larger tie capacity (100–1000). For the purposes of this argument, the ABM is simplified in that this is the *only* way in which the GM differs from other agents. "He" has the same potential life span, is not treated any differently by others and makes his ties by exactly the same decision process. (Although because he has much more tie capacity, he actually forms many more ties in a single time period on average.) Following the same measurement process as before, we compare the average size of the GC for 4000 time periods before the birth of the GM and 4000 time periods after his death. Again, we find no significant difference between the average giant component size in the two periods with the whole simulation being run ten times (0.64518 and 0.64769). What we observe (with images of the simulation runs available on request from the authors for reasons of space and to avoid colour reproduction) is that the GC increases dramatically in size during the lifetime of the GM but then immediately reverts to its previous size on his death. In retrospect, it is easy to understand what has happened here. By making many friends at once in several time periods, the GM immediately makes himself very central to the network and at the same time makes it very vulnerable to his death. Although all the other agents continue to make ties randomly, these are not directed at sustaining the large component created by the central GM so his death always causes it to fragment.

Now in a sense this is not obviously interesting from a historical perspective. I doubt anyone would claim that the world is no different with kings than without them or that people of exceptional abilities have no impact even when they are exercising them. But what is much more interesting is what this stylised ABM reveals about the nature of the GM debate. It appears that it is not enough for there just to be a GM, he also has to leave something tangible behind[17]. What makes the GM great, as far as history goes, is not just that he won the battle but that the battle resulted in a treaty that changed who owned which colonies subsequently (in other words some institutional or wider behavioural change). Historical studies are thus

---

[17] In this case, the GM sometimes visibly disrupts the robustness of the network afterwards although it gradually recovers to the previous level by "social forces" (the habitual actions of the mass). We could tentatively link this to the succession problem in politics [27]. A crucial role in a network is not beneficial for society unless you can pass it on to someone adequately competent. This issue motivates the third "institutional" variant of the ABM presented here.

(as we have already suggested) most likely to find a use for ABM that can cope with this situation of ongoing structuration (rather than those which simply unfold the implications of fixed rules over time).

Thus, even though a GM in the sense presented here is neither historical nor plausible, this very stylised ABM still allows us to see clearly what we might mean by changing the course of history. In the context of a theory of evolving networks we can ask whether (and how) a GM could produce a lasting change in network structure such that the world is truly different thereafter than if they had never lived.

Further, thinking about the problem in this way enables us to dig more deeply into the rather nebulous idea of individual agency versus "social forces" that seems to underpin the GM debate. By assumption, in the ABM, all agents make their network formation decisions in the same way. The GM is only different in being able to form many more ties than the ordinary agent. We might expect, therefore, that the GM *couldn't* have a lasting effect. Social forces in this context mean the great mass of people with small tie capacities being born, choosing their ties at random and then dying. But, of course, these ties are not strictly random. They depend on both chooser and chosen having capacity and this in turn depends on births and deaths. If the GM has the capacity to change the overall structure of the system (for example by creating a very centralised GC) then this may change the choices open to new agents in perpetuity or at least for very long periods. The question then becomes what effect (or effects) does the GM need to have for this to occur? In the ABM discussed here, clearly forming lots of random ties on your own account is not enough. But we have started to move the GM debate from a contest between abstract philosophical positions to an exploration of different ABM assumptions that might, at least in principle, take account of data.

But this insight takes us further in our thinking in an important way. What would be impressive about this situation from a historical point of view is that even though the greatness of the GM was only manifested in his lifetime (and left behind no tradition of changed behaviour like Jesus or formal institution like the UK Civil Service after Trevelyan) it is still possible for an imprint of that greatness to be left behind (in this case in the equilibrium structure of dynamic social networks.) This makes what we mean by changing the course of history better defined and also more interesting in certain ways. In a sense, after reflection on this case, it would be much less surprising if someone did this by leaving behind Krupp or the Catholic Church as stable bureaucratic entities that can perpetuate themselves. However, the idea that someone can, through activities only in their own lifetime and leaving no trace in the behaviour of others, leave their mark on a system that is constantly changing through individual demography and decision is a historically intriguing one, unlikely to be identified or articulated clearly without ABM (even using a ridiculously stylised example).

This discussion leads to a second variant of the ABM in which, after a certain point (time period 16000), one agent starts to make their network ties by an innovative decision process (choosing the least connected agent rather than choosing at random). Once invented, this innovative "practice" can spread through the population in three ways. Agents can "catch it" with some probability from those who already have it (observation), they can be born with it (reflecting socialisation by an innovating parent) or they have a very small chance to discover it without reference to network or socialisation (perhaps reflecting other forms of transmission like books.) The results are interesting. In the first place, unlike the last two experiments, there *is* a significant difference before and after the innovation (0.64836 and 0.462418). However, the result is recognisably negative. This is because, although on an individual level, choosing the least connected agent to make a tie to may seem like a good way to increase the size of the GC (by "incorporating" isolates), it actually results in a centralised network that is much more vulnerable to the death of some agents. Thus we see, even in a stylised example, the potentially counter-intuitive conflict between the logic of individual choices and aggregate social outcomes [28][18].

The previous discussions have been leading up to the third variant ABM, a very simple form of institutional growth. As we have seen, a GM cannot be great (in this ABM at least) just on the basis of his own actions. He must also, somehow, change behaviour, networks or institutions. Individual decisions, spreading through the population may actually harm the outcome they are intended to promote (in this case GC size.) How then can social order come about over long periods without assuming high levels of rationality or knowledge? This variant ABM provides one possible answer. In it, after a certain date, there is an innovation, but of a different kind to that already discussed. An agent decides that it will, before its death, seek a replacement for its position in the social network. In this case, other agents will consider whether replacing the existing agent or just making the usual random connection will lead to better connectedness. For simplicity the agent that wants to be replaced does not apply any selection criteria. Furthermore, there is no attempt to ensure that the replacement agent has surplus capacity to take on the ties of the dying agent. Instead, it just sheds some existing ties randomly to ensure that maximum tie capacity is maintained. As before, the social innovation of wanting to be replaced can spread through the population. This gives rise to a situation where an institution consists of a set of nodes that all want to be replaced and the connections between them. Provided replacement actually occurs, this institution will reproduce itself over time. The final ingredient in increasing order is the assumption that if a node wants to be replaced but nobody wants to replace it (because that node is very poorly connected for example) then replacement will simply fail. This gives rise to an evolutionary process in which there is a continuous random variation in the population of nodes that want to replace themselves (which might include the innovative practice dying out again) but a selective retention of those that happen to be organised in ways that ensure that others want to continue replacing them.

---

[18]As with all variants reported here, this is the result of ten simulation runs, averaging the giant component size over 4000 time periods (to allow for noise) at two points in the simulation (before and after the "event" whether that be innovation or a GM.) The only difference in this condition was that the measurement periods were 12000–16000 and 24000–28000 time periods. This was to allow time for the innovation to diffuse since that is a more protracted process than the life of a GM.

(Cliques—network structures where everyone is connected to everyone else—are a good example. Once they form randomly, agents have a strong incentive to replace because they instantly gain access to a large number of ties and everyone else in the clique is part of the same virtuous circle.) Using the same measurement procedures as before, there is again a significant difference between the GC size before and after the innovation of institution formation (0.65751 and 0.70427). However, unlike the case where agents started to choose the least connected partners individually rather than choosing at random, this social innovation has positive effects for the connectedness of the network[19]. Furthermore, these arise on the basis of individual decisions based on very limited information that gradually self-organise into a mutually sustaining institution.

4   Conclusion

Because the purpose of this chapter is to present case studies of ABM to start a dialogue about the potential of ABM for history (both in developing models that *will* appeal to historians and avoiding those that *won't*), there is a real danger that a conclusion will simply recapitulate the chapter itself. There is however one set of linked ideas that can usefully be presented as concluding this argument at this stage.

The hope has been to present ABM of three social processes that might be historical in the view of historians and which might be hard to represent and understand without ABM (a genuine social innovation, possible roles of unique individuals and the growth of a stylised institution.) There are at least three ways in which these ideas could be developed, all of them positive for dialogue between history and ABM. The first is simply additions to the list of potentially historical processes that could then be modelled. (If these examples are not, or are not exhaustively, what is meant by history then what do we need to add?[20]) The second is just to follow these ideas through in more detail and explore their consequences. For example, we have shown the effects of single social innovations in a simulated society that has (by design) reached a steady state purely so we can present clear differences in GC size. But, in fact, societies may never reach a steady state and face more or less constant social innovation. Is an ABM the only way we can possibly characterise such profoundly dynamic systems? Finally, and this was hinted at in the discussion of the GM, we have so far treated each historical process as if were separate, purely for simplicity of exposition. In fact, of course, it is highly likely that these effects combine. Our GM had no lasting impact. But could he have done so if he had also introduced a social innovation in behaviour even if, without him, that

---

[19]Note that these effects are unintended. An agent wants to be replaced. This *happens* to increase the size of the GC. As with selection of least connected agents, it need not have this effect.

[20]As well as advancing modelling, this dialogue could thus help to clarify theoretical thinking in history in a way that narrative theorising might not.

behaviour could not have spread (or perhaps even worked?) Almost regardless of the actual findings of our ABM, we hope that we have both raised some cautionary notes for the style of existing "historical" ABM and suggested some new avenues for "more historical" approaches that might start a more effective dialogue with historians.

References

1. Anderson, A., Bryan, P., Cannon, C., Day, B., Jeffrey, J.: An experiment in combat simulation. the battle of Cambrai, 1917. J. Interdiscip. Hist. **1**, 229–247 (1972)
2. Clements, R., Hughes, R.: Mathematical modelling of a mediaeval battle: the battle of Agincourt, 1415. Math. Comput. Simul. **64**, 259–269 (2004)
3. Komlos, J., Artzrouni, M.: Mathematical investigations of the escape from the Malthusian Trap. Math. Popul. Stud. **2**, 269–287 (1990)
4. Diamond, J.: Life with the artificial Anasazi. Nature **419**, 567–569 (2002)
5. Düring, M.: The potential of agent-based modelling for historical research. In: Youngman, P., Hadzikadic, M. (eds.) Complexity and the Human Experience: Modeling Complexity in the Humanities and Social Sciences, pp. 121–140. Pan Stanford, Singapore (2014)
6. Barth, F.: On the study of social change. Am. Anthropol. **69**, 661–669 (1967)
7. Graham, S.: Networks, agent-based models and the antonine itineraries: implications for roman archeology. J. Mediterr. Archaeol. **19**, 45–64 (2006)
8. Janssen, M.: Understanding artificial Anasazi. J. Artif. Soc. Soc. Simul. **12**(2009) http://jasss.soc.surrey.ac.uk/12/4/13.html
9. Murgatroyd, P., Craenen, B., Theodoropoulos, G., Gaffney, V., Haldon, J.: Modelling medieval military logistics: an agent-based simulation of a Byzantine army on the march. Comput. Math. Organ. Theor. **18**, 1–19 (2012)
10. Newman, C.: The Kentucky Dark Patch Knight Riders' rebellion. N Engl. J Polit. Sci. **1**, 3–28 (2006)
11. Small, C.: Finding an invisible history: a computer simulation experiment (in virtual polynesia). J. Artif. Soc. Soc. Simul. **2**(1999). http://jasss.soc.surrey.ac.uk/2/3/6.html
12. Carpenter, C., Sattenspiel, L.: The design and use of an agent-based model to simulate the 1918 influenza epidemic at Norway House, Manitoba. Am. J. Hum. Biol. **21**, 290–300 (2009)
13. Hill, R., Champagne, L., Price, J.: Using agent-based simulation and game theory to examine the WWII Bay of Biscay U-boat campaign. J. Def. Model. Simul. Appl. Methodol. Technol. **1**, 99–109 (2004)
14. Mithen, S.: Simulating prehistoric hunter-gatherer societies. In: Gilbert, N., Doran, J. (eds.) Simulating Societies: The Computer Simulation of Social Phenomena, pp. 165–193. UCL Press, London (1994)
15. Doran, J., Palmer, M., Gilbert, N., Mellars, P.: The EOS Project: modelling upper paleolithic social change. In: Gilbert, N., Doran, J. (eds.) Simulating Societies: The Computer Simulation of Social Phenomena, pp. 195–221. UCL Press, London (1994)
16. Ferguson, N. (ed.): Virtual history: alternatives and counterfactuals. Basic Books, New York (1999)
17. Evans, R.: Altered pasts: counterfactuals in history. Little Brown, London
18. Tucker, A.: Our knowledge of the past: a philosophy of historiography. Cambridge University Press, Cambridge (2004)
19. Garavaglia, C., Malerba, F., Orsenigo, L., Pezzoni, M.: A simulation model of the evolution of the pharmaceutical industry: a history-friendly model. J. Artif. Soc. Soc. Simulat. **16**(2013) http://jasss.soc.surrey.ac.uk/16/4/5.html

20. Chattoe-Brown, E., Edmonds, B.: Modelling evolutionary mechanisms in social systems. In: Edmonds, B., Meyer, R. (eds.) Simulating Social Complexity, pp. 455–495. Springer-Verlag, Berlin (2013)
21. Ewert, U., Roehl, M., Uhrmacher, A.: Hunger and market dynamics in pre-modern communities: insights into the effects of market intervention from a multi-agent model. Hist. Soc. Res. **32**, 122–150 (2007)
22. Snijders, T., Baerveldt, C.: Multilevel network study of the effects of delinquent behavior on friendship evolution. J. Math. Sociol. **27**, 123–151 (2003)
23. McPherson, M., Smith-Lovin, L., Brashears, M.: Social isolation in America: changes in core discussion networks over two decades. Am. Sociol. Rev. **71**, 353–375 (2006)
24. Wilensky, U.: NetLogo. Center for Connected Learning and Computer-Based Modeling, Northwestern University, Evanston, IL (1999). http://ccl.northwestern.edu/netlogo
25. Wilensky, U.: NetLogo giant component model. Center for Connected Learning and Computer-Based Modeling, Northwestern University, Evanston, IL (2005). http://ccl.northwestern.edu/netlogo/models/GiantComponent
26. Carneiro, R.: Herbert Spencer as an anthropologist. J. Libert. Stud. **5**, 153–210 (1981)
27. Neher, C.: Political succession in Thailand. Asian Survey **32**, 585–605 (1992)
28. Merton, R.: The unanticipated consequences of purposive social action. Am. Sociol. Rev. **1**, 894–904 (1936)

Evolutionary Cooperation in a Multi-agent Society

Marjolein de Vries and Pieter Spronck

Abstract According to the evolutionary ethics theory, humans have developed a moral sense by means of natural selection. A "moral sense" supposes some form of cooperation, but it is not immediately clear how "natural selection" could lead to cooperation. In this research, we present a multi-agent society model which includes a cooperative act in order to investigate whether evolution theory can explain to what extent cooperative behavior supports the emergence of a society. Our research shows that cooperation supports the emergence of a society in particular when taking into account individual differences.

Keywords Evolutionary ethics • Evolutionary algorithms • Multi-agent societies • Simulation • Cooperation • Individual differences

1 Introduction

For many years, philosophers, biologists, and psychologists have been studying how the human race got instilled with a moral sense [1]. A moral sense can be defined as "a motive feeling which fueled intentions to perform altruistic acts" [1]. It drives humans to exhibit moral behavior such as performing acts of cooperation. According to the evolutionary ethics theory, humans have developed a moral sense by means of evolution and natural selection [1]. Natural selection implies competition and thus can be argued not to lead to cooperation [2], which has made the existence of cooperation in human behavior a challenge to explain [3]. Moreover, according to the Social Exchange theory, humans require some cost–benefit mechanism in order to successfully engage in a cooperative act [4].

While the ideas behind evolutionary ethics are appealing, it is difficult to determine their value experimentally. Performing experiments with organisms, such as humans, on evolutionary ethics is hard, as evolution takes place over hundreds of thousands of years and experiments on ethics with humans are

M. de Vries (✉) • P. Spronck
Tilburg center for Cognition and Communication, Tilburg University, Tilburg, The Netherlands
e-mail: mdv@marjoleindevries.com; p.spronck@uvt.nl

© Springer International Publishing AG 2017
W. Jager et al. (eds.), *Advances in Social Simulation 2015*, Advances in Intelligent Systems and Computing 528, DOI 10.1007/978-3-319-47253-9_6

unethical themselves. However, simulations may provide a means to perform such experiments. In particular, the artificial life approach using a multi-agent society may be suitable to perform such experiments, as agent-based models are an effective tool for modeling the emergence of social phenomena [5].

Previous research on evolutionary ethics by Spronck and Berendsen [6], based on the work of Epstein and Axtell [7] and Mascaro [8], shows that artificial multi-agent societies may provide new insights in evolutionary ethics theory. Nonetheless, no specific act of cooperation was present in their computational model, even though social and ethical behavior implies some form of cooperation.

In our research we present a multi-agent society model based on the model of Spronck and Berendsen [6], in which agents are able to perform a specific cooperative act. With our model, we aim to investigate to what extent evolution theory can explain the emergence of cooperation in the formation of a society. Because research has shown that individual differences strongly affect evolutionary outcomes [9], we will differentiate between regular cooperation and cooperation in which individual differences are taken into account.

The outline of this chapter is as follows: Sect. 2 provides the theoretical background needed for the setup of the computational model, which is described in Sect. 3. Section 4 presents our results. In Sect. 5, our results are discussed. Section 6 formulates the conclusion of this chapter.

2 Theoretical Framework

Section 2.1 describes the research area of Evolutionary Ethics. Section 2.2 outlines the concept of cooperation.

2.1 Evolutionary Ethics

In "The Origin of Species," Darwin [10] describes a new way of looking at our contemporary behavior from an evolutionary approach. According to Darwin, one of the most important features of species is self-reproduction, a system of positive and negative feedback which we refer to as natural selection [11]. Individuals having any advantage over others, and thus having a better fitness, have the best chance of surviving and procreating. Darwin referred to this process as "survival of the fittest" [12]. From a sociobiological approach, natural selection can explain patterns of social behavior of humans and other species [11]. According to the evolutionary ethics theory, natural selection has caused humans to have a moral sense, i.e., "a motive feeling which fueled intentions to perform altruistic acts" [1]. This moral sense drives humans to behave morally, such as performing acts of cooperation.

Natural selection implies competition, as every organism is designed to enhance its own fitness at the expense of the fitness of others [2]. Contrariwise, humans are amongst the most communal and socially dependent organisms alive on our planet [1]. The aspect which distinguishes us from most other species is that we cooperate even with individuals which are not genetically closely related in order to achieve common goals [13]. Moreover, evolutionary ethicists state that if members of the same species execute a process of cooperation instead of struggling with each other, more personal benefit can be achieved [14]. Evolutionary biologists have been fascinated by this phenomenon for several decades [2].

A research method used for evolutionary experiments is the artificial life method. Artificial Life as a research field attempts to resemble biological phenomena by means of computer simulations in order to create theoretical grounding for biological theories [15]. Artificial societies are inhabited by autonomous agents, which are capable of exhibiting emergent behavior. Emergent behavior is behavior which is not programmed into the agents by the designer, but instead is created using a process of self-adaptation. This resembles the assumed automatic emergence of moral behavior within humans [16]. De Weerd, Verbrugge and Verheij, for example, show that agent-based models are suitable for investigating emerging social phenomena [5].

In order to research ethical behavior using artificial agent societies, a specific definition of which actions are called "ethical" is needed. According to Jaffe [17], a distinction between the benefit to society 'S' and the compound benefit to the individual "I" needs to be made. Jaffe [17] classifies actions as either:

"True altruism", if $S \geq 0$ and $I < 0$
"Social investment", if $S \geq 0$ and $I \geq 0$
"Destructive egoism", if $S < 0$ and $I \geq 0$
"Destructive behavior", if $S < 0$ and $I < 0$

The common moral sense of a society can be defined as the average of executed actions in a particular period in a society [6]. A 'moral' society would predominantly perform actions of true altruism and social investment.

Earlier research done on ethics in agent societies have been performed by Epstein and Axtell [7], who experimented with an artificial society built by defining simple local rules. Their research shows that these simple rules can lead to complex emergent behavior in an agent society. Mascaro [8] proposed using genetic algorithms to let the behavioral rules evolve over time, resembling the evolution process described by Darwin. Mascaro also presented basic environments in which agents are able to perform a finite number of actions.

Spronck and Berendsen [6] based their model for researching the emergence of moral behavior on the work of Epstein and Axtell [7] and Mascaro [8]. In their society model, agents are able to perform four actions: wander, forage, steal, and share. They found some specific configurations in which agents would steal very little, and would even share for a small percentage of their actions. However, a specific act of cooperation was missing from this model, even though social and moral behavior implies cooperative acts. Therefore, in our research, a cooperative action will be added to the model of Spronck and Berendsen [6].

2.2   Cooperation

To design a cooperative action, we looked at earlier research on cooperation from an evolutionary perspective. Such research has been done on the (iterated) Prisoner's Dilemma [18]. However, it is hard to draw conclusions from the Prisoner's Dilemma, as it is only a model of a simple problem with a finite number of options [6]. Moreover, it only models the behavior of two agents, and not the behavior of a society. Therefore, a different proposal for a design of a cooperative act should be made.

In the interest of designing an act of cooperation for agents in a society, a representation of the concept of cooperation is needed. A cooperative social act includes some individual costs, but also supplies an amount of collective group benefits which is divided equally over all the cooperating members [16]. A cooperative act enables not only an agent's own objectives, but also the objectives of other agents [19]. This kind of cooperation is closely related to Social Exchange, which is defined as "cooperation between two or more individuals for mutual benefit" [4]. During an exchange, an individual is generally obligated to pay a cost in order to be able to receive a benefit. This requires humans to have specialized cognitive mechanisms for deciding whether the benefits of a potential exchange outweigh its costs.

To be able to fully exploit the benefits of cooperation, it might help if there exist individual differences in a society. Research has shown that individual differences strongly affect evolutionary outcomes such as survival and mating success [9]. Diversity creates value and benefit for team outcomes [20]. Introducing individual differences will have an effect on the outcome of a group process such as cooperation.

3   Society Model

In this section, we describe the design of our society model. The society model is based on the model of Spronck and Berendsen [6]. In Sect. 3.1, the general setup of the environment is explained. Section 3.2 describes the design of the agents and their attributes. Section 3.3 construes the set of rules each agent possesses and Sect. 3.4 illustrates the mating process. Section 3.5 states by what means we have measured the effect of a cooperative act in a multi-agent society. Extensive details of the model are given by de Vries [21].

3.1   The Environment

The society is situated in an environment which consists of a two-dimensional 30×30 grid with correspondingly 900 cells. The environment is bounded by the edges, which means agents close to the boundary have a limited environment.

At the start of the experiment, the environment contains 200 agents and 9 cells contain nutrition. During a round, each of the 900 cells can contain nothing, an agent, nutrition or both an agent and nutrition.

A "round" is the unit of time which is used in the society model. Each run in the model lasts 20,000 rounds. In every round, an agent executes one action. At the beginning of each round, new nutrition is be added to the environment. Moreover, every agent is allowed to start the mating process (if the agent meets some specific requirements), which creates a new agent. The mating process counts as one action, which means that after mating an agent is not able to execute another action anymore. The order in which agents are allowed to perform actions is randomly decided. At the end of the round, the health and age of each agent changes, which could result in agents dying when their health or age is respectively too low or too high

In order to create a dependency and a possible urge to cooperate between agents two nutrition types are available in the environment: food and water. Half of the cells in the environment are assigned to contain water and the other half to contain food. On a turn, agents are only able to forage one type of nutrition. Assigning the cells to one of the nutrition types is done randomly.

3.2 Agent Attributes

Each agent has an **age**. This is the number of rounds the agent has been alive. The age of an agent at birth is 0, and an agent dies when it reaches the age of 100. Moreover, agents have a **health** attribute represented as an integer. An agent dies when its health reaches zero or less. An agent created at the beginning of a run has a health of 100. An agent created by the mating process inherits 25 % of both its parents' health. The health attribute is used as fitness measure. Agents lose 5 U of health at the end of each round. Furthermore, agents have a field of **vision**, which determine how many cells an agent can take into account when making decisions. All agents can look in four main directories (north, east, south, and west) and have a vision of three cells.

Furthermore, agents have a **food basket** and a **water basket**, of which its value represents the amount of units of the corresponding nutrition type. When agents receive nutrition, the acquired units are added up to the corresponding baskets. At the end of a round, agents are able to digest nutrition from their baskets and increase their health. An agent can only consume water and food in equal amounts, thereby implying a need for both nutrition types. Agents created at the beginning of a run have a value of zero for each basket, and agents created by the mating process inherit 25 % of each of the baskets of both parents.

3.3   Agent Rules

Agents have a rule set of seven rules. Rules consist of zero, one or two tests with a specified action, thus having the following structure: $(Test \wedge Test) \rightarrow Action$. Agents created at the beginning of a run get a randomly made rule set. Agents created by the mating process inherit rules from their parents (Sect. 3.4).

Each test begins with specifying an agent on which the test is performed. A choice is made among (1) the agent itself, (2) the closest agent, (3) the furthest agent, (4) the healthiest agent, (5) the weakest agent, (6) the agent with the most food, (7) the agent with the least food, (8) the agent with the most water and (9) the agent with the least water. Then, an attribute is chosen from (1) health, (2) age, (3) food availability, (4) water availability, (5) food basket, and (6) water basket. Moreover, a mathematical operator is selected from smaller than, larger than, equals and wildcard. If a test contains a wildcard, it always yields "True." Next, a whole number is selected.

Finally, an action is executed by an agent if both tests are applicable. The following actions are available, where cooperation is not available in the baseline experiment.

Wander. The agent moves to a random cell within its vision. In the case that the new cell contains nutrition, the agent puts it in the corresponding basket with a maximum of 30 U.

Forage Food. The goal for the agent with this action is to search for food within its vision. The agent then moves to the cell with food and puts it in the food basket with a maximum of 30 U.

Forage Water. The goal for the agent with this action is the same as Forage Food, only this action is aimed at water instead of food.

Steal Food. The agent steals an amount of 15 U from the food basket of the specified agent. The amount of units is based on the model of Spronck and Berendsen [6]. If the specified agent does not have 15 or more units in its food basket, the maximum quantity possible is stolen. The agent generally aims the steal action at the agent specified in the second test. If that agent is the agent itself, the aim of the action is the first specified agent or otherwise a random agent within the active agent's vision.

Steal Water. The goal for the agent with this action is the same as Steal Food, only this action is aimed at water instead of food. The target agent is specified in the same way as for the Steal Food action.

Share Food. With this action, the agent gives away 25 % of its basket with food to the specified agent. The target agent is specified in the same way as for the Steal Food action.

Share Water. With this action, the agent gives away 25 % of its basket with water to the specified agent. The target agent is specified in the same way as for the Steal Food action.

Cooperate. Agents have to pay a cost of x units of their fullest basket. Agents are only allowed to engage in the cooperative act if they have enough units in their basket to pay the cost. These costs are removed from the environment, which means they are lost to the population of agents. After paying the costs, agents are randomly assigned to either forage food or forage water. Then, agents have to hand in the units of food or water they just have foraged. At the end of the round, the total nutrition collected is divided equally amongst all the participating agents. The cost that an agent pays to perform the cooperative action is thus the entry fee of x units of their fullest basket. The benefit of the cooperative act is that the agent is guaranteed to receive both food and water, instead of at most one type of nutrition. This act satisfies the description of cooperation made by Gintis [16] and by Doran, Franklin, Jennings, and Norman [19].

According to Jaffe's dimensions [17], we can classify wandering, foraging and cooperating as *social investment*, stealing as *destructive egoism* and sharing as *true altruism*.

In a second experiment, individual differences are introduced into the society. To reflect the concept of individual differences, each agent is instilled with a preference for foraging water or food. When the agent has a preference for foraging nutrition y, the agent receives $\frac{3}{2}$ times its foraged units of the type y and $\frac{1}{2}$ times its foraged units of the other nutrition type. Agents at the start of a run get a randomly assigned preference. Agents created by the mating process have a 50 % chance to receive each of their parents' preferences, with a 2 % chance of being mutated. While engaging in the cooperative act, agents are now assigned to forage the nutrition type which has their preference.

An example rule is *(closest agent age > 28) and (agent itself health < 40) then forage*. A child inherits its rule sets by means of a uniform crossover operator which operates on the different tests in the rule, where each rule has a 2 % chance of having a test being mutated.

Rules are ordered on complexity, and agents start with the most complex rule when executing their rules. Agents go through their rules according to their complexity until one of the rules apply, then the according actions is executed. The specified agent for the steal and share actions is generally the agent specified in the second test of the rule. If the specified agent is the agent itself, the specified agent of the first rule is selected and otherwise a random agent within an agent's vision is selected. If all rules do not apply, the agent executes the action *foraging* as a way of using an instinct.

3.4 Mating

An agent tries to mate with another agent before a round starts. If an agent already has mated with another agent, it is not able to mate again in the same round. Mating is only possible if an agent has an age of at least 18, a minimum health of 50 and at least 30 U of nutrition in each of its two baskets. The agent chooses randomly from suitable agents in his vision.

A child agent is created by the mating process and placed on a cell nearby its parents if possible. A child inherits 25 % of each of its parents' health and 25 % of each of its parents' baskets. A child inherits a combination of the rules of its parents by means of a uniform crossover operator. Moreover, each rule has a 2 % chance of having a test part being mutated.

3.5 Measurements

Statistics are gathered during the runs in order to investigate the common moral sense of a society. Demographics of the society, statistics of the executed actions by the agents, and information about the executed rules are taken into account when answering the research questions.

We define the common moral sense as the averages of the moral evaluation of the actions in the rule set. This means that the common moral sense exists of certain percentages of true altruism, social investment, destructive egoism, and destructive behavior. An ethical moral sense would consist of mostly true altruism and social investment. Because it is only interesting to look at agents who are strong enough to survive and to mate, we only take the statistics of the agents who survived at least to the mating age. Moreover, only the executed actions by the agents in the last 5000 rounds (of 20,000) are used for determining the common moral sense.

4   Results

Section 4.1 describes the baseline behavior. Section 4.2 describes the experiment with a cooperative act, and the experiment in which individual differences are also introduced. The results of the forage, steal and share actions for both nutrition types are grouped together for the purpose of clarity (e.g., "forage" is the sum of "forage food" and "forage water").

4.1 Baseline Behavior

The simulation was executed 20 times, which results in 20 runs where each run consists of 20,000 rounds. Table 1 shows the percentage of executed actions, number of mature agents and health of mature agents averaged over 20 runs and the standard deviation.

Table 1 shows that the baseline behavior consists of mostly foraging and stealing. Therefore, we can conclude that the common moral sense of the baseline society is mainly egoistic, because the agents only use actions which are helping themselves. The results of the baseline behavior in this research are quite similar to the results of the baseline experiments of Spronck and Berendsen [6].

Table 1 Percentage of executed actions and descriptive statistics in the baseline behavior

	Average	St. dev.
Wander	0.13	0.07
Forage	53.55	11.82
Steal	46.15	11.76
Share	0.27	0.23
Number of mature agents	151.73	12.69
Health of mature agents	95.80	4.99

Table 2 Percentage of executed actions and descriptive statistics in the cooperative experiment, standard deviation between brackets

	Cost 0	Cost 1	Cost 2	Cost 3	Cost 4
Wander	0.16 (0.05)	0.14 (0.10)	0.14 (0.10)	0.16 (0.11)	0.13 (0.10)
Forage	17.20 (27.22)	30.33 (27.44)	40.82 (24.32)	45.50 (14.12)	58.08 (12.95)
Steal	41.39 (9.52)	47.20 (10.22)	41.90 (9.54)	48.66 (14.12)	41.53 (12.88)
Share	0.58 (0.50)	0.29 (0.20)	0.34 (0.34)	0.23 (0.15)	0.17 (0.13)
Cooperate	40.85 (20.85)	22.19 (22.90)	16.95 (20.20)	5.59 (10.32)	0.19 (0.18)
Nr of mature agents	143.09 (4.62)	134.00 (3.89)	129.27 (4.03)	133.33 (4.40)	131.72 (4.37)
Health of mature agents	106.57 (8.25)	99.95 (10.52)	97.02 (9.36)	93.84 (4.92)	90.69 (3.53)
Compound benefit	7.63 (0.73)	8.53 (0.63)	8.42 (1.01)	8.91 (0.84)	n/a (n/a)

4.2 Cooperation Experiments

The simulation of the society with the introduction of a cooperative act is executed with different costs for engaging in the cooperative act, where the cost range from 0 to 4 U. For each cost value, 20 runs were executed. Table 2 shows the percentage of executed actions, number of mature agents, health of mature agents and the compound benefit (the sum of food and water units) received from the cooperative act, all averaged over 20 runs. With a cost of 4 U, the cooperative was hardly executed anymore. Note that a cooperative act with a cost of 0 U cannot be called a true cooperative act, because it does not require any cost for engaging in the cooperation, while by definition a cooperative act must have a cost associated with it.

Table 2 shows that the cooperative act is executed less and the execution of the foraging act increases when the cost increases. Clearly the benefits of participating in the cooperative act do not always outweigh its costs. No statistically significant linear dependence of the mean of compound benefit on the cost was detected. Table 2 also shows that the average number of agents decreases with the introduction of the cooperative act when comparing to the baseline behavior. We assume that this is a consequence of the fact that the environment loses nutrition (which is paid as a cost for the cooperative act), thereby having less nutrition to nourish the agents in the society. Even when the cooperative act is no longer executed according to Table 2, there might still be agents which execute it, but which are not counted in Table 2 as they do not reach the age of maturity.

Table 3 Percentage of executed actions and descriptive statistics in the individual differences experiment, standard deviation between brackets

	Cost 0	Cost 1	Cost 2	Cost 3	Cost 4
Wander	0.16 (0.09)	0.16 (0.06)	0.18 (0.07)	0.16 (0.07)	0.16 (0.06)
Forage	4.98 (10.17)	3.07 (7.95)	2.05 (3.46)	6.58 (14.52)	12.33 (16.40)
Steal	36.36 (18.57)	38.57 (11.66)	41.51 (7.37)	44.84 (8.17)	48.95 (7.50)
Share	0.85 (0.85)	0.49 (0.27)	0.79 (1.50)	0.86 (1.87)	0.52 (0.53)
Cooperate	58.00 (25.44)	57.98 (15.15)	55.69 (8.93)	47.73 (12.76)	38.22 (16.99)
Number of mature agents	170.10 (23.90)	158.67 (11.59)	148.25 (9.33)	134.18 (10.19)	126.06 (9.00)
Health of mature agents	111.93 (6.64)	112.42 (4.43)	113.42 (2.05)	111.58 (4.80)	109.91 (6.66)
Compound benefit	8.25 (2.35)	9.71 (1.43)	10.71 (1.16)	11.14 (1.24)	11.70 (1.52)

The common moral sense is quite similar to the baseline experiment. Nevertheless, agents do choose using the cooperative act over using the individual action "foraging" when the costs are low, which results in somewhat more social behavior than in the baseline experiment.

Table 3 shows the results, averaged over 20 runs, when providing agents with a foraging preference as a reflection of the concept of individual differences. Table 3 shows that the cooperative act is now chosen considerably more than any of the other acts, in all tests. The benefits received from the cooperative act now outweigh the costs more convincingly, as also can been seen in the increase in benefit comparing with the previous cooperation experiment. With the use of linear regression, we can see that the cost significantly predicts the compound benefit, $\beta = 0.83$, $t(3) = 6.84$, $p < .01$. The cost also explains a significant proportion of variance in the health scores, $R^2 = 0.94$, $F(1, 3) = 46.74$, $p < .01$. Table 3 also shows that the introduction of the foraging preference attribute results in a larger mature population and an increase in the average health.

Furthermore, the common moral sense of the society with a cost of 0, 1, or 2 U contains of more *social investment* than the common moral sense in the baseline experiment. Consequently, we can conclude that the introduction of a cooperative act with a cost of 1 U and the foraging preference attribute results in a superior society with the largest reduction of *destructive egoism*. In the light of the dimensions provided by Jaffe (2004), we can define its common moral sense as 61.05 % *social investment*, 38.57 % *destructive egoism*, and 0.49 % *true altruism*. Therefore, this experiment shows that with the introduction of a new action and attribute, the common moral sense can become significantly less antisocial.

5 General Discussion

Our research, which uses computer science to investigate questions in social sciences, supports the Social Exchange theory, as the cost–benefit behavior of the agents is evident in all our experiments. Because agents have to pay a cost for the cooperative act, agents need to receive a benefit which outweighs the cost, i.e., a benefit which is enough to maintain health. Therefore, the compound benefit should increase as the cost increases, as more compound benefit is needed to make up for the paid cost. We can see that the benefit is significantly predicted by the cost. Future research could implement actions with more complex cost–benefit calculations

Moreover, the results of our research support the positive effect of individual differences on team outcomes [9], as the individual differences experiments yielded an increase in benefit and a morally superior society.

The research also introduces a new way of representing a cooperative act. Our society model provides an alternative to modeling cooperation by means of the (iterated) Prisoner's dilemma [18]. Moreover, many agent-based models on cooperation consist of agents having a shared goal on which they have to work together [19]. However, in our research, agents do not have a shared goal but rather individual goals on which they can achieve more effectively with the help of others.

In this chapter, the results for the actions concerning both nutrition types are grouped together. When looking at the distinctive results for food and water [21], it can be seen that in some cases for example more food was foraged than water but more water was stolen than food. One explanation could be that agents need a minimal amount of food and water in order to reach their maximum age. For this purpose, it does not matter if an agent has more of one nutrition type than the other, as long as the lowest one is sufficient. Therefore, it is possible that the evolutionary process results in a society in which the two nutrition types are not foraged equally. This does not prevent a healthy society from emerging.

We succeeded in evolving a common moral sense that leans towards mainly social behavior, even with our simple agent model. Naturally, our society model is in no way comparable with human society. Many subtle influences of a human society are left out in the computational society, but could be included in future work. Future research could investigate the effect of other actions, attributes and settings.

Finally, we remark that when experimenting with agent models, one has to be careful not to "build in" the conclusions that one seeks. In our experiments, the evolved behavior is the result of only an agent's ability to survive and procreate, which is only dependent on age and health. Therefore, no preference for a specific act is built in our model.

6   Conclusion

The goal of this chapter is to investigate to what extent evolution theory can explain the emergence of cooperation in the formation of a society. Our research shows that cooperation supports the emergence of a society, in particular when taking into account individual differences. With the introduction of individual differences, benefits from engaging in cooperation increased. The positive effect of individual differences on the society can be predicted by the Social Exchange theory [4], as individual differences lead to higher benefits from cooperation and thus more agents engaging in cooperative acts. Our research thereby supports the importance of individual differences for group outcomes [9].

References

1. Richards, R.J.: A defense of evolutionary ethics. Biol. Philos. **1**, 265–293 (1986)
2. Nowak, M.A.: Five rules for the evolution of cooperation. Science **314**(5805), 1560–1563 (2006)
3. Axelrod, R., Hamilton, W.D.: The evolution of cooperation. Science **211**(4489), 1390–1396 (1981)
4. Cosmides, L.: The logic of social exchange: has natural selection shaped how humans reason? Studies with the Wason selection task. Cognition **31**(3), 187–276 (1989)
5. de Weerd, H., Verburgge, R., Verheij, B.: Agent-based models for higher-order theory of mind. Adv. Intell. Syst. Comput. **229**, 213–224 (2014)
6. Spronck, P., Berendsen, B.: Evolutionary ethics in agent societies. Int. J. Soc. Robotics **1**(3), 223–232 (2009)
7. Epstein, J.M., Axtell, R.: Growing Artificial Societies. MIT Press, Cambridge (1996)
8. Mascaro, S.: Evolutionary ethics. Master's Thesis, School of Computer Science and Software Engineering, Monash University, Victoria, Australia (2001)
9. Buss, D.M.: How can evolutionary psychology successfully explain personality and individual differences? Perspect. Psych. Sci. **4**(4), 359–366 (2009)
10. Darwin, C.: On the origin of species. John Murray, London (1859)
11. Tooby, J., Cosmides, L.: Conceptual foundations of evolutionary psychology. In: Buss, D.M. (ed.) The Handbook of Evolutionary Psychology, pp. 5–67. John Wiley & Sons, Hoboken (2005)
12. Paul, D.B.: The selection of the "Survival of the Fittest". J. Hist. Biol. **21**(3), 411–424 (1988)
13. Kurzban, R., Neuberg, S.: Managing ingroup and outgroup relationships. In: Buss, D.M. (ed.) The Handbook of Evolutionary Psychology, pp. 653–675. John Wiley & Sons, Hoboken (2005)
14. Ruse, M.: Evolutionary ethics: a phoenix arisen. Zygon **21**(1), 95–112 (1986)
15. Pfeifer, R., Scheier, C.: Understanding Intelligence. The MIT Press, London (2001)
16. Gintis, H.: The Hitchhiker's Guide to altruism: gene-culture coevolution, and the internalization of norms. J. Theor. Biol. **220**(4), 407–418 (2003)
17. Jaffe, K.: Altruism, altruistic punishment and social investment. Acta Biotheoretica **52**(3), 155–172 (2004)
18. Mascaro, S., Korb, K.B., Nicholson, A.E., Woodberry, O.: Evolving Ethics: The New Science of Good and Evil. (2010). Retrieved from http://www.csse.monash.edu.au/~korb/chap1.pdf
19. Doran, J.E., Franklin, S., Jennings, N.R., Norman, T.J.: On cooperation in multi-agent systems. Knowl. Eng. Rev. **12**(3), 309–314 (1997)

20. Mannix, E., Neale, M.A.: What differences make a difference? Psychol. Sci. Publ. Interest. **6**(2), 31–55 (2005)
21. Vries, M. J. de: The introduction of a cooperative act into an evolutionary ethics agent society. Bachelor's thesis, Tilburg University, The Netherlands (2014). http://www.marjoleindevries.com/bachelors-thesis

Design of an Empirical Agent-Based Model to Explore Rural Household Food Security Within a Developing Country Context

Samantha Dobbie and Stefano Balbi

Abstract Food security is an unresolved issue, especially so among developing countries. Models of food security usually focus on the farming components (availability) and neglect to fully incorporate its multi-dimensional nature: including access, utilization and stability (of food). This article proposes the design of an empirical agent-based model representing Malawian smallholders, which operationalizes the FAO framework on food security at the household level. While the general structure can be customized and replicated for other contexts, the agents' characterization is based on national survey data. Preliminary results suggest that non-agricultural workers are more food secure. However important feedbacks with the natural (ecosystem services) and the economic system (local/international market) are foreseen but not fully implemented.

Keywords Agent based • Farming households • Food security • Social-ecological systems

1 Introduction

Food security remains a deep-seated issue, particularly within the developing world. Here, complex social, ecological and political factors act to undermine the achievement of food security—defined by [26] as a state in which *"all people at all times have physical and economic access to sufficient, safe and nutritious food to meet their dietary needs and food preference for an active and healthy life"*. This paper uses Malawi as a case study. Over 90 % of the rural population are engaged in subsistence farming [16]. Agriculture continues to be rain-fed, leaving smallholders vulnerable to climatic shocks [23]. High population densities lead to small plot sizes,

S. Dobbie (✉)
University of Southampton, Southampton SO17 1BJ, UK
e-mail: s.dobbie@soton.ac.uk

S. Balbi
Basque Centre for Climate Change (BC3), 48008 Bilbao Bizkaia, Spain
e-mail: stefano.balbi@bc3research.org

with households cultivating less than 1 hectare of land on average, while poor soil quality further compounds food insecurity [16].

Within smallholder systems such as these, decisions made by farming households can have significant impacts upon household food security. Cropping plan decisions, for instance, in which farmers must allocate land to selected crops may influence future food availability, access, utilisation and stability [9]. Taken together, these four dimensions of food security represent the four pillars as described by FAO [10]. A recent review by Burchi and De Muro [6], however, found current approaches to evaluate food security failed to sufficiently take into account its multi-dimensional nature. In particular food security is also a problem of human behaviour that derives from the capabilities of social agents at risk [8], including health and education, and from the interaction between the natural environment and the larger social environment that often dictates the range of options available [4].

Under these assumptions agent-based modeling (ABM) emerges as a valuable tool to better explore the complexity of smallholder behaviour and food security. ABMs are computationally intense and micro-detailed simulations where many heterogeneous agents can interact at multiple temporal and spatial scales [2]. Agents interact within an environment through predisposed rules, behaviour at the system level is an emergent property of collective behaviour at the local level [11].

A key strength of ABM lies in its ability to account for the uniqueness of individuals and the interactions between them and their environment [1]. An underlying assumption of many ABMs dealing with economic elements is rational behaviour. Endowed with clear preferences and all available information, a rational agent will always elect the optimum solution with no associated cost [14]. However, farming households are seldom efficient maximizers [4].

In this study we depart from the notion of rational agents by employing fast and frugal heuristics. Heuristics are regarded to be fast if they can be computed in a small amount of time and frugal if they can be ascertained with little information [12]. They represent simple 'rules of thumb' used in the decision-making process [13]. Agents adhering to fast and frugal decision-making can be modeled using decision trees and coded with simple if-then rules (See Fig. 4). Moreover the existent ABM literature on this topic has mainly emphasized the smallholders' farming components and given limited attention to the off-farm behavioural elements, thus missing a crucial component of food security especially in view of the changing climate [28].

Overall, it is proposed that this study will design a simple empirical ABM of household food security, in line with the up-to-date qualitative food security research, using Southern Malawi as a case study. The ABM is implemented in NetLogo 5.1 [31]. In the next section we provide an overview of the model development process. The problem of food security is conceptualized for the purpose of modeling rural households within Southern Malawi. The simulation design is also described according to accepted protocols. In the third section we anticipate some results but we mainly present current calibration and validation challenges. In the final section we describe the added value of this research and its expected continuation.

2 Model Development

Model development comprised of four key stages (Fig. 1). Firstly, the aim of the model was clarified. In the second stage a conceptual model of household food security was developed using available literature, qualitative and quantitative data. This fed directly into the third stage in which the ABM of food security was constructed. The fourth stage employed expert knowledge to validate the ABM. A link between the validation stage and further development of the conceptual model highlights the iterative nature of the model development process. The remainder of this section will describe each stage of model development in greater detail. The model purpose is stated as part of the ODD protocol (Sect. 2.2).

2.1 Conceptual Model Development

Early approaches to the analysis of food security emphasized food availability [27]. The 1996 world summit plan of action, however, promoted the multi-dimensional nature of food security which led to the creation of the four pillars, namely: availability, access, utilisation and stability [10].

In the context of the four pillars of food security, food availability can be regarded as the availability of sufficient quantities of quality food from domestic production and imports including food aid. Food access, on the other hand, reflects the ability of individuals and/or households to acquire appropriate foods for a nutritious diet. This is determined by access to adequate resources or entitlements, where entitlements constitute the set of commodity bundles over which a household or individual can establish command [24]. The ability of a household and/or individual to establish command over a set of commodity bundles is in turn dependent upon the political, economic and social backdrop of the community in which they reside [24]. Food utilisation refers to the use of food to achieve a state of nutritional wellbeing in

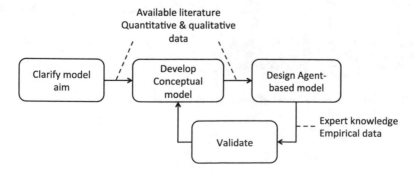

Fig. 1 Iterative stages of model development

which all physiological needs are satisfied. The ability to utilize food is dependent upon a number of factors including access to clean water, energy, good sanitation, sufficient healthcare and an adequate diet [18]. Finally, food stability reflects the temporal dimension of food security. Ecological, political, economic and social shocks may act to destabilize food security on a temporary or long-lasting basis [10].

In developing an empirical ABM of food security we recognized the difficulty in operationalising the traditional four pillars framework by means of a quantitative model. The overlapping nature of each of the four dimensions poses a particular issue. For this reason we adapted the framework to form a well-defined hierarchy, where boundaries are artificially designed across the four dimensions (Fig. 2). Availability of food is necessary for food security, but is not sufficient to guarantee access without accounting for its stability. Similarly, food access is also required for food security, but not sufficient to ensure adequate utilisation.

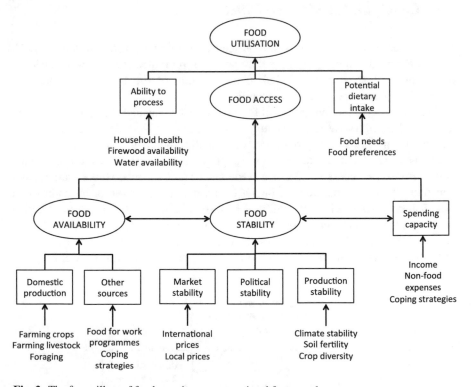

Fig. 2 The four pillars of food security—an operational framework

2.2   Agent-Based Model Construction

The design and communication of ABMs is challenging especially for empirical models. It is therefore important to refer to an existing standard [20]. In this article we use multiple narratives and graphical protocols like Overview, Design concepts, and Details (ODD), Unified Modeling Language (UML) and pseudo-code. First, a description of the ABM is presented according to the ODD protocol [15].

Purpose

The purpose of this ABM is to simulate the behaviour of households within a village and observe the emerging properties of the system in terms of food security. The overall aim of the model is to better represent the multi-dimensional nature of food security at the level of rural households.

Entities, State Variables and Scales

Households are the main entity and are distinguished into three types (1) farmers, (2) agricultural labourers and (3) non-agricultural workers. The main attributes of households include human, physical, natural and financial capital. Households are connected through a social network, which is modeled as a small world [30]. Individuals belong to households and are defined by their gender, education, age and bodyweight. Four different environments are considered, including forest, farm plots, dimba plots[1] and rivers. Key attributes of the different environments include: area, fertility and ownership. Households can own both dimba and farm plots, while forests and rivers remain communal. Other entities are captured in the simplified UML class diagram (Fig. 3). The model runs at the village level and is non-spatial. It is assumed that within a month, each human agent is able to traverse the entire village boundary.

Process Overview and Scheduling

The model works on a monthly time step. It begins by defining the month of the year and the corresponding agricultural and fishing seasons. Basic needs in the form of food, water and fuel are then calculated for each of the households. Household labour is allocated between productive activities, including firewood collection, water collection, on-farm agricultural activities, off-farm agricultural activities and off-farm non-agricultural work. At the beginning of the agricultural

---

[1]Farm plots used for winter cultivations primarily based on the residual moisture of areas bordering streams and rivers.

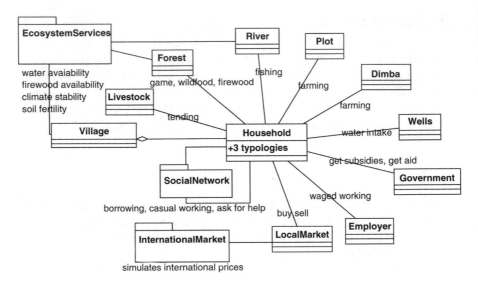

Fig. 3 Simplified UML class diagram

season, households decide how to allocate their farm plots with different crops. Eleven different combinations of basic grains, annual roots, permanent roots, nuts and pulses, fruits, vegetables and cash crops are possible. Decisions are constrained by land, labour, subsistence, input and knowledge requirements. In the months that follow, households are able to adjust land allocation decisions based on labour availability. In addition to farming, households may also tend to livestock, forage for wild and indigenous foods, go fishing, hunt game and carry out off-farm activities such as casual labour (ganyu) and non-agricultural employment. Towards the end of the time step, the food security status of a household is evaluated. Spending capacity is calculated as the difference between income from productive activities and expected non-food expenses. If the availability of calories from self-production and market access is insufficient at this point, households undertake coping strategies. These include: participation in government led 'food for work' programs, sale of livestock and borrowing food and money from the social network, which represents a key interaction between households. Utilisation of available calories is dependent upon a household's ability to access and process food. At the end of the time step the village food security status can be derived from various indicators, summarized for each household typology, including: proportion of food deficient households, mean daily food energy consumption per capita, average proportion of food energy from staples, diet diversity, income and access to food at the market.

Design Concept

Agents are driven by their need to achieve food security in terms of calories, diversity of diet and cultural preferences. Livelihood strategies are based on their typology (i.e. farmers, agricultural labourers and non-agricultural workers). They are aware of their aggregate labour force, their assets and land endowment and of their social ties as modeled in the social network. Emergent phenomena include food security and the state of natural resource stocks at the village level. These are not solely resulting from mere aggregation given that households interact with other households in their social network by sharing resources. However, households are reactive and do not formulate predictions.

Initialization

Individual, household and environmental entities are initialized with survey data. To represent a hypothetical village within Southern Malawi, data corresponding to a hundred unique households is drawn from a sample of the IHS3 dataset ($n = 2492$). The same sample data set was used to construct the initial household typology [7]. For each household a number of attributes are calculated from the survey data. These include: household size, financial capital, livelihood assets and type (farmer, agricultural labourer or non-agricultural worker). Attributes of individuals and farm patches corresponding to the selected households are also initialized using survey data. Gender, age, education level, household status and health of individuals are drawn from the IHS3 dataset. As the bodyweight of individuals is only recorded for children up to the age of 5 years in the IHS3, additional datasets for Nigeria and Malawi are used as an approximation. A single variable, plot area is assigned to patches using data from the IHS3.

In total there are 480 individuals, belonging to 100 households with 195 patches. The majority of households are farmers (64 %), followed by non-agricultural workers (23 %) and agricultural labourers (13 %). On average households comprise of 5 individuals and own just 0.3 ha of land. However, variation can be seen between the different household types (Table 1).

Input and Submodels

Key submodels include: households, crops and livestock. Parameter values build upon primary data and expert knowledge. Where possible they are defined as range-values to introduce some stochasticity.

A static representation of the modeled system entities is proposed in Fig. 3. Two external modules are foreseen for modeling ecosystem services and market prices. Both are envisaged to capture a bidirectional relationship. In the first case, the flow of benefits from the natural environment to the households is fundamental to take into account key variables of food security like water and firewood availability

Table 1 Summary of attribute values for model entities following initialization

Attribute	Full sample		Farmers		Agricultural labourers		Non-agricultural workers	
n	100		64		13		23	
	Mean	SD	Mean	SD	Mean	SD	Mean	SD
Household size	4.8	2.19	4.8	2.29	5.2	2.15	4.6	1.97
Financial capital (USD)	60.4	34.9	21.9	28.9	20.4	20.3	190.3	723.2
No. of cattle owned	0.1	0.42	0.1	0.53	0	0	0	0
No. of poultry owned	2.3	4.68	1.9	3.68	1.4	2.60	3.9	7.22
No. of farm patches owned	2.0	1.18	2.0	1.06	1.5	0.88	2.1	1.59
Total area of land owned (ha)	0.3	0.27	0.4	0.26	0.4	0.25	0.3	0.33

among the others. At the same time an increased demand for firewood can drive deforestation at an unsustainable rate making the ability to process food more challenging. In the second case, it is important to consider the market effect to households' consumption patterns in particular for substituting local productions to commodities produced elsewhere. The market can also induce price shocks at the local level lowering the households' spending capacity and thus food access.

The dynamics of the model are specified in ninety procedures. While the full list and brief description is available online,[2] here we emphasize two procedures that define labour (see pseudo-code below) and land allocation (Fig. 4). In the first case, total labour availability (in hours per month) is calculated for the household. This is based upon the household typology (hh-type), number of working adults, gender and health. The proportion of time spent by the household on different productive activities is then determined according to household typology using averages calculated from IHS3 data.

Pseudo-code for the labour allocation procedure

```
PROGRAM AllocateLabour
FOR each individual
  IF (householdType = "Farmer")
    IF (age > 15 and age <65 and gender = "M" and illness = "F")
    indLabour = 80;
    ELSE IF (age > 15 and age <65 and gender = "M" and illness = "T")
    indLabour = 78;
    ELSE IF (age > 9 and age < 65 and gender = "F" and illness = "F")
    indLabour = 100;
    ELSE IF (age > 9 and age < 65 and gender = "F" and illness = "T")
    indLabour = 88;
    // continue for all household types
  END IF
END FOR
```

[2]Documentation of the model procedures is available here: http://tinyurl.com/nhmx9at.

```
FOR each household
  tLabour = SUM indLabour of individuals;
  IF (householdType = "Farmer" and month = "Jan")
  fwLabour = 0.06 * tLabour; // labour for firewood collection
  wcLabour = 0.15 * tLabour; // labour for water collection
  fhLabour = 0.72 * tLabour; // labour for on-farm agri. activities
  foLabour = 0.05 * tLabour; // labour for odd-farm agri. activities
  naLabour = 0.03 * tLabour; // labour for off-farm non agri. work
  // continue for all household types and each month of the year
  END IF
END FOR
END
```

While labour allocation is statically derived from household survey data, land allocation is dynamically contingent upon key variables. The procedure in which households set long term land allocation decisions occurs at the start of the growing season. Using simple decision trees calibrated for each household typology, for each patch of land owned, households choose between eleven combinations of 6 different crop types. Decisions are constrained by land, labour, subsistence, input and knowledge requirements.

As described in Fig. 4, if the requirements are met, the household goes on to allocate land according to the 'farm-vegetables', 'farm-food-crops', 'farm-cash-crops' or 'plant-fruit-trees' sub-procedures. If labour requirements aren't met, households can use the 'seek-labour' sub-procedure to hire labour from their social network, otherwise the land is left to fallow using the 'fallow' sub-procedure.

Figure 5 describes the logic following the decision to plant food crops. Every household typology has a different probability to pick a certain crop pattern among basic grains only, nuts and pulses, annual roots, and a combination (intercropping) of grains and nuts and pulses. Households will adjust their labour and land availability accordingly, while accounting for an increased level of subsidence.

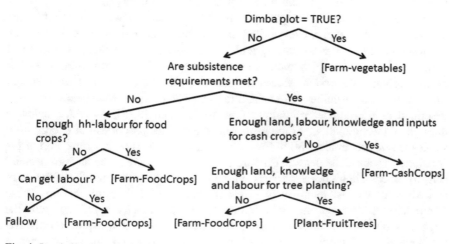

Fig. 4 Land allocation decision tree

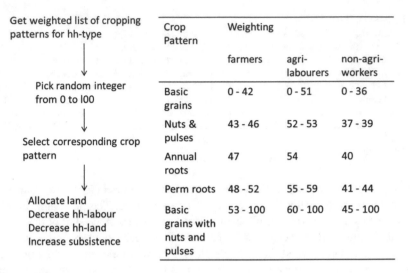

Fig. 5 Farm-FoodCrops decision logic

3 Model Analysis and Preliminary Results

Validation of ABMs is a crucial stage of the development process. However, the inherently complex nature of ABMs poses a significant challenge to validation efforts [19]. A recent review by Bert et al. [5] identified two distinct, but complementary approaches to model validation. These comprise: (1) validation of model components and processes and (2) empirical validation. For the former, face validation is employed to ensure that model mechanisms and properties correspond to those of the real world [22]. While for the latter, simulation outputs are validated against empirical data representing one or more stylized facts [5].

Preliminary simulation outputs for the ABM described within this study are summarized within Fig. 6. Non-agricultural workers appeared the most food secure when compared to farmers and agricultural labourers. On average, over 80 % of agricultural labourers and farmers were food energy deficient, compared to 71.1 % of non-agricultural workers. Values for diet diversity and the proportion of energy from staples were similar across all three household types. Mean diet diversity, described as the number of unique food types eaten [0, 8], remained below 4 throughout the simulation. At the same time the average proportion of energy from staples was approximately 80 % for each household. The monthly time step of the simulation enabled the seasonal nature of food security to be taken into account (Fig. 6). Overall, households were most food secure between the months of February and May, as the vast majority of crops are harvested during this period. These preliminary results are in line with traditional statistical analysis [25] and with the ongoing global urbanization process [3]. According to [3] the aspiration of rural smallholders to abandon subsistence farming for more secure income producing activities has driven urbanization in poor countries.

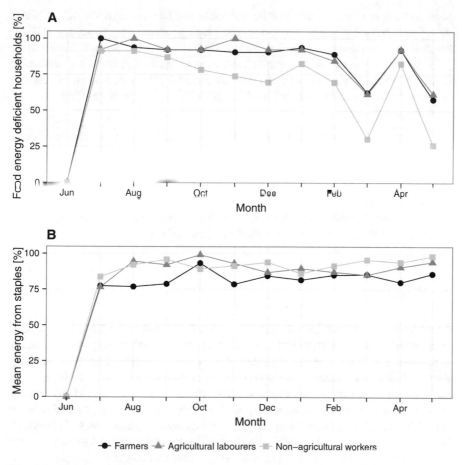

Fig. 6 Preliminary results from a single model run. (**a**) Energy deficient households. (**b**) Food energy from staples

Thorough calibration of certain model procedures, however, is required to verify results. Agricultural yield, for example, is much lower than expected with an annual average of just 20.4 kg/ha for farm plots and 9.6 kg/ha for dimba plots. Low crop yields such as these may be explained by the large proportion of land left to fallow. Currently, over the course of an average simulation 33.7 % of agricultural land is left to fallow. Survey data corresponding to the study area of interest, however, suggests that just 1 % of land was left to fallow in 2010–2011 [21]. We are therefore envisioning a modified land allocation procedure (Fig. 4) to account for the fact that high population densities and sustained parcelization of land limit the practice of fallow in rural Malawi.

In addition to the aforementioned calibration needs, a workshop was held at the end of January 2015 in Southampton to validate the structure of the ABM using expert knowledge (See Acknowledgments), this will lead to further model fine-tuning. Decision trees were tested using a role-playing game in which participants took on the role of one of the three household types: farmers, agricultural labourers or non-agricultural workers. The use of role-playing games to construct and validate the behavioural rules of agents has been widely documented for empirical ABMs [17]. In this study, by taking a participatory approach to validation we were able to engage experts from a range of domains including the social sciences, natural sciences and economics. A number of suggested improvements resulted from the workshop, including:

- To stagger land allocation decisions of households to limit the number of farm plots left to fallow.
- To allow households to allocate crops to larger farmer patches first, rather than at random.
- To include the possibility to rent land.
- To improve interactions between households by refining the scheduling of food sharing and money borrowing within the social network.

4 Discussion

Overall, this paper describes the development of an empirical ABM of household food security in a developing country context. Further calibration and validation is required before the simulation outputs can be used to evaluate the food security status of households accurately. However, in its current form the model does address the multi-dimensional nature of food security. In contrast to existing ABMs that act to address food security, we have proposed a modeling framework that adheres to the FAO four pillars of food security (availability, access, utilization and stability). By characterizing distinct household types within the study area of interest, namely farmers, agricultural labourers and non-agricultural workers it is also possible to take into account the heterogeneous nature of rural households. Furthermore, the use of decision trees acts as a departure from the notion of rational agents, by employing fast and frugal heuristics.

The design and communication of ABMs can pose significant challenges. There is a tradeoff between model simplicity, generality and truth [19]. Indeed, a recent survey by Waldherr and Wijermans [29] found common criticism on social simulation models being too complex, too simple, not theory based, not realistic, a black box, etc. In order to overcome these issues, the ABM of food security introduced in this paper was described using multiple standard narratives and graphical protocols, including: ODD, UML and pseudo-code. Even so we recognize that these efforts are unsatisfactory for the sake of analytical replicability [20]. The entire code, developed with NetLogo 5.1, will be published once the authors are satisfied with the model. At the same time the general modeling framework, as presented in this article, is of greater value in view of modeling food security in compliance with the up-to-date food security literature.

This study emphasizes the development of models of complex social-ecological systems as an iterative process. Future work will build upon the characterization of household behaviour, the environment and their interactions.

Once the model has been vigorously tested techniques from exploratory modeling analysis (EMA) will be used to further explore model uncertainty and enable the development of robust policies. Although preliminary results seem to suggest that moving away from subsistence agriculture improves food security, we are particularly interested to explore the role of ecosystem services, on the one hand, and the role of food and commodities market, on the other hand. This is the focus of our future research. It is hoped that future applications of the ABM may enable both scientists and decision makers to better understand the complex, dynamic nature of household food security within a developing country context.

Acknowledgements This work was supported by an EPSRC Doctoral Training Centre grant (EP/G03690X/1). It was also inspired by an ESPA funded research project: Attaining Sustainable Services from Ecosystems (ASSETS), which aims to "explicitly quantify the linkages between ecosystem services that affect—and are affected by—food security and nutritional health for the rural poor at the forest-agricultural interface" (http://espa-assets.org/). The participatory rural appraisal (PRA) carried out by the University of Southampton and LEAD-SEA, Malawi in the Zomba and Machinga districts of Malawi (6 villages of Chilwa East and West areas) was subjectively interpreted by the authors to build the model general framework which was subsequently assessed by project experts. The authors wish to thank the advice and support of: Kate Schreckenberg, Carlos Torres Vitolas (PRA experts), Nyovani Madise, Dalitso Kafumbata, Patrick Linkongwe (local experts), and Simon Willcock, Ferdinando Villa and James Dyke (modeling experts).

References

1. An, L.: Modeling human decisions in coupled human and natural systems: review of agent-based models. Ecol. Model. **229**, 25–36 (2012)
2. Balbi, S., Giupponi, C.: Agent-based modelling of socio-ecosystems: a methodology for the analysis of adaptation to climate change. Int. J. Agents Technol. Syst. **2**(4), 17–38 (2010)
3. World Bank: Global Monitoring Report 2013: Rural-Urban Dynamics and the Millennium Development Goals. World Bank, Washington (2013). doi:10.1596/978-0-8213-9806-7
4. Barlett, P.F.: Agricultural Decision Making: Anthropological Contributions to Rural Development. Academic, New York (1984)
5. Bert, F.E., Rovere, S.L., Macal, C.M., North, M.J., Podestá, G.P.: Lessons from a comprehensive validation of an agent based-model: the experience of the pampas model of Argentinean agricultural systems. Ecol. Model. **273**, 284–298 (2014)
6. Burchi, F., De Muro, P.: From food availability to nutritional capabilities: advancing food security analysis. Food Policy **60**, 10–19 (2016)
7. Dobbie, S., Dyke, J.G., Schreckenberg, K.: Unpacking diversity: typology creation and livelihoods analysis to support food security policy in rural Southern Malawi. Food Security (in review)
8. Dreze, J., Sen, A.K., et al.: Hunger and Public Action. Clarendon, Oxford (1991)
9. Dury, J., Schaller, N., Garcia, F., Reynaud, A., Bergez, J.E.: Models to support cropping plan and crop rotation decisions. A review. Agron. Sustain. Dev. **32**(2), 567–580 (2012)

10. FAO: Voluntary guidelines to support the progressive realization of the right to adequate food in the context of national food security (2005). http://www.fao.org/docrep/meeting/009/y9825e/y9825e00.htm. Accessed 11 April 2013
11. Farmer, J.D., Foley, D.: The economy needs agent-based modelling. Nature **460**(7256), 685–686 (2009)
12. Gigerenzer, G., Gaissmaier, W.: Heuristic decision making. Annu. Rev. Psychol. **62**, 451–482 (2011)
13. Gigerenzer, G., Goldstein, D.G.: The recognition heuristic: a decade of research. Judgm. Decis. Mak. **6**(1), 100–121 (2011)
14. Gigerenzer, G., Todd, P.M.: Fast and frugal heuristics: the adaptive toolbox. In: Simple Heuristics that Make Us Smart, pp. 3–34. Oxford University Press, Oxford (1999)
15. Grimm, V., Berger, U., Bastiansen, F., Eliassen, S., Ginot, V., Giske, J., Goss-Custard, J., Grand, T., Heinz, S.K., Huse, G., et al.: A standard protocol for describing individual-based and agent-based models. Ecol. Model. **198**(1), 115–126 (2006)
16. IFAD: Republic of Malawi sustainable agriculture production programme (sapp), programme design report (2011). http://www.ifad.org/operations/projects/design/104/malawi.pdf. Accessed 11 April 2013
17. Le Page, C., Becu, N., Bommel, P., Bousquet, F.: Participatory agent-based simulation for renewable resource management: the role of the cormas simulation platform to nurture a community of practice. J. Artif. Soc. Soc. Simul. **15**(1), 10 (2012)
18. Misselhorn, A., Aggarwal, P., Ericksen, P., Gregory, P., Horn-Phathanothai, L., Ingram, J., Wiebe, K.: A vision for attaining food security. Curr. Opin. Environ. Sustain. **4**(1), 7–17 (2012)
19. Moss, S.: Alternative approaches to the empirical validation of agent-based models. J. Artif. Soc. Soc. Simul. **11**(1), 5 (2008)
20. Müller, B., Balbi, S., Buchmann, C.M., De Sousa, L., Dressler, G., Groeneveld, J., Klassert, C.J., Le, Q.B., Millington, J.D., Nolzen, H., et al.: Standardised and transparent model descriptions for agent-based models: current status and prospects. Environ. Model. Softw. **55**, 156–163 (2014)
21. NSO: Integrated household survey 2010/11 (ihs3) (2012)
22. Rand, W., Rust, R.T.: Agent-based modeling in marketing: guidelines for rigor. Int. J. Res. Mark. **28**(3), 181–193 (2011)
23. Sahley, C., Groelsema, B., Marchione, T., Nelson, D.: The Governance Dimensions of Food Security in Malawi, vol. 20. USAID, Washington, DC (2005)
24. Sen, A.: Development: which way now? Econ. J. **93**(372), 745–762 (1983)
25. Smith, L.C., Subandoro, A.: Measuring Food Security Using Household Expenditure Surveys, vol. 3. International Food Policy Research Institute, Washington, DC (2007)
26. The Rome declaration on world food security. Popul. Dev. Rev. **22**(4), 807–809 (1996). http://www.jstor.org/stable/2137827
27. UN: Report of the World Food Conference, Rome 5–16 November 1974. New York (1975)
28. van Wijk, M., Rufino, M., Enahoro, D., Parsons, D., Silvestri, S., Valdivia, R., Herrero, M.: Farm household models to analyse food security in a changing climate: a review. Glob. Food Secur. **3**(2), 77–84 (2014)
29. Waldherr, A., Wijermans, N.: Communicating social simulation models to sceptical minds. J. Artif. Soc. Soc. Simul. **16**(4), 13 (2013)
30. Watts, D.J., Strogatz, S.H.: Collective dynamics of "small-world" networks. Nature **393**(6684), 440–442 (1998)
31. Wilensky, U.: NetLogo (1999). http://ccl.northwestern.edu/netlogo/

Comparing Income Replacement Rate by Prefecture in Japanese Pension System

Nisuo Du and Tadahiko Murata

Abstract In this study, we examine an income replacement rate of each prefecture in Japanese public pension system using an agent-based simulation. OECD Glossary of Statistical Terms defines the income replacement rate as a ratio of an individual's pension (P) and the average income (I), that is, P/I. On the basis the statistics of each prefecture, we calculate the amount of pension, wage structure, and marriage behavior using the agent-based model. From simulation results, we try to see circumstances of current pensioners in each prefecture based on the population change, the income replacement rate, and the marital state relationship.

Keywords Income replacement rate • Pension system • Agent-based social simulation

1 Introduction

Japanese pension system employs the "pay-as-you-go" system. Under the pension system, the current workers pay the fund of pension for the current recipients. Although yearly amount of pension is adjusted in conjunction with the commodity price, the yearly amount of pension stays higher than the original level of the yearly amount of pension because of a special measure of a commodity price slide, the special measure of a commodity price slide comes into force in 2015. Paying higher amount of pension will cause the fund shortage for younger generations. In addition, the prediction of Japanese demographic for coming years indicates that Japan will experience a declining birth rate and an aging population. The change in the demographic structure will cause different contribution in the pension system among generations. In 2004, the Japanese government enhanced the sustainability of the public pension system and reviewed a reform of the public pension system [1].

N. Du
Graduate School of Informatics, Kansai University, Takatsuki, Japan
e-mail: maosuo131@gmail.com

T. Murata (✉)
Faculty of Informatics, Kansai University, Takatsuki, Japan
e-mail: murata@kansai-u.ac.jp

© Springer International Publishing AG 2017
W. Jager et al. (eds.), *Advances in Social Simulation 2015*, Advances in Intelligent Systems and Computing 528, DOI 10.1007/978-3-319-47253-9_8

Since global population aging in the twenty-first century, the pension financing has been becoming the most serious problem in developed countries. In order to verify the sustainability of the pension system under the problem of aging population, many researchers have carried out numerous analytic studies using simulation. Boldrin et al. [2] calculated the pension expenditure percentage of GDP in European countries. Viehweger and Jagalski [3] employed a system dynamics model to analyze the German public senile social security program in 2002. Their simulation results focused on the macro effects, such as population structure change and pension financing. In addition, Kapteyn et al. [4] and Fehr et al. [5] focused on the interaction between retirement decision and pension finance. The Japanese public pension system was analyzed through estimating the sum of the benefits and insurance payments of individuals in Hirata et al. [6] using simulation.

There are three types of micro-simulation model for examining Japanese pension system. They are household information analysis model (INAHSIM) by Inagaki [7–10], a pension analysis model using the dynamic micro simulation techniques (PENMOD) by Shiraishi [11] and another micro-simulation model, that is CAM-MOD by Chen and Murata [12].

Most of Japanese social security programs including pension system have been designed on assumptions of model cases of average household. However, as pointed out by a recent study [13], all households in Japanese society can no longer be classified as model cases because industrial structures change and people tend to delay marriage. Traditionally, there are two typical household types in Japanese society. In the first-type household, the husband is a fulltime employee. He works for the company after graduation from school until retirement and his wife's occupation is limited such as a part-time worker or a fulltime housewife. After husband's retirement, they live on pension benefits. In the other type household, the husband is a self-employer. He takes over the family business, and his wife assists him as a family worker. There is no specific age of retirement for self-employees. After they reach an advanced age, they leave their family business for their children and get supported from them while living on pension benefits. Those two household model cases can stand with assumptions such as "every fulltime employees can work until retirement unless they choose to leave," "all persons can get married," "couples never divorce," "family business can last forever," just to list a few. More and more people are involved in irregular employment, and do not have traditional marriage. For example, the divorce rate is increasing. Therefore, it is more than obvious that the pension system designed based on limited model cases does not fit the reality anymore.

By performing micro-simulation with many agents, we are able to discuss the impact on each individual in some topics such as social security, and tax policies. For example, a new tax policy has a possibility of expanding the income gap among citizens using the simulation based on the life cycle model [14].

In previous research, Du and Murata [15] showed transitions of population trends, household structure, household type, employment status, and wage structure using a micro-simulation method. However, the previous research [15] only provided the analysis over a nation. In this study, we develop simulation models using parameters of each prefecture such as wage growth rate, average wages, bonuses, age-specific population, age-specific couple population, insured population of 5-year-old division, and beneficiary population. From simulation results, we examine population fluctuations, income replacement rate, and marital state in the prefecture level. Among them, OECD Glossary of Statistical Terms defines the income replacement rate as a ratio of an individual's pension (P) and the average income (I), that is, P/I. Using agent-based model, we are able to compare the income replacement rate by marital status in each prefecture. We can see that the income replacement rates differ from prefecture to prefecture. That will cause a difference in the quality of life by prefectures.

2 Japanese Pension System

Figure 1 shows the outline of the current Japanese Pension System. The Japanese pension system consists of public pension and private pension. All residents aged between 20 and 65 join compulsorily the first tier National Pension (NP). In the second-tier, employees in the private sector join compulsorily the Employees' Pension Insurance (EPI), and civil servants in governments should join Mutual Aid Associations (MAAs). These are included in Public Pension. Private pensions include corporate pension funds and private savings plans on a voluntary basis. To make simulation model simpler, we mainly target at retirement pension of NP and EPI (MAAs is regarded as part of EPI in simulations because of almost the same nature of pension scheme).

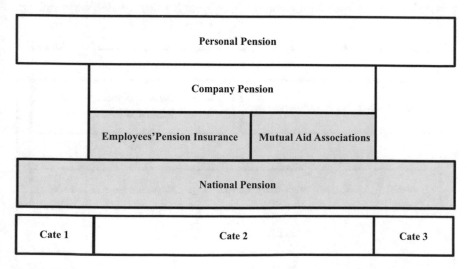

Fig. 1 Pension system

The NP divides subscribers into three categories (Cate 1, Cate 2, and Cate 3 in Fig. 1). In any category, those who are from 20 to 65 in age should pay insurance premium. The first category consists of self-employed persons and their families, students, and unemployed persons. They are referred to as the first insured persons. The first insured persons need to pay flat rate insurance premiums (the monthly insurance premium was 14,660 yen in 2009, and it becomes 15,250 yen in 2014). The second category consists of private company employees and civil servants. They are referred to as the second insured persons. The second insured persons need to pay earnings-related insurance premiums (the insurance rates was 15.704 % in 2005, and it becomes 17.474 % in 2014). The third category are spouses of the second insured persons, they don't work, and are dependent on the second insured persons. They are referred to as the third insured persons. The third insured persons' don't need to pay their insurance premiums individually because their premium is included in the payment of their spouses. As for pension benefits, depending on the number of payments of basic pension in the first tier of the system, all subscribers to meet the requirement can receive the basic pension. The second insured persons benefit from pension based on the ratio of payment premiums.

3 Simulation Model

In this study, we employ CAMMOD that is a model proposed in our previous study [15]. We show a schematic diagram of the model in Fig. 2. Our model has agents whose attributes are based on demographic statistics, economic parameters, and pension finance. Agents are created according to statistical data of age-specific population and age-specific pensioners. Each agent has age, gender, wage records,

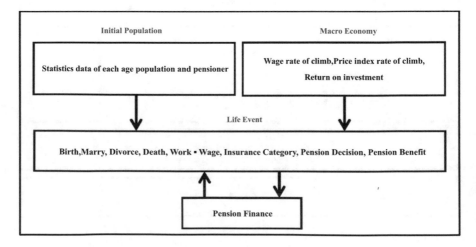

Fig. 2 CAMMOD in [15]

insurance payment record, and insurance benefit record. As for economic indicators, we employ a rate of wage increase, rate of price increase, and return on investment in the macro economy. As for life events of each agent, it starts to have a job stochastically based on the age-sex-specific working rate when it becomes Age 20. Agents at work earn wages that vary stochastically according to the average wage in that year and age-sex-specific wage structure rate. After they reach 65 years old, they start to receive pension benefit until they die. New agents are generated according to the birth rate, and their death is determined by the annual survival rate. The pension finance calculates premium payment according to insurance type of agents and pension benefit. About premium payment, agents at work pay pension premiums according to their insurance type and their wage in that year. About pension benefit, agents receive pension benefit calculated from their payment records.

3.1 Demographic

The initial population is generated according to a demographic data by 1-year-age group of 2004 of each prefecture (when there is no demographic data by 1-year-age group but by 5-year-age group in a certain prefecture, we equally allocate to each age group from 5-year-age group). Simulation of demographic change contains predictions about the number of future birth and the number of death. As for the birth population of each prefecture, we calculate it by multiplying the percentage of the population of each prefecture to the birth population of the country that is estimated by an intermediate scenario by the National Institute of Population and Social Security Research in 2006 [16]. As for the mortality, we determine the mortality rate of the agent by the 20th life table issued from the Ministry of Health, Labor and Welfare [17].

3.2 Insured and Beneficiary

As shown in Sect. 2, each Japanese should belong to one of three types of insurance. First insured persons receive their pension only from the national pension of first tier. Second insured persons receive from the national pension and their welfare pension. Third insured persons receive from the national pension. As for the movement of insured persons from one category to another, it is caused by changing their jobs or getting married. In order to assign the insurance category to each agent, we employ the statistical data with the number of insured and beneficiary in each prefecture [18].

3.3 Individual Wages and Employment

The wage of each agent is generated according to the average wage distribution of the wage structure rate with age groups. In our simulation, the amount of wage increases according to the wage growth rate. We employ the wage structure rate that is estimated by the Ministry of Health, Labour and Welfare (MHLW) in the Basic Survey on Wage Structure of each prefecture [19]. Although the real wage structure changes gradually, we handle it as a constant in our simulation. In addition, agents at work depend on the age-specific employment rate estimated by age-sex-specific working rate of census of Ministry of Internal Affairs and Communications (MIAC) [20].

3.4 Marriage and Divorce

The population of annual marriage is calculated according to age-specific unmarried rate for men and women. Agents who get married are determined in consideration of their occupation and the age difference between couples. As for the marriage rate, during the years from 2005 to 2010, we employ the marriage rate for 2005 to 2010, after 2011, we employ the marriage rate in 2010. Since each prefecture does not collect the marriage rate, we employ the first marriage rate in the age-specific unmarried data from the National Institute of Population and Social Security Research announced report (20–69-year-old, and 70 years of age or older, 5-year-old separated) [21]. That means the same marriage rate is employed in every prefecture. About the occupation-specific marriage rate, we employ the data of the marriage of the first insured persons and the second insured persons that Ministry of Internal Affairs and Communications has announced in 2007 Employment Status Survey [20]. As for the divorce rate, we employ the divorce rate in the age-specific married data from the National Institute of Population and Social Security Research announced report [22].

4 Simulation Results

4.1 Income Replacement Rate

We show the income replacement rate of five prefectures in Japan: Hokkaido, Tokyo, Shizuoka, Kyoto, and Fukuoka. Figures 3 and 4 show average income replacement rate over 100 trials by household marital status in 2015 and 2050, respectively. In Figs. 3, 4, and 5, "m" and "f" mean men and women, respectively. "1," "2," "3" mean first insured person, the second insured person, and the third insured person, respectively. As shown in Figs. 3 and 4, Hokkaido and Fukuoka have

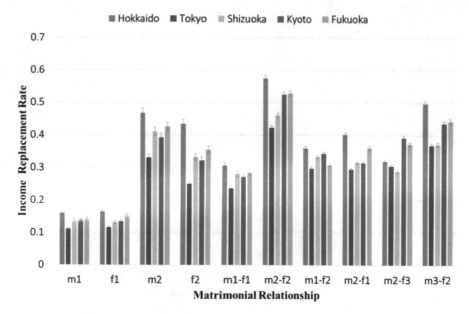

Fig. 3 Changes of income rate (2015)

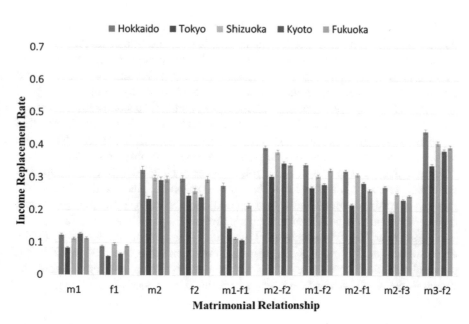

Fig. 4 Changes of income rate (2050)

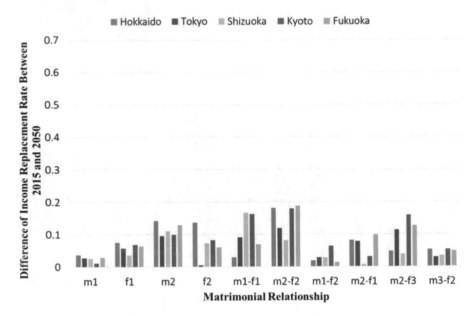

Fig. 5 Difference of income replacement rate between 2015 and 2050

the high-income replacement rates than the other prefectures. On the other hand, the income replacement rate is low in Tokyo, Kyoto, and Shizuoka. Although the average wage and the wage growth rate of each region are different, the uniform pension premium should be paid by contributors. This is because the wage growth rate is larger in urban areas than rural areas. In addition, by comparing Figs. 3 and 4, we can see that the average income replacement rate becomes lower in every household type in 2050 when it is compared to 2015. In Fig. 5, we can observe that the average income replacement rate of households is much lower in 2050 for the households with the second insured person. We can also observe that the average income replacement rate of m2-f2 is much lower in Hokkaido, Tokyo, Kyoto, and Fukuoka. And we can observe that the average income replacement rate of m1-f1 is much lower in Shizuoka and Kyoto. In addition, we can see the average income replacement rate of m1-f1 is lower than single household such as m2 or f2.

In Figs. 3 and 4, we show standard deviation of income replacement, respectively. As shown in Fig. 3, we can see all standard deviation are below 0.04 in 2015. All standard deviations become reduced with the decrease of income replacement. As shown in Fig. 4, all standard deviation will below 0.02 in 2050.

4.2 Population Trends

In order to consider the difference of the income replacement rate by regions, it is important to estimate population in each prefecture. In this chapter, we employ regions with the following characteristics: metropolitan regions, regional urban center, and regions away from metropolitan regions. In our simulation, we employ the data of the nine prefectures in Japan: Hokkaido, Aomori, Tokyo, Kanagawa, Shizuoka, Kyoto, Nara, Fukuoka, and Kagoshima [23–31]. The population trend of each prefecture is shown in Figs. 6, 7, and 8. The figures of population trend of each prefecture show simulation results up to 2120 along with the statistical data up to 2014. In Figs. 6, 7, and 8 we show population trend of each prefecture data up 2010-2055 by the National Institute of Population and Social Security Research (IPSS) [32].

As shown in Fig. 6, the population of statistical data changes higher than the simulation results in metropolitan regions and regions close to the metropolitan regions (Tokyo, Kanagawa, Fukuoka). In Fig. 7, regional urban centers (Hokkaido, Kyoto, Kagoshima) show a small difference between the simulation results and the statistical data. On the other hand, the simulation result of Shizuoka is below the statistical data in Fig. 8. The simulation result and the statistical data are almost identical in Nara, and Aomori. We can see similar drop tendency in our simulation results and IPSS results.

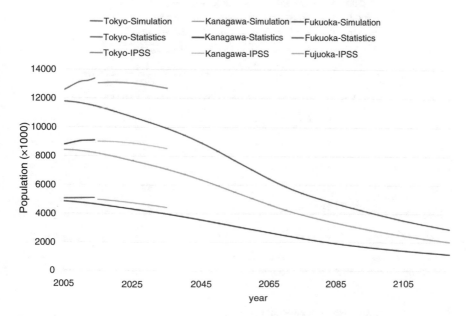

Fig. 6 Metropolitan regions and regions close to the metropolitan region

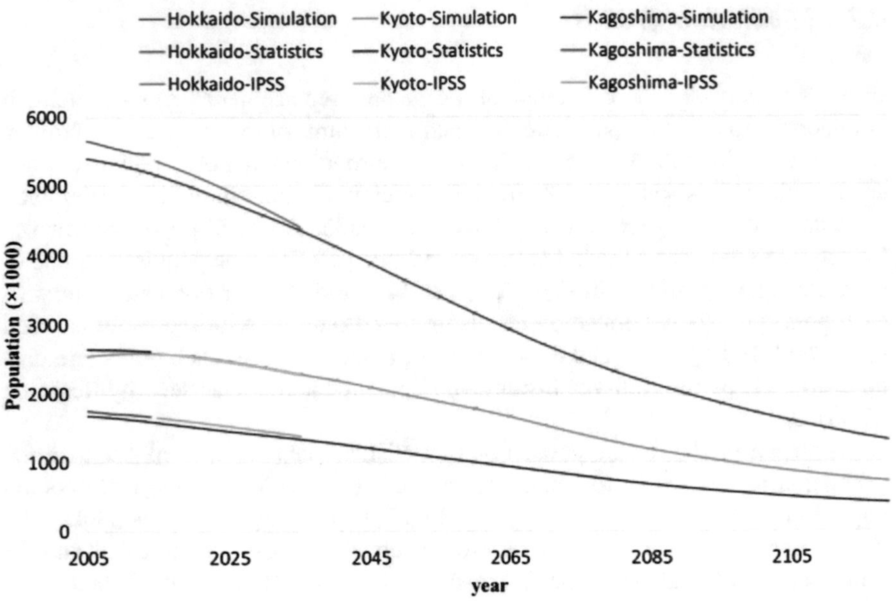

Fig. 7 Regional urban center

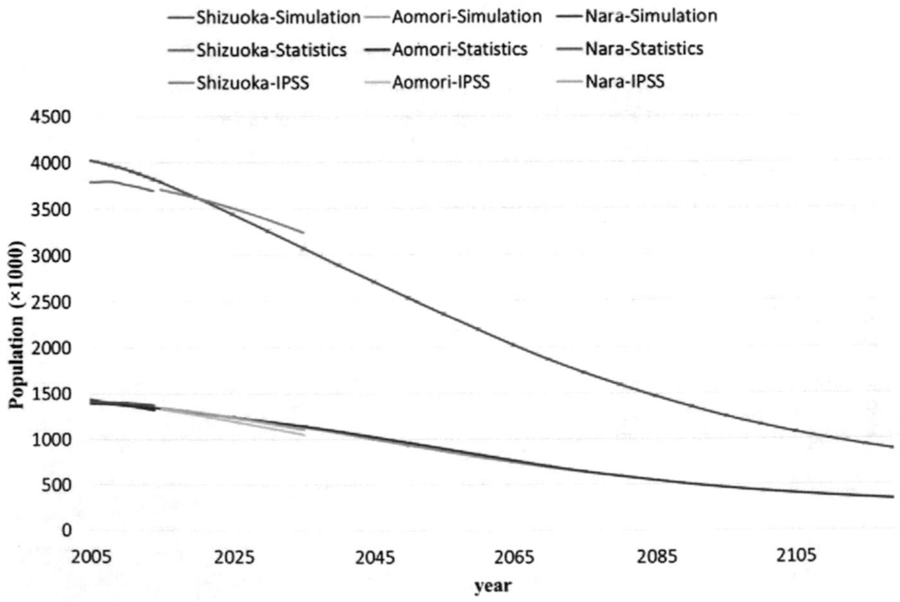

Fig. 8 Regions away from the metropolitan region

In our simulation, the characteristic of population changes varies in three types as shown in Fig. 6, 7, and 8: Metropolitan region, regional urban centers, and regions away from metropolitan region. This difference may come from the difference of the birth rate employed in each prefecture. In this chapter, we employ the same birth rate of the national statistics. In fact, many young people are rearing child in urban areas such as metropolitan region or regions close to metropolitan regions. In those regions, the population increases more than the average increase rate. That is observed in Fig. 6. On the other hand, young generation reduces in regions away from metropolitan regions such as Shizuoka.

We show the standard deviation of population change in the nine prefectures in Table 1. As shown in Table 1, we can see that values of all standard deviation are very small when compared to population. And all standard deviations tend to growth until 2030 or 2060. After that, all standard deviations will gradually become smaller. There are two reasons why standard deviation changes small. The first reason is the constant value of the birth population. Birth population in every year is fixed value in our simulation. The second reason is a small change in population. Every prefecture of population base tends to decrease over time. That is to say, kinds of ratio do not have overlap effect in our simulation.

5 Conclusion

In this chapter, we examine the income replacement rate of each prefecture in the pay-as-you-go Japanese public pension system using an agent-based simulation. The simulation period is over 115 years from 2005 to 2120. We show the income replacement rate of marital status and population trends of several prefectures. We can see that the income replacement is higher in rural areas than urban areas. And the average income replacement rates in 2050 become lower than 2015. As for the population trends, we see some differences of birth rate when we employ the national birth rate. As for the future, population fluctuations must also be taken into consideration due to population transference between each prefecture. With the improved estimation, we try to examine the value of public pension in each prefecture by comparing the income replacement rate of each prefecture. In addition, we try to employ different parameters, and then will try to discuss ways of maintaining income replacement in different prefectures in Japan.

Table 1 Standard deviation of population (1000×)

Year	Tokyo	Kanagawa	Fukuoka	Hokkaido	Kyoto	Kagoshima	Shizuoka	Aomori	Nara
2005	9	8	6	8	3	4	6	3	3
2006	13	12	9	10	4	7	9	4	4
2007	16	15	12	12	6	8	11	5	5
2008	18	17	14	12	6	10	12	7	6
2009	18	18	13	13	7	10	13	7	5
2010	19	19	14	13	8	9	14	7	6
2011	22	19	16	14	9	9	14	7	7
2012	24	19	17	14	9	10	15	7	8
2013	23	20	17	15	9	9	15	8	8
2014	22	20	18	17	10	10	16	8	8
2015	23	20	17	18	11	11	16	10	8
2020	30	20	17	20	13	11	18	10	9
2025	29	26	17	18	13	11	18	10	10
2030	31	26	18	22	16	11	17	9	10
2035	32	27	23	21	14	11	18	10	11
2040	35	29	21	21	14	11	16	11	10
2045	31	26	21	22	14	12	14	12	11
2050	34	27	19	20	14	10	16	13	11
2055	33	24	22	20	15	9	16	11	11
2060	30	23	21	18	16	10	16	11	11
2065	31	27	19	17	15	9	14	11	10
2070	29	26	19	14	12	9	14	9	9
2075	26	22	18	17	12	8	15	9	9
2080	23	22	17	17	12	10	15	9	9
2085	22	23	16	15	11	9	12	9	9
2090	26	21	15	15	11	8	13	10	8
2095	25	19	13	16	11	9	13	8	8
2100	21	19	15	17	9	9	13	7	8
2105	22	16	13	16	9	8	12	7	7
2110	21	16	13	14	9	8	11	6	7
2115	21	16	13	13	9	8	11	6	7
2119	19	16	13	12	8	7	11	7	7

References

1. Ministry of Health, Labour and Welfare, The 2004 Actuarial Valuation of the Employees' Pension Insurance and the National Pension, http://www.mhlw.go.jp/topics/nenkin/zaisei/zaisei/04/index.html. (2006) (in Japanese)
2. Boldrin, M., Jose, J., Dolado, Jimeno, J.F., Peracchi, F.: The future of pensions in Europe. Econ. Pol. **14**, 287–320 (1999)
3. Viehweger, B, Jagalski, T.: The reformed pension system in germany – a system dynamics model for the next 50 years. 21st System Dynamics Conference Proceeding, no. 191, 10 p. in CDROM (2003)

4. Kapteyn, A, de Vos, K.: Simulation of pension reforms in The Netherlands. In: Gruber J., Wise D.A. (eds.) Social security programs and retirement around the world: fiscal implications of reform. (University of Chicago Press), Chapter 8, pp. 327–349. (2007)
5. Fehr, H., Kallweit, M., Kindermann, F.: Pension reform with variable retirement age—a simulation analysis for Germany. Netspar Discussion Paper No. 02/2010-013, pp. 1–33, (2010)
6. Hirata, T., Sakamoto, K., Ueda, M.: A simulation of the policy of public pension finance in Japan. Kawasaki J. Med. Welfare **13**(2), 127–136 (2008)
7. Inagaki, S.: Projections of the Japanese socioeconomic structure using a microsimulation model (INAHSIM), IPSS Discussion Paper Series, No. 2005-03, pp. 1–37. (2005)
8. Inagaki, S.: Effect of proposals for pension reform on the income distribution of the elderly in Japan. The Second General Conference of the International Microsimulation Association, Ottawa, June 8–10, 20 p. (2009)
9. Inagaki, S.: The effects of proposals for basic pension reform on the income distribution of the elderly in Japan. Rev Socionetwork Strategies **4**, 1–16 (2010)
10. Inagaki, S.: Simulating policy alternatives for public pension in Japan. The Third General Conference of the International Microsimulation Association, Stockholm, Sweden, 9 June, pp. 129–144. (2011)
11. Shiraishi, K.: The use of microsimulation models for pension analysis in Japan. PIE/CIS Discussion Paper, No. 409, pp. 1–50, (2008) (in Japanese)
12. Chen, Z., Murata, T.: Examination of possible progress of japanese pension system using an agent-based model. Proceedings of the 2nd International Symposium on Aware Computing, 6 p. (2010)
13. Yamada, M.: The Time of working poor reconstruction of social safety net . Bungeishunju (Tokyo, Japan), (2009) (in Japanese)
14. Hashimoto, K.: Inequality of consumption tax and the measure to ease it. Kaikei Kensa Kenkyu **41**, 35–53 (2010) (in Japanese)
15. Du, N., Murata, T.: Study on income replacement ratio in pension system using agent simulation. Proceedings of SICE 7th Symposium of Technical Committee on Social Systems, pp. 21–26, (2014) (in Japanese)
16. National Institute of Population and Social Security Research, estimated future population in Japan. http://www.ipss.go.jp/syoushika/tohkei/suikei07/suikei.html. (2006) (in Japanese)
17. Ministry of Health, Labour and Welfare.: The 20th Abridged Life Table. http://www.mhlw.go.jp/toukei/saikin/hw/life/20th/index.html. (2007) (in Japanese)
18. Social Insurance Agency Japan, Survey of Status of Public Pension, http://www.mhlw.go.jp/topics/bukyoku/nenkin/nenkin/toukei/dl/h16a.pdf. (2004) (in Japanese)
19. Ministry of Health, Labour and Welfare.: Basic statistical survey of japanese wage structure. http://www.e-stat.go.jp/SG1/estat/NewList.do?tid=000001011429. (2010)
20. Ministry of Health, Labour and Welfare. Demographic statistics. http://www.e-stat.go.jp/SG1/estat/GL08020103.do?_toGL08020103_&listID=000001101925&disp=Other&requestSender=dsearch (2007)
21. First Marriage Rates by Age of Groom and Bride to Never-Married Persons, Nuptiality, National Institute of Population and Social Security Research, http://www.ipss.go.jp/p-info/e/psj2012/PSJ2012.asp, 2012.
22. Divorce Rates by Age and Sex to Currently Married Persons, Nuptiality, National Institute of Population and Social Security Research. http://www.ipss.go.jp/p-info/e/psj2012/PSJ2012.asp. (2012)
23. Tokyo Statistical Data, Population Statistical Data.: http://www.toukei.metro.tokyo.jp/jsuikei/js-index4.htm (in Japanese)
24. Hokkaido Web Site, Population Statistical Data.: http://www.pref.hokkaido.lg.jp/ss/tuk/900brr/index2.htm (in Japanese)
25. Aomori Prefecture Web Site, Population Statistical Data.: http://www.pref.aomori.lg.jp (in Japanese)
26. Kanagawa Web Site, Population Statistical Data.: http://www.pref.kanagawa.jp/life/6/26/140/ (in Japanese)

27. Shizuoka Prefecture Web Site, Population Statistical Data.: http://toukei.pref.shizuoka.jp/chosa/02-030/ (in Japanese)
28. Kyoto Prefecture Web Site, Population Statistical Data.: http://www.pref.kyoto.jp/tokei/yearly/tokeisyo/tsname/tsg0207.html (in Japanese)
29. Nara Prefecture Web Site, Population Statistical Data.: http://www.pref.nara.jp/dd.aspx?menuid=6437 (in Japanese)
30. Fukuoka Prefecture Web Site, Population Statistical Data.: http://www.pref.fukuoka.lg.jp/dataweb/search-1-1619.html (in Japanese)
31. Kagoshima Prefecture Web Site, Population Statistical Data.: http://www.pref.kagoshima.jp/tokei/bunya/jinko/suikei/ (in Japanese)
32. Population & Household Projection, National Institute of Population and Social Security Research.: http://www.ipss.go.jp/pp-fuken/j/fuken2007/t-page.asp. (2007) (in Japanese)

Hybrid Simulation Approach for Technological Innovation Policy Making in Developing Countries

Maryam Ebrahimi

Abstract The aim of this study is to create a hybrid simulation model which is a combination of systems dynamic (SD) and agent-based modeling (ABM). It analyzes the market share of redesigned and independent designed technologies compared to the acquired ones. For this purpose, supply chains of technology suppliers and firms that trade on their own developed technologies through applying SD are modeled. In case of ABM, consumers' decisions are influenced by marketing activities, word-of-mouth between consumers, and work experience of the companies. Delivery time is a key variable that affects the performance of each company. Additionally, some policies are proposed regarding the significant impacts of marketing on absorbing consumers, collaboration between development and manufacturing on the production rate, and resources on time to innovate. The key finding is that by improving marketing, collaboration, and resource management, market share of new developed technology will be improved.

Keywords Agent-based modeling • Systems dynamic • Technological innovation policy planning

1 Introduction

It is assumed that the transfer of technology helps improve the technological abilities of the technology-recipient country. These abilities include investment ability, operational ability, and dynamic learning ability [1]. As argued by [2], the emergence of technological ability presented as a continuum expanded from the purchase of equipment by an acquirer (in principle, formation of financial exchanges with no technology transfer) to total technology transfer giving the acquirer equal technological cooperation with the owner. Along this continuum, four levels are designated in a technology ability ladder: (1) assembly or turnkey operations, (2) adaptation and localization of components, (3) product redesign,

M. Ebrahimi (✉)
Postdoctoral Fellow of the Alexander von Humboldt Foundation – Georg Forster Research, Information Systems Management, University of Bayreuth, Bayreuth, Germany
e-mail: mar.ebrahimi@gmail.com

© Springer International Publishing AG 2017   109
W. Jager et al. (eds.), *Advances in Social Simulation 2015*, Advances in Intelligent Systems and Computing 528, DOI 10.1007/978-3-319-47253-9_9

and (4) independent design of products. The first two levels are likely existed in the developing countries. Nevertheless, if they are not going to remain a follower in their efforts to compete in internationally, the developing countries will have to conduct their technology policies more and more at the third and fourth levels. This should take into consideration that gaining first-mover advantage according to the redesigned or independently adapted technology will not be easy for these countries.

Whereas the owner may have no tendency to manifest all about a technology, the acquirer desire to obtain as much understanding as possible to decrease dependency on the supplier [3]. Buying a new machine or adopting a turnkey project does not contribute an acquirer with a new technology as a replacement for technology transfer [2]. In addition, about identifying technology adapted to local conditions, automation does not necessarily result to more effective manufacturing since greater compliance with a set of procedures needs higher levels of know-how [4].

Besides, the move towards technology development affects economic development of nations. Despite the importance of innovation within each country, developing countries due to the lack of knowledge or high costs prefer to import technologies. While sometimes inappropriate or outdated technologies are selected because of political, legal, and cultural differences between various countries. That is why developing countries in addition to the concern of failures to utilize effectively transferred technologies consider the development of essential technologies in order to provide industry needs.

The purpose of this study is to create a hybrid simulation model which is a combination of systems dynamic (SD) and agent-based modeling (ABM) by using Anylogic. The subject of this research is to examine market share of redesigned and independent designed technologies compared to the acquired ones in developing countries. Because of choosing improper technologies and some political constraints, they attempt to strengthen research and development centers. As stated by [5], R&D organizations usually are the main agents in technology development, thus public research institutions and research policies in developing countries have recently gone through major transformations. For instance, they have been transforming its direction from library and laboratory research into technology development through scaling up from bench scale to industrial scale.

Companies in these countries are generally classified into two categories: suppliers of technologies who are responsible solely as the technology sales representatives, and firms with research and development who attempt to produce their own developed technologies. Each company has its own supply chain that delivers technologies to the end consumers. It is clear that there is delivery time difference between suppliers and the firms that produce their own developed technologies. In this study for the aim of comparing the market share of each technology, developed within the country and imported technology, supply chain for technologies suppliers and developers with the help of SD are modeled in addition to modeling the consumer market through using ABM. Then, models of supply chain and consumer market are integrated, simulated, and finally the results and policies are presented and discussed.

2 Methods

Supply chains are complicated dynamical systems triggered by customer demands. Proper management decisions impact on efficiency of supply chains, which are often based on intuition and experience [6]. SD in supply chain is used to develop an integrated inventory model of supply chain for deteriorating items among a supplier, a producer, and a buyer in [7]. The method of SD is applied to model supply chain information sharing by [8]. Fuzzy estimations and SD are adopted for improving supply chains in [9]. The importance of customer orientation for evaluation and improvement of supply chain process supported by an integrated usage of discrete-event simulations models and SD models is explained by [10]. SD methodology as a modeling and analysis tool to tackle strategic issues for food supply chains is used in [11]. Applying ABM, the impact of alternative production–sales policies on the diffusion of a new generic product and the generated profit is analyzed in [12]. ABM in order to integrate demand and supply for Smart Meters is applied by [13]. According to [14], SD allows a systematic examination of a complex system, supply chains of the companies, and ABM helps the exploration of individual-level theories, consumers' decisions.

3 Hybrid Simulation Model

3.1 Supply Chains: SD

Delivery time is a key variable that affects the performance. A fast response time usually means the customer gets a better impression of the company. The ability to deal with inquiries and fulfill orders quickly means the company is able to serve more customers, resulting in higher profit. If customers have to wait for responses or products, they may cancel orders and go to other companies. A product which hits the market too late cannot be successful. For developing a new technology as a product, it is necessary to harmonize all development stages—only in this way the product development time can be reduced [15]. Supply chain for firms that trade on their own developed products considers technological innovation as the transformation of an idea into a new technology as a product which should be commercialized to sale to a customer. Technology here means a field of technologies which is the working area of the suppliers and developers. To represent this supply chain, three main stock variables including current technological innovation, finished technological innovation, and finished technology production are applied.

This supply chain works as follows (see Fig. 1):

- The technology can be purchased by a consumer just from a stock of finished technology production, initially holding a certain amount of technology.

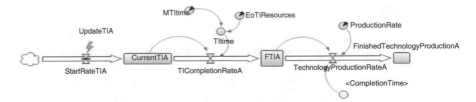

Fig. 1 Supply chain of firms that trade on their own developed products

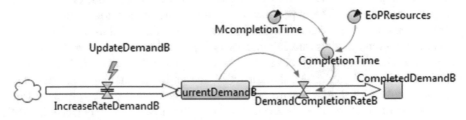

Fig. 2 Supply chain of suppliers

- The technology is initiated to be developed based on the start rate of techno-logical innovation per week, and this rate may vary because it can be adjusted according to the demand which is known to the firm.
- The current technological innovation will be completed within technological innovation time which is dependent on the sufficiency of resources and the average time to innovate which is supposed to be 48 weeks.
- The finished technological innovation will be produced according to the produc-tion rate.

Adding an event nearby the flow StartRateTIA, there will be a link between the AB and SD models. The action of the event is specified by:

$$\text{StartRateTIA} = (\text{consumers.NWantA ()} + \text{consumers.NWantAny ()}$$
$$+ \text{consumers.NUseAny ())} * 0.7; \tag{1}$$

Thus, at the beginning of each week the StartRateTIA will be modified according to the number of consumers who are willing to buy the product. NWantA() shows con-sumer population demanding TechnologyA, NwantAny means the customers who are dissatisfied of long time waiting for buying, and NuseAny shows dissatisfaction of customers due to dysfunctionality. In addition, just 70 % of orders are accepted after feasibility study analysis.

Supply chain of suppliers works as follows (see Fig. 2):

- The technology can be acquired by a consumer only from the stock of completed demand, initially holding a certain amount of technology.

- The technology is supplied based on the increase rate of demands per week, and this rate may vary because it can be adjusted according to the demand which is known to the firm.
- The current demands will be completed within completion time which is dependent on the sufficiency of resources and the average time to supply the technology which is assumed to be 8 weeks.

Therefore, two main stock variables including current demands and finished demands are applied which are related through demands completion rate as a flow variable.

Adding an event nearby the flow IncreaseRateDemandB, there will be a link between the AB and SD models. The action of the event is defined by:

$$\text{IncreaseRateDemandB} = \text{consumers.NWantB()} + \text{consumers.NWantAny()}$$
$$+ \text{consumers.NUseAny()}; \qquad (2)$$

Thus, at the beginning of each week the IncreaseRateDemandB will be modified according to the number of consumers who are willing to buy the product. NWantB() shows consumer population demanding TechnologyB.

In the models, resources effect on the time needed to fulfill technological innovation and also technology supplying. Resources include all assets, capabilities, organizational processes, firm attributes, information, knowledge, etc. controlled by a firm that enable the firm to conceive of and implement strategies that improve its efficiency and effectiveness [16].

3.2   Consumer Market: ABM

In this study, consumers are modeled as agents who make adoption or purchase decisions based on a word of mouth influences from local interactions with other agents (consumers). In order to model in agent-based way, the model of consumer market of technologies is developed in multiple phases with following order:

1. Concerning just one technology.
2. Considering supply chain for the selected technology.
3. Adding another technology (Fig. 3).

In the model of consumer market seven states (top down): PotentialCustomer, WantTechnologyA, WantTechnologyB, UseTechnologyA, UseTechnologyB, WantAnything, and UseAnything. Transitions from PotentialCustomer to WantA are AdoptionA, MarketingEffectivenessA, and PreviousExperienceCoA. In case of adoption, it is assumed that agents start talking to each other. The transition will be taken periodically and on each occurrence the agent-user will send a message to a random other agent saying that technology A is good. If another agent is a potential customer (i.e. is in the state PotentialCustomer), he will react to such a message by

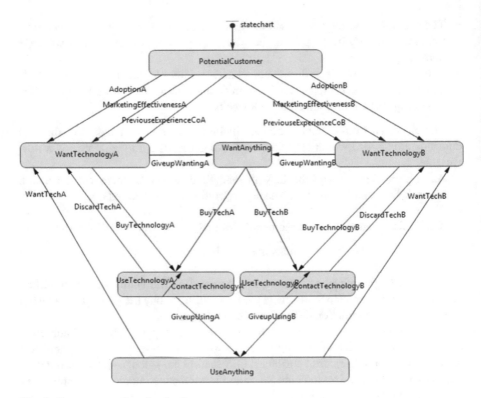

Fig. 3 Consumer market of technology

changing its state to WantTechnologyA. ContactTechnologyA is triggered by rate and its action is: send "Buy TechnologyA!," Random; and adoption is triggered by message which is: "Buy TechnologyA. Transition named as MarketingEffectivenessA will model the effect of marketing in case of requirement identification and advertisement. In addition, the transition of PreviousExperienceCoA is defined which shows work experience of the company influencing on its upcoming sales. These transitions can be presented in the same way for technology B.

The transition BuyTechnologyA is triggered by type condition. In this stage, AB and SD can be interacted. In this way that in the state of WantTechnologyA a consumer constantly monitors the stock of FinishedTechnologyProductionA; when the stock contains at least one package of technology, the transition is taken and, as a result, one package of technology A is deducted from the stock. BuyTechnologyA is triggered by condition which is: get_Main().FinishedTechnologyProductionA>=1 and the action is: get_Main().FinishedTechnologyProductionA−−;. Regarding BuyTechnologyB, in the state of WantTechnologyB a consumer constantly monitors the stock of CompletedDemandB and when the stock contains at least one package of technology, the transition is taken and one package of TechnologyB is deducted from the stock. BuyTechnologyB is triggered by condition which is: get_Main().CompletedDemandB >= 1 and the action is: get_Main().CompletedDemandB−;.

As the technology is discarded after a certain period, technology life time, and the user needs to buy a replacement, it is necessary to define a transition from UseTechnologyA to WantTechnologyA triggered by timeout with value uniform (96, 144). This transition is presented in the same way for technologyB. In this study, customer dissatisfaction has two meanings. Firstly, if a consumer waits too long for a particular technology, he becomes impatient and will buy whatever is available (A or B), so technology switching is possible. In this regard, there are two timeout transitions from WantTechnologyA and WantTechnologyB to a new state WantAnything. From that state, in turn, there will be two alternative condition-type transitions to UseTechnologyA and UseTechnologyB triggered by the purchase of the corresponding product. Secondly, UseAnything state shows if a customer after utilizing a particular technology for reasons related to functionality or quality could not keep on using the product, he will buy the accessible one. In this case, there are transitions named as GiveupUsingA and GiveupUsingB triggered by rates in addition to WantTechA and WantTechB triggered by timeouts.

4 Results

As is clear from the results in the Table 1, although the market share of technology A is growing slightly, but the supplier's market share is almost total. This means that with the assumed conditions, creating a new technology approach will not be successful. Many completed innovations do not reach to the production stage due to the lack of a precise definition of the needs of potential consumers and communication problems between technological innovation and production. Due to the cognitive and process difference, and interdependencies of their performance, unwittingly and unintentionally, they interpret other side activities potentially or actually injurious to their interests. Generally, innovation processes are time-consuming that due to resource constraints, the time will increase further. Thus, lack of attention to

Table 1 Results of SD and ABM before applying policies

Weeks		48	96	144	192	240
Technologies	CurrentDemandB	7.98	38.03	50.64	15.41	21.52
	CurrentTIA	21.51	99.39	155.27	101.76	92.70
	FTIA	12.83	56.62	169.83	287.33	367.50
customers	FinishedTechnologyProduction	0.15	0.97	2.101	5.80	12.18
	NWantA	0	0	0	0	0
	NWantAny	0	0	0	0	0
	NUseAny	1	4	6	3	2
	NWantB	0	0	0	0	0
	NUseA	1	0	1	3	5
	NUseB	7	86	88	94	93

marketing, understanding of market requirements as well as advertising, and only devotion to innovation, lack of effective communication and integration between technological innovation and production, and resource constraints cause failure in developing and manufacturing new technologies.

Regarding marketing policy, market share of technology A increases significantly. Because marketing impacts on purchase requests of potential consumers and also the production rate of developed technology. Marketing influences on the number of production of developed technology which can be seen in the Table 2 clearly. In this policy, the marketing effectiveness and production rate are assumed to be increased ten times.

In case of collaboration policy, the production rate is assumed to be increased two times. The production and technology development are in great need of being integrated [17]. In this regard, instead of dividing the company into departments, the way to coordinate development and manufacturing activities should be considered. This shows positive effect on market share of developed technology (see Table 3).

Resolving the resource constraints of technological innovation about 30 % enhances the market share of technology A and also the completed technological

Table 2 Results of SD and ABM after applying marketing policy

Weeks		48	96	144	192	240
Technologies	CurrentDemandB	24.78	25.52	37.67	31.5	40.46
	CurrentTIA	200.52	133.27	136.3	136.99	182.15
	FTIA	120.22	219.74	271.42	317.12	381.36
	FinishedTechnologyProduction	2.65	46.68	111.27	181.78	273.98
Customers	NWantA	0	0	0	0	0
	NWantAny	0	0	0	0	0
	NUseAny	4	4	5	6	6
	NWantB	0	0	0	0	0
	NUseA	11	16	19	17	16
	NUseB	70	77	75	77	78

Table 3 Results of SD and ABM after applying collaboration policy

Weeks		48	96	144	192	240
Technologies	CurrentDemandB	22.86	55.69	25.09	12.67	31.17
	CurrentTIA	28.40	157.81	122.55	87.23	136.80
	FTIA	5.25	94.06	220.15	308.04	398.18
	FinishedTechnologyProduction	0.14	1.42	7.70	20.02	37.91
Customers	NWantA	1	0	0	0	0
	NWantAny	0	0	0	0	0
	NUseAny	4	8	4	2	3
	NWantB	0	0	0	0	0
	NUseA	1	2	3	5	5
	NUseB	62	82	91	93	92

Table 4 Results of SD and ABM after applying resource management policy

Weeks		48	96	144	192	240
Technologies	CurrentDemandB	19.46	17.31	40.06	37.67	64.76
	CurrentTIA	58.35	60.22	144.04	153.94	184.31
	FTIA	24.20	79.63	172.29	308.14	443.47
	FinishedTechnologyProduction	0.24	0.75	0.163	4.025	13.72
Customers	NWantA	2	0	2	0	0
	NWantAny	0	0	0	0	0
	NUseAny	2	4	4	4	11
	NWantB	0	0	0	0	0
	NUseA	0	1	5	5	3
	NUseB	68	88	80	91	86

Table 5 Results of SD and ABM after applying all the policies

Weeks		48	96	144	192	240
Technologies	CurrentDemandB	15.72	11.55	31.67	59.39	83.64
	CurrentTIA	176.8	123.56	129.15	155.36	298.38
	FTIA	107.87	202.33	242.91	275.00	397.49
	FinishedTechnologyProduction	1.73	52.11	122.43	198.14	293.62
Customers	NWantA	0	0	0	0	0
	NWantAny	0	0	0	0	0
	NUseAny	3	2	3	10	8
	NWantB	0	0	0	0	0
	NUseA	13	16	18	20	26
	NUseB	72	81	79	70	66

innovation. Availability of sufficient human resources, equipment, raw materials, and budget causes technological innovation to be finished at scheduled time which is equal to 48 weeks (see Table 4).

By applying all policies simultaneously, the market share of technology A grows considerably with the assumption that the supposed conditions of supplied technology are constant. Comparing results of Tables 1 and 5 shows eighty percent increase in the market share of technology A by the 240th week. Following this approach makes the conditions of developed technologies and research and development centers in developing countries improved.

5 Conclusion

In this study a hybrid simulation model with combination of SD and ABM was modeled and simulated by using Anylogic in order to analyze the market share of redesigned and independent designed technologies compared to the acquired ones. SD was used to model supply chains of technology suppliers and technology developers. ABM was adopted to analyze the consumers market. Then, models of supply chain and consumer market were integrated, simulated and finally the results and policies were presented and discussed. Results showed: the significant impact of marketing on absorbing customer, increasing the production rate, and market share; the problem of collaboration between technology development and its production, and the influence of improved collaboration on production rate and market share; the impact of resolving resources constraints on time to innovate, completion of technological innovation, and market share. The results show that the market share of developed technology is more sensitive to the marketing. A proposed model is useful and it can be judged to be valid for the purpose for which it is constructed.

Thus following activities should be done to improve the conditions of the firm that trade their own technology:

• Participative resource planning and allocating.
• Partnership with other companies to overcome the resource scarcity.
• Participative planning and scheduling between marketing, technological innovation, and manufacturing to have shared goals and values.
• Marketing should provide new ideas, consumers' requirements, and market analysis to the technological innovation and production.
• Technological innovation should perform according to the market's needs.
• Production should collaborate to produce market's needs.
• Marketing, technological innovation, and manufacturing should collaborate in designing user and service manuals.
• Suitable mechanisms should be implemented to provide effective communication.
• It should be provided some services for customers such as training and answering the questions.

References

1. Kumar, V., Kumar, U., Persaud, A.: Building technological capability through importing technology: the case of Indonesian manufacturing industry. J. Technol. Transfer. **24**, 81–96 (1999)
2. Leonard-Barton, D.: Wellsprings of knowledge. Harvard Business School Press, Boston (1995)
3. Marcus, G.P.: Present and emergent identities: requirements for ethnographies of late twentieth-century modernity worldwide. In: Lash, S., Friedman, J. (eds.) Modernity and identity. Blackwell, Oxford (1992)
4. Bohn, R.E.: Measuring and managing technological knowledge. Sloan Manage. Rev. **36**, 61–73 (1994)
5. Ekboir, J.M.: Research and technology policies in innovation systems: zero tillage in Brazil. Res. Pol. **32**, 573–586 (2003)
6. Sarimveis, H., Patrinos, P., Tarantilis, C.D., Kiranoudis, C.T.: Dynamic modeling and control of supply chain systems: a review. Comput. Oper. Res. **35**, 3530–3561 (2008)

7. Lee, C.F., Chung, C.P.: An inventory model for deteriorating items in a supply chain with system dynamics analysis. Procedia Soc. Behav. Sci. **40**, 41–51 (2012)
8. Feng, Y.: System dynamics modeling for supply chain information sharing. Phys. Procedia. **25**, 1463–1469 (2012)
9. Campuzano, F., Mula, J., Peidro, D.: Fuzzy estimations and system dynamics for improving supply chains. Fuzzy Set. Syst. **161**, 1530–1542 (2010)
10. Reiner, G.: Customer-oriented improvement and evaluation of supply chain processes supported by simulation models. Int. J. Prod. Econ. **96**, 381–395 (2005)
11. Georgiadis, P., Vlachos, D., Iakovou, E.: A system dynamics modeling framework for the strategic supply chain management of food chains. J. Food Eng. **70**, 351–364 (2005)
12. Amini, M., Wakolbinger, T., Racer, M., Nejad, M.G.: Alternative supply chain production–sales policies for new product diffusion: an agent-based modeling and simulation approach. Eur. J. Oper. Res. **216**, 301–311 (2012)
13. Rixen, M., Weigand, J.: Agent-based simulation of policy induced diffusion of smart meters. Technol. Forecast. Soc. Change **85**, 153–167 (2014)
14. Rand, W., Rust, R.T.: Agent-based modeling in marketing: guidelines for rigor. Int. J. Res. Market. **28**, 181–193 (2011)
15. Kusar, J., Duhovnik, J., Grum, J., Starbek, M.: How to reduce new product development time. Robot. Comput. Integr. Manuf. **20**, 1–15 (2004)
16. Barney, J.: Firm resources and sustained competitive advantage. J. Manag. **7**, 99–120 (1991)
17. Drejer, A.: Integrating product and technology development. Eur. J. Innov. Manag. **3**, 125–136 (2000)

Modelling Contextual Decision-Making in Dilemma Games

Harko Verhagen, Corinna Elsenbroich, and Kurt Fällström

Abstract Social dilemmas such as the Prisoner's Dilemma and the Tragedy of the Commons have attracted widespread interest in several social sciences and humanities including economics, sociology and philosophy. Different frameworks of human decision-making produce different answers to these dilemmas. Common for most real-world analyses of the dilemmas is finding that behaviour and choices depend on the decision context. Thus an all-in-one solution such as the rational choice model is untenable. Rather, a framework for agent-based social simulation of real-world behaviour should start by recognising different modes of decision-making. This paper presents such a framework and an initial evaluation of its results in two cases, (1) a repeated prisoner's dilemma tournament playing against a set of well-known base models, and (2) a Tragedy of the Commons simulation.

Keywords Agent-based modelling • Social dilemmas • Context • Action theory • Social ontology

1 Introduction

One set of interesting social phenomena are collective dilemmas, such as the Tragedy of the Commons [1]. Although often discussed in rather abstract terms, collective dilemmas exist in every shared kitchen, every community project, every collective endeavour or service. Collective dilemmas are of interest as in the real world they seem often to resolve despite their dilemma structure. This is similar to the finding of the high cooperation in Prisoners' Dilemma games in experiments with real people, contrary to the predictions of game theory. Similar to the resolution

H. Verhagen (✉) • K. Fällström
Department of Computer and Systems Sciences, Stockholm University, P.O. Box 7003, Stockholm 164 07, Sweden
e-mail: verhagen@dsv.su.se; Kurt.Fallstrom@cybercom.com

C. Elsenbroich
Centre for Research in Social Simulation, University of Surrey, Guildford, UK
e-mail: c.elsenbroich@surrey.ac.uk

© Springer International Publishing AG 2017

121

W. Jager et al. (eds.), *Advances in Social Simulation 2015*, Advances in Intelligent Systems and Computing 528, DOI 10.1007/978-3-319-47253-9_10

of this empirical incongruity, there are several approaches to analyse collective dilemmas, such as invoking institutions [2], norms of fairness [3, 4], collective identity and group belonging [5] or collective reasoning [6, 7].

This plethora of solutions seems to suggest that there is no one-size-fits-all solution to human decision-making. A meta-framework systematising this variety of decision-making is the Computational Action Framework for Computational Agents (CAFCA). CAFCA is a two dimensional framework of contexts, where each dimension has three elements, a social dimension constituted by the individual, social and collective and the reasoning dimension consisting of automatic, strategic and normative reasoning [8].

2 Collective Strategies in a Prisoner's Dilemma Tournament

In the 1950s the idea of a Prisoner's Dilemma (PD) was developed to discuss decision situations in which the outcome for one actor are dependent upon the choices of another actor. More specifically, the choices are either defect or cooperate with the other actor. In its simplest form the PD is a game described by the payoff matrix in Table 1.

In a multiplayer Prisoner's Dilemma setting, a collectivist strategy can be seen in different ways. Our starting point to determine the switch from an individualistic to a collective mode of decision-making is the size of the coalition k (where coalition means the subjectively experienced group of peers). If an agent thinks the coalition is big enough to make it worthwhile for the collective, it will start cooperating. We concentrate on three collective strategies for making make decisions based on a collectivity value that is updated after each round of the game. These strategies are labelled the individual strategy, the memory strategy, and the neighbourhood strategy respectively.

Initially, the agents are scattered randomly on a grid and the collectivity value (the relative collective mindedness of the agent, a real between 0 and 1) is distributed randomly across the agent population. A threshold for unconditional cooperation is determined (to be compared to the collectivity measure). Three modes of behaviour change are implemented:

Table 1 General Prisoner's Dilemma pay-off matrix (T = temptation pay-off, R = reward pay-off, P = punishment pay-off, and S = sucker pay-off) satisfying that T>R>P>S

	Cooperate	Defect
Cooperate	R, R	S, T
Defect	T, S	P, P

1. Individual Payoff: k is extrapolated using one's own last payoff as an estimate of k. There is a choice between global and local dynamics for updating the collectivity level (and for the cases where memory is relevant the memory is updated as well). In the local case the average payoff within the game's radius is considered. In the global case it is the agent's own last-round-payoff. The payoffs are also used to generate dynamics for changing the collectivity of the agents. If an agent has defected in the last round and the respective payoff is lower than the reward payoff, its collective commitment goes up by 0.01. If an agent has cooperated and the respective payoff is higher than the reward or lower or equal to punishment, its collectivity goes down by 0.01.

   (a) Collectivity $\leq$ cooperation threshold then the agent defects.
   (b) Collectivity $>$ cooperation threshold, then the agent.

      • Cooperates if it's last round payoff is $\geq$ Sucker pay-off.
      • Defects otherwise

2. Memory: k is extrapolated from an agent's memory, experiencing defection above a threshold makes agents defect. Memory is a list consisting of 1s and 0s. In every round the last item in the list is deleted and a 1 or 0 appended in the front, depending on whether the experience was positive (1) or negative (0). A positive experience is one in which the last round payoff is $\geq$ to the Reward payoff. Memory can be constructed globally or locally similar to the collectivity dynamic.

   (a) Collectivity $>$ cooperation threshold, the agent cooperates.
   (b) Collectivity $\leq$ cooperation threshold, the agent.

      • Cooperates if the number of positive interactions $>=$ memory threshold.
      • Defects otherwise.

3. Neighbourhood Evaluation: k is extrapolated from the average neighbourhood payoff.

   (a) Collectivity $>$ the cooperation threshold, the agent.

      • Cooperates if the neighbourhood payoff $>$ Punishment.
      • Defects otherwise.

4. Collectivity $\leq$ the cooperation threshold, the agent defects.

The model was implemented in Repast and pitches a set of decision-making strategies against each other, comparing the average score achieved in each round. The strategies compared are Tit-for-Tat (TfT), Always Defect, Always Cooperate & Random Choice as examples of individualistic strategies and Individual, Memory and Neighbourhood as three implementations of collective decision-making. The individualist strategies were compared to the NetLogo implementation of the iterated PD [9] and displayed the same behaviour. Four experiments were performed to compare the collective strategies against TfT.

Table 2 Average scores in iterated Prisoner's Dilemma situations for three collective strategies in comparison to TfT

| | Collective strategies | | | | TfT |
	Individual	Memory	Neighbourhood	Combined	
Experiment 1	2.12	1.96	2.64	2.24	2.21
Experiment 2	2.05	2.09	2.29	2.14	2.32
Experiment 3	2.06	2.03	2.05	2.05	2.26
Experiment 4	2.04	2.11	2.36	2.16	2.14

Simulations ran for 200 steps and we average over 4 runs (we ran 5 runs and removed the outlier as in the original Axelrod tournament [10, 11]). We also used the same pay-off values (T=5, R=3, P=1 and S=0). The main results were that whilst some collective strategies (Individual and Memory) perform worse than TfT, the neighbourhood-based collective strategy performs similar or outperforms TfT. Table 2 below shows that the results for the different strategies are relatively similar and that the combined model resembles the results of the TfT model.

3 Team Reasoning on the Commons

Team reasoning is an extension to game theory which allows keeping the idea of utility maximisation but changes the agent the way utility calculation is applied to from the individual to the group or collective. Team reasoning explicitly allows for both, individual and collective utility and the main question is when agents switch from one mode to another. The simplest theory for switching put forward is that of Bacharach [6]. According to Bacharach people automatically switch between individual and collective reasoning when the collective solution to the situation— or "a game"—is strongly Pareto dominant. It is not a Nash equilibrium but Pareto Optimality that people are looking out for.

In [12] a model of the Tragedy of the Commons is presented which implements a variety of "psychological dispositions" such as cooperativeness, fairness, reciprocity, conformity, and risk aversion. These can be seen as implementations of various normative decision mechanisms. Due to space restrictions we will not discuss the results in detail here but rather present an implementation of an operationalisation of this switch of between individual and collective utility maximisation.

When Team Reasoning is switched on, an agent in the model compares the expected utility from a selfish and a cooperative action and if the payoff of the latter is greater than the former, the action reward is set to −1, making it less likely for the agent to add a cow to the pasture. The model was run over the parameter space in Table 3 below.

Table 3 Parameter space for experimentation

Variables	Values
Herdsman	6
Regrowth-rate	0.0005
Cow-grazing	1
Cow financial benefit	150
Learning factor	0.25
Initial number of cows	50
Team-reasoning	On, Off
Selfishness	0, 0.25, 0.5, 0.75, 1
Cooperativeness	0, 0.25, 0.5, 0.75, 1

Most of the variables were kept constant. We investigated the influence of team-reasoning on levels of sustainability, inequality and efficiency. Sustainability was assessed by the number of runs until the system runs out of grass. Inequality was measured by the Gini-coefficient [13] plus an absolute measurement of herdsmen with 2 cows or less (an arbitrary minimum level). The Gini coefficient is the most commonly used measure of inequality among a population, expressing the statistical of resources between 0 (absolute equality) and 1 (absolute inequality). Efficiency was assessed by comparing the number of cows and the levels of grass. The variables that were varied were selfishness and cooperativeness. Each combination was run ten times.

Results in [12] shows (a) that selfish scenarios are not sustainable, and (b) explores several psychological amendments to the payoff function, showing that some lead to sustainable outcomes. The selfish scenario compares the financial benefit of adding a cow with the cost adding the cow. The cooperative scenario compares the groups' average financial benefit with the cost of this action in comparison to the current situation.

We interpret the cooperative scenario as utility maximisation for the group, equivalent to the individual utility maximisation of the selfish case. Team reasoning is simply implemented by comparing the individual utility and the cooperative utility. Varying selfishness and cooperativeness determines to which extent the final decision is informed by their respective calculation.

The first result is that team reasoning simulations are sustainable even if the levels of selfishness are high. Figure 1 shows that in the individualistic model runs the Gini coefficient is low due to unsustainability of the Commons (top of Fig. 1). In the collectivistic case, the Commons is sustainable resulting in a higher Gini coefficient than in the individualistic case except in the bottom right corner (bottom Fig. 1). Thus team reasoning is a viable option to use as a decision-making strategy in commons dilemmas.

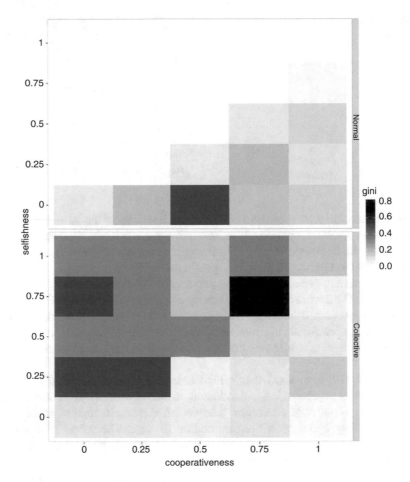

Fig. 1 Simulation results displaying the Gini-coefficient value variations in the individualistic (*top*) and collectivist (*bottom*) case respectively. *X*-axis = cooperativeness, *Y*-axis = selfishness

4 Conclusions

We presented two implementations of kinds of collective reasoning into models of social dilemma, one the iterated PD, the other the Tragedy of the Commons. The main purpose of the simulations was to show how alternatives to the classical rational choice decision-making can be used to model the empirical phenomenon of social dilemma being resolved. The first model explored whether collective strategies can compete in payoff terms to the winning strategy Tit-for-Tat in the original tournament exploration. The answer was that overall collective strategies perform similarly to TfT and that the neighbourhood focussed collective strategy overall outperforms TfT. The second model explored whether an explicit implementation of team reasoning can be used to explain the resolution of collective resource dilemmas such as the Tragedy of the Commons. The experiments showed that team reasoning indeed outperforms simple selfish decision strategies and has in addition positive consequences for the equity of society, without relevant reduction in profits. Future work is to fully explore the extension to the Tragedy of the Commons model and implement other versions of team reasoning which are more dependent on group features.

References

1. Harding, G.: The tragedy of the commons. Science 162(3859), 1243–1248 (1968)
2. Ostrom, E.: Governing the commons: the evolution of institutions for collective action. Cambridge University Press, Cambridge, NY (1990)
3. Fehr, R.E., Schmidt, K.M.: A theory of fairness, competition, and cooperation. Q J Econ 114(3), 817–868 (1999)
4. Eek, D., Biel, A.: The interplay between greed, efficiency, and fairness in public-goods dilemmas. Soc Justice Res 16(3), 195–215 (2003)
5. de Boer, J.: Collective intention, social identity, and rational choice. J Econ Methodol 15(2), 169–184 (2008)
6. Bacharach, M.: Interactive team reasoning: a contribution to the theory of cooperation. Res Econ 53, 117–147 (1999)
7. Gold, N., Sugden, R. (eds.): Michael Bacharach: beyond individual choice: teams and frames in game theory. Princeton University Press, Princeton, NJ (2006)
8. Elsenbroich, C., Verhagen, H.: The simplicity of complex agents—a contextual action framework for computational agents. J. Mind Soc. 15(1), 131–143 (2016). doi:10.1007/s11299-015-0183-y
9. Wilensky, U.: NetLogo. http://ccl.northwestern.edu/netlogo/. Center for Connected Learning and Computer-Based Modeling, Northwestern Institute on Complex Systems, Northwestern University, Evanston, IL (1999)
10. Axelrod, R.: Effective choice in the Prisoner's Dilemma. J Conflict Resolut 24, 3–25 (1980)
11. Axelrod, R.: More effective choice in the Prisoner's Pilemma. J Conflict Resolut 24, 379–403 (1980)
12. Schindler, J.: Rethinking the tragedy of the commons: the integration of socio- psychological dispositions. J Artif Soc Soc Simulat 15(1), 4 (2012)
13. Gini, C.: Variabilitá e mutabilita. Reprinted In: Pizetti, E., Salvemini, T. (eds.) Memorie di metodologica statistica Rome: Libreria Eredi Virgilio Veschi (1912)

Preliminary Results from an Agent-Based Model of the Daily Commute in Aberdeen and Aberdeenshire, UK

Jiaqi Ge and Gary Polhill

Abstract Rapid economic and population growth have posed challenges to Aberdeen City and Shire in UK. Some social policies can potentially be helpful to alleviate traffic congestion and help people maintain a healthy work–life balance. In this initial model, we study the impact of flexi-time work arrangement and the construction of a new bypass on average daily commute time and CO_2 emissions. We find that both flexi-time scheme and the new bypass will effectively reduce average daily commute time. Introducing a 30-min flexi-time range will reduce daily commute time by 6.5 min on average. However, further increasing flexi-time range will produce smaller saving in commute time. The new bypass will also reduce daily commute time, but only by one minute on average. As for environmental impact, introducing a 30-min flexi-time range will decrease CO_2 emissions by 7 %. Not only that, it also flattens the peak emission at rush hour. The bypass, on the other hand, will increase CO_2 emissions by roughly 2 %.

Keywords Daily commute • Flexible work scheme • Road infrastructure

1 Introduction and Motivation

This chapter develops an agent-based model of daily commute in Aberdeen City and the surrounding area in Scotland, UK. The area of study is home to more than 300,000 people, and has seen strong economic growth in recent years. In the 10 years since 2001, gross value added in Aberdeen City and Shire has increased by 72 %, compared with 51 % in Scotland and 48 % in the UK [1]. Due largely to the North Sea oil industry, Aberdeen City has the highest income in Scotland and the second highest income in the UK, just after inner London. It also has one of the lowest unemployment rates in the UK, at 1.5 % in 2013. Oil and energy sector plays an important role in the local economy. Out of the 18 main companies in Aberdeen

J. Ge (✉) • G. Polhill
The James Hutton Institute, Aberdeen AB15 8QH, UK
e-mail: Jiaqi.Ge@hutton.ac.uk; Gary.Polhill@hutton.ac.uk

© Springer International Publishing AG 2017
W. Jager et al. (eds.), *Advances in Social Simulation 2015*, Advances in Intelligent Systems and Computing 528, DOI 10.1007/978-3-319-47253-9_11

City and Shire in 2014, 16 are in oil and energy sector. In 2013, enterprises in oil and energy sectors accounted for 67 % of all turnovers and 17 % of employment in Aberdeen City.

As a result of the growing economy, population in Aberdeen City has been increasing in nine consecutive years, from 205,710 in 2004 to 224,970 in 2012, an increase of 9.4 %. Between 2011 and 2012, Aberdeen's population grew by 2510, of which 1878 are net migration into the city. Prosperity and population growth has led to an increased demand for accommodation and transport infrastructure. Prices of residential properties and public transport in Aberdeen City are already among the highest in the UK, and they are still growing. Between 2012 and 2013, residential property prices had increased by 9.9 % in Aberdeen City, compared with 0.7 % in Glasgow and −2.5 % in Edinburgh during the same period. Aberdeen City also has the highest car ownership rate in the country. More than 71 % of residents in Aberdeen own at least one car, compare with 59 % in Edinburgh and 49 % in Glasgow. In Aberdeenshire, 83 % of people own at least one car, and more than 40 % of people own two or more cars.

Rapid population and economic growth have posed challenges to existing transport network. As the roads become more congested, the daily commute to and from work takes longer. Adding to the challenge is that Aberdeen, like many other UK cities, is on the sea and expansion of the city is thereby constrained. For commuters, being stuck in congested traffic everyday could potentially become a major source of stress. More time spent on road also means less time for everything else, which could also affect people's overall life satisfaction. Moreover, driving in congested traffic is not only annoying to the commuters but also harmful to the environment due to the increased CO_2 emission and concentration.

This paper develops an agent-based model of daily commute in Aberdeen City and the surrounding area. We set up the environment so that it represents the actual landscape and existing road network in the area. We also set up commuters and employers to reflect the geographic distribution of businesses and homes in the area. By tailoring the model to the area of study, we hope that the model can better capture the reality and provide policy advice more relevant to the local community.

Using the agent-based model, we study the impacts of various treatments on the overall commute patterns and the transportation system. One of the treatments is flexi-time work scheme. The work scheme allows employers to arrive at work at a time of their choice within a range. For example, employers may choose to arrive at work anytime between 7:30 a.m. to 9:30 a.m. and, depending on the time of arrival, leave work between 4 p.m. to 6 p.m. Another treatment is the construction of a new bypass on the outskirts of Aberdeen City. The new bypass is expected to carry more than 43,000 vehicles each day once finished and help alleviate traffic in and out of Aberdeen City in rush hours. It is currently under construction. The agent-based model provides a framework with which to study the impact of treatments like flexi-time scheme and the new bypass on average commute time and total CO_2 emissions.

Preliminary results show that both the flexi-time scheme and the bypass will reduce average daily commute time. We find that the introduction of a 30-min flexi-time range will lead to a significant reduction in commute time in comparison with no flexibility at all. However, further increasing flexi-time range from 30 min will have a decreasing effect on saving in commute time. In terms of CO_2 emissions, introducing the flexi-time system will reduce not only the total CO_2 emission but also the concentration of CO_2 emissions at peak hours. The bypass, on the other hand, will slightly increase total CO_2 emissions.

The chapter proceeds as follows. In the next section we will give a review of literature on microscopic transport simulation. Then we will present the agent-based model using the overview, design concepts and details (ODD) protocol. After that we will present primary results. Finally there is the conclusion.

2   Relevant Literature

2.1   Traffic Cellular Automata (TCA) Models

Traffic cellular automata (TCA) models are dynamic computer simulations of many running vehicles that form traffic flows. Space in TCA is discretized into small cells, as is time into short time steps. Due to their flexibility and efficiency, TCA models are powerful tools to simulate realistic microscopic traffic flow. They are able to reproduce common traffic flows such as free flow, transition flow and congested flow [2].

Typical TCA models use the car-following rule to model vehicle movement, where a vehicle follows the car in front of it while trying to maintain a desired space gap. Most TCA models have single lanes and do not allow lane changing or overtaking, though there are also models that have a two-lane roadway [3]. Some TCA models are deterministic, meaning there is no stochastic term in the equations that determine the movement of vehicles. Deterministic TCA models therefore rule out any spontaneous formation of traffic jams except for those stemming from initial conditions. Stochastic or probabilistic cellular automata models, on the other hand, allow for spontaneous emergence of traffic jams. Some TCA models are based on simple topology such as a ring or straight line traffic [4] while others use a grid or complex city network [5].

Like TCA models, this chapter also adopts the car-following rule for vehicle movement. However, unlike TCA models, space is modelled using a coarser grain where each patch is 100 m × 100 m, compared with 7.5 m-wide cells in most TAC models [6]. Space is modelled in a coarser grain because we want to simulate an area of roughly 2000 km^2 and it is the daily commuting patterns and its impact of on various aspects of the society that we are most interested in. Our main focus is not to simulate car movements as realistically and accurately as possible, but to have vehicles drive in a reasonable way within a larger dynamic system that includes not just the vehicles, but also decision-making individuals, the landscape, the road network and social scenarios.

Individual vehicles or drivers in TCA models are largely treated as automated elements with little intelligence. Many TCA models simulate vehicles driving from point A to B, which is essential and sufficient for the purposes of the studies, but leaves little room for the vehicles to "think" and to do more complicated tasks such as route choosing. There is also little concern given to the landscape beyond the roads, such as the distribution of residential areas near the road or the distribution of businesses. In TCA models, vehicles are injected into one end of the road and exit from the other without a purpose, and traffic volume is a control factor outside the model. All these are fine for the purposes of the studies. However, this chapter goes beyond the scope of TCA models while adopting from TCA models movement rules and simulation techniques.

2.2 Driver Route Choice

As previously stated, TCA models have drivers follow the car in front of them without making an explicit route choice decision. In contrast, extensive literature has been devoted to the route choice behaviour of drivers. In route choice models, drivers are decision-making agents with preferences, habits and decision-making rules. They can also negotiate, learn and react to new information.

Studies of driver behaviour differ by their assumptions regarding driver rationality, drivers' capacity to process information and driver heterogeneity. Most equilibrium analyses of transportation systems are based on the assumptions that drivers are rational, homogeneous, have perfect information and infinite information processing capabilities. A classic example is Wardrop et al. [7]. However researchers are starting to acknowledge the limitations of such assumptions. Nakayama et al. [8] countered the assumption of rational driver and argued that drivers' rationality is bounded due to cognitive limitations. Instead, a driver uses simple rules to choose a route. Moreover, drivers are heterogeneous in their preferences and perceptions. Likewise, Ramming [9] showed that drivers failed to use full information to choose a route to minimize distance or time. Rather, their behaviour seems habitual. Arslan and Khisty [10] developed a heuristic way for handling fuzzy perceptions in route choice behaviour. Fuzzy "if-then" rules are adopted to represent a driver's route choice rules to capture driver preferences.

Another central issue is how drivers respond to information, such as how they respond to real-time or pre-trip travel time information, such as under the Advanced Traveller Information System (ATIS). Mahmassani and Liu [11] studied drivers' route switch decision under ATIS, using data from a laboratory experiment. The authors concluded that drivers will only switch route if expected travel time reduction exceeds a certain threshold. Abdel-Aty and Abdalla [12] conducted laboratory experiments with a travel simulator and found that remaining travel time and familiarity with the device are two factors significantly influencing route-switch response to ATIS. As for navigation information and route guidance, Adler [13] uses laboratory experiments to investigated the effects of route guidance and traffic advisories on route choice behaviour. The author found that in-vehicle navigation information is valuable to unfamiliar drivers in the short term, but its value diminishes as drivers become more familiar with the road network.

Laboratory and field experiments are often employed to explore different aspects of driver behaviour and route choice under various circumstances. For example, Bogers et al. [14] looked at the impacts of various aspects such as learning, risk attitude under uncertainty, habit and information on route choice behaviour using two laboratory experiments carried out using an interactive travel simulator. Selten et al. [15] conducted laboratory experiments to study driver's behaviour as a result of his daily experience on the road. The experiments involve repeated (200 times) interactive choices between two routes with and without feedback on travel time. They found that feedback on travel time has a significant impact on route-choosing behaviour. Zhang and Levinson [16] conducted a field experiment in which 100 drivers were asked to drive four alternative routes in Twin City in Minnesota, USA, given real-time congestion information with different degrees of accuracy. They found that travellers are willing to pay up to $1 per trip for pre-trip travel time information.

2.3   Agent-Based Models

Agent-based modelling is a research approach that models multiple individual agents in a common interacting environment. Like cellular automata models, agent-based models are typically implemented in computer software, but unlike cellular automata models, which focus on the states of the cells, agent-based models focus on the agents and their behaviour and interactions within the environment, which could be made up of cells. For example, agents could be decision-making human beings with heterogeneous preferences and the capability to process information and to learn. In other words, agents in agent-based models can do far more than simply update states in response to their neighbours as cells do in most cellular automata models. This advantage of agent-based models enables researchers to explore more sophisticated route choice behaviour while maintaining the model's ability to simulate emerging traffic flow patterns from individual vehicle behaviour.

Dia [17] develops an agent-based approach to modelling individual driver behaviour given real-time information. Drivers' route choice behaviour information is collected in a behavioural survey conducted on a congested commuting corridor in Brisbane, Australia. This behaviour information was then used to estimate driver preference parameters, which was then used in an agent-based route-choice model implemented within a commercially available microscopic traffic simulation tool.

Adler et al. [18] developed a multi-agent model to investigate the impact of cooperative traffic management and route guidance. In the model, agents from different parties negotiate to seek a more efficient route allocation across time and space, and the result is better network performance and increased driver satisfaction.

Bazzan et al. [19] proposed to use agent-based models to incorporate more realistic behaviour of drivers as a way to improve the accuracy of the existing ATIS. Klügl and Bazzan [20] developed an agent-based traffic simulation in which agents use a simple heuristics adaptation model to choose route. The authors then used experimental data to validate the heuristics behaviour rules. A traffic control system that perceives drivers decisions and return a travel time forecast is also developed to study its impact on the entire system.

This chapter develops an agent-based model of daily commute in Aberdeen. The model simulates commuters who live across the region and go to work every day via the existing road network in Aberdeen. Rather than a stylized environment, the landscape, the road network and the location of homes and business are obtained from GIS data. We will simulate measures such as a flexi-time scheme and the addition of a new bypass and study their impact on average commute time and CO_2 emissions. The scope of this chapter therefore goes beyond the typical microscopic traffic simulation. We will focus on how individual commuters have shaped the patterns of daily commuting in Aberdeen, and how we can achieve a more desirable outcome by changing policies and infrastructure.

3   The Agent-Based Model

In this section, we use Grimm et al. [21, 22] overview, design concepts and details (ODD) protocol to describe the model.

3.1   Purpose

The purpose of the model is to simulate daily commute in Aberdeen and to study factors that affect commute patterns, average commute time and total CO_2 emission from transportation.

3.2   Entities, State Variables and Scales

Main active entities in the model are daily commuters, who are people living and working in Aberdeen City and surrounding area. Each commuter has a car and specific home and work locations. Commuters drive to and from work every day at a time of their choice within the limits imposed by the flexi-time system. Environment entities include the shape and distribution of road network, location of businesses and homes and social and business policies such as speed limit and flexi-time scheme. The state variables include CO_2 emissions at any point of time and commute time for each individual commuter.

3.3   Process Overview and Scheduling

The model is developed in NetLogo. Time is discrete. Each time step represents about three seconds in real time. Every morning, individual commuters set off for work by car at a time of their choice, which depends on the flexi-time range and the distance from home to work. In the same way, commuters drive home from work in the afternoon. The time they leave work depends on the time they arrive at work in the morning.

Commuters can adopt different approaches to choose the route to and from work. They may choose the shortest route, which stays the same over time, or the fastest route, which will change over time depending on what the other commuters do. Every time a commuter drives on a section of a road, she stores in her memory how long it takes her to drive through the section. The commuter can later use that information to choose the fastest route to work according to her memory.

3.4   Initialization and Calibration

The landscape and road network of Aberdeen City and Shire form the environment in which commuters drive and interact. We simulate the landscapes and road network using Ordnance Survey GIS data. There are six types of road: main road type A, trunk main road type A, main road type B, minor road, urban road and urban trunk road. Each type has a different speed limit. We can also include the new bypass currently being built. Note, however, that we assume a junction wherever two roads intersect (i.e. no bridges—which will not always be the case), and the model does not include traffic flow controls such as traffic lights and roundabouts.

Each commuter is assigned a home and a work address using Ordnance Survey address data. Work addresses are found by identifying unique postcodes with a valid operation name such as "Aberdeenshire Council" or "Balmoral Group Ltd." There are 718 work locations identified in this way. On the other hand, home addresses are drawn from addresses that do not occupy a unique postcode and has no operation name. We randomly draw addresses from all valid home addresses in the data and assign it to each commuter. Therefore the spatial distribution of homes and businesses in the model environment matches the actual geographic distribution in the area.

We do not know the relative size of the 718 identified employers in the area, so we assume they are of the same size and assign them equal probability to be selected as workplace for people. As can be seen in Fig. 1, businesses are highly clustered spatially, which may alleviate the problem caused by assigning the same weight to each business. Another assumption we make here is the independence between a commuter's work and home location, because we do not have information on real individuals' home and work locations. In reality, however, people consider their workplace location when choosing where to live. There are of course other

Fig. 1 Roads, homes (*blue*)
and businesses (*red*) in
Aberdeen area with the
yellow road represents the
new AWPR bypass (*yellow*)

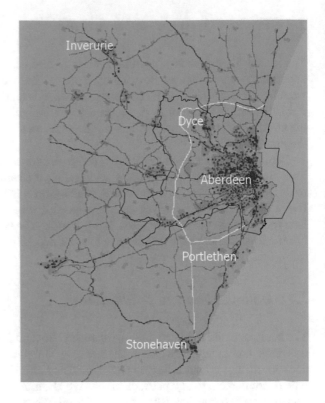

factors affecting people's choice of home, such as the building itself, housing price, access to services, the reputation of local schools, work locations of other household members etc., which are also beyond the scope of the study. We may see the situation modelled here as the worst-case scenario where people do not optimize distance between home and work.

3.5   Sub-models

Driving and Route Choosing Behaviour

Driving is modelled by a simplified car following rule. A vehicle speeds up and drives at the speed limit until it is close enough to the car ahead, and then it slows down and follows the car ahead, as is typical in TCA models. To avoid over-complications, traffic is single-lane and overtaking prohibited. The acceleration and deceleration rate at which the vehicle speeds up and slows down are fixed and the same for all cars.

As for route choosing behaviour, commuters can simply stick to the route they used last time, or they can use the experience they obtain on the road to search for fastest routes. When the commuters are first created in the simulation, they have no experience on the road so they all choose the shortest route available using Dijkstra's algorithm [23]. Once the commuters have been on the road, they remember the congestion level on each section of the road network. Later they will retrieve from their memory the information of expected time needed to drive through each road section, and choose the fastest route accordingly. Not every commuter would do this calculation and update the route in each period. In each period, some commuters go through this route-updating process and the rest stick to the old route. The model only simulates commuter traffic. No other traffic (e.g. freight) is simulated.

CO_2 Emissions Model

The CO_2 emission model we used is a statistical model of vehicle emissions developed by Cappiello et al. [24]. The authors estimated a statistical model of CO_2 emissions for two categories of light-duty vehicles in the U.S.: category 7 and 9. Category 7 includes cars under Tier 0 (less than Euro 1) emission standard with less than 50,000 miles accumulated; whereas Category 9 includes cars under Tier 1 (roughly equal to Euro 1) emission standard and more than 50,000 miles accumulated. Tailpipe emissions are estimated as a function of speed (v) and change in speed (a). The estimated function is then validated on actual traffic data. In the model, we will use the estimated function to calculate real-time tailpipe CO_2 emissions from each vehicle.

3.6 Treatment Factors and Experiment Design

Treatment factors in the model include the range of the flexible working time (flexi-time) scheme and whether there is a new bypass. The flexi-time scheme allows employees to arrive at work at any time within a range. The larger the range, the more flexibility the scheme has. The new bypass (the AWPR bypass) can be turned on and off. The two functional values of the model are CO_2 emissions and average commute time of all commuters. Table 1 summarizes the treatment factors and experiment design.

Table 1 Treatment factors and experiment design

Treatment factor	Parameter space	Experiment design
Bypass	True, false	True, false
Flexi-time	[0] min	0, 30, 60, 90, 120 min

4   Preliminary Results

The simulation is developed and run in Netlogo. Every morning and afternoon, cars are driving on roads to and from work. Different types of roads have different speed limits. During each run, the average commute time and total tailpipe CO_2 emissions are calculated. Each run consists of 60 consecutive days. For every parameter combination, we run 10 replications, each with a different random seed. Figure 3 below is based on data of the 60th day in the first replication. Table 2 shows the mean and standard error of commute time and CO_2 emission under different flexi-time ranges with and without bypass.

4.1   Flexi-Time

We find that first introducing a 30-min flexi-time when there is no flexi-time before reduces average daily commute time by 6.5 min. However, further increasing the flexi-time range from 30 min will lead to much smaller reduction in commute

Table 2 Mean and standard error of mean of commute time and CO_2 emission under different flexi-time ranges, with and without bypass

		Commute time (min)		Total CO_2 emission	
Flexi-time	Bypass	Mean	Std. err.[a]	Mean	Std. err.
0	False	34.37	0.20	8073.75	56.34
	True	33.21	0.16	8234.88	46.25
30	False	27.83	0.16	7528.22	43.98
	True	26.81	0.18	7695.88	64.25
60	False	27.83	0.16	7516.77	40.93
	True	26.81	0.18	7667.83	36.55
90	False	26.77	0.21	7506.33	42.64
	True	25.60	0.14	7643.03	48.02
120	False	26.60	0.15	7487.19	40.94
	True	25.58	0.14	7674.69	52.88

[a]Mean difference, standard error of mean difference, same for Tables 3 and 4

Table 3 Change in commute time and CO_2 emission from increasing flexi-time

	Commute time (in min)			Total CO_2 emission		
Flexi-time (bypass = false)	Mean	Std. err.	t	Mean	Std. err.	t
0→30	−6.54	0.26	−25.51	−545.54	71.47	−7.63
30→60	−0.83	0.22	−3.70	−11.45	60.08	−0.19
60→90	−0.23	0.27	−0.85	−10.44	59.10	−0.18
90→120	−0.16	0.26	−0.61	−19.15	59.11	−0.32

Table 4 Change in commute time and CO_2 emission from adding new bypass

Bypass (flexi-time)	Commute time (in min)			Total CO_2 emission		
	Mean	Std. err.	t statistics	Mean	Std. err.	t statistics
False→True (f=0)	−1.16	0.26	−4.50	161.13	72.89	2.21
False→True (f=30)	−1.02	0.24	−4.30	167.66	77.86	2.15
False→True (f=60)	−1.08	0.20	−5.41	151.06	54.87	2.75
False→True (f=90)	−1.17	0.26	−4.57	136.7	64.21	2.13
False→True (f=120)	−1.02	0.21	−4.91	187.51	66.88	2.80

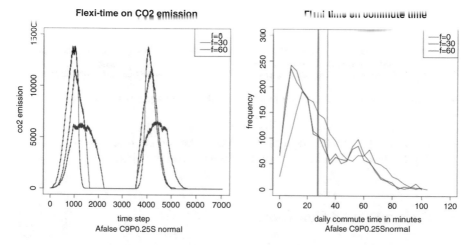

Fig. 2 Impact of flexi-time range from one of the ten replications with no bypass

time. The impact on commute time becomes non-significant as flexi-time increases. As for CO_2 emission, increasing flexi-time from 0 to 30 min reduces total CO_2 emissions by about 7 %. It does not only reduce total CO_2 emissions but also lowers the concentration of CO_2 emissions at rush hours. However, as flexi-time increases, the environmental benefit diminishes and becomes Roads, homes (blue) and businesses (red) in Aberdeen area with the yellow road represents the new bypass non-significant. Figure 2 show the impact of flexi-time ranging from 0, 30 and 60 min with no bypass. The distribution of commute time is shown on the left, with vertical lines showing the means; CO_2 emissions on the final day are shown on the right.

4.2 Bypass

We find that having the new bypass reduces average commute time by roughly a minute per day, regardless of flexi-time. It also slightly increases total CO_2 emissions by about 2 %. Figure 3 shows the impact of new bypass with flexi-time

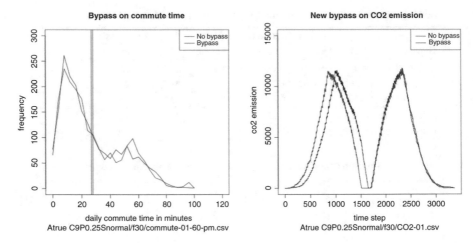

Fig. 3 Impact of new bypass from one of the runs with flexi-time equals 30 min

equals 30 min. Commute time distribution is shown on the left, with vertical lines showing the means; total CO_2 emissions are shown on the right.

5   Discussion and Conclusion

We find that both flexi-time scheme and the new bypass will effectively reduce average daily commute time. Under the situation where employers are allowed no flexibility at all, introducing a 30-min flexi-time will result in a large time saving of 6.5 min in average commute time, compared with roughly one minute reduction in average commute time from the bypass. However, once a moderate amount of flexibility is in place, further increasing flexi-time range will produce a much smaller saving in commute time. The benefit from allowing more flexi-time diminishes quickly as more flexi-time is already in place. Savings in commute time become non-significant as current flexi-time increases. The new bypass, on the other hand, will always reduce commute time by around one minute, no matter how much flexi-time is currently allowed.

Zero flexibility is rather an extreme case: most employers would probably tolerate being 5 min late, for example, employees may implement a form of "negative" flexi-time in aiming to arrive at work early to avoid variability in traffic, several employers in the area already implement a flexi-time scheme, and different employers have different expected starting times. Though in future work we plan to examine the threshold in flexibility at which the significant gains in commute time are observed, it would also be interesting to study how the effects of such flexibility could be achieved equivalently by some of the other points raised above.

As for environmental impact, under zero flexi-time, introducing a 30-min flexi-time range will decrease CO_2 emissions by 7%. Not only that, it also flattens the concentration of emissions at peak hours. But again, the environmental benefit from flexi-time quickly diminishes as current flexi-time increases. The bypass, on the other hand, will increase CO_2 emissions by roughly 2%. Though not confirmed by studying data from the model, noting from equations [1] and [2] that tailpipe emissions are a cubic function of speed, we suspect this may be because of the increased speed that the (dual carriageway) bypass would permit. Were this to be the case, it would be interesting to simulate a case where a speed limit less than that of a typical dual carriageway in the UK were imposed on the bypass, to see if this would mitigate the expected increase in CO_2 emissions.

Many other aspects of the model are yet to be explored. For example, the effect of individual decision-making on overall dynamics, through a parameter determining the percentage of people who update their route by referring to memory; reducing the speed limit in rural roads; how different types of vehicle affect CO_2 emissions; and how does urban sprawl, or more generally, the geographic distribution of home and businesses affect the overall system dynamics. These are the subject of likely future research.

Other options for improving the model include heterogeneous driving style and vehicle capabilities (particularly acceleration), implementing road junctions properly, simulating business-to-business traffic throughout the working day, freight and including other domestic driving activities, such as the "school run," shopping trips and leisure activities. The model would also benefit from allowing commuters to choose other transportation options (buses, bicycles, walking), and exploring options for increased flexibility, such as working from home, or using "business hubs" near where people live rather than necessarily requiring all employees to be in the same building throughout the working day.

References

1. Aberdeen City Council.: Behind the Granite Aberdeen key facts 2014. http://www.aberdeencity.gov.uk/tourism_visitor_attractions/tourists_visitors/statistics/stt_home.asp. (2014)
2. Lárraga, M., del Río, J., Mehta, A.: Two effective temperatures in traffic flow models: analogies with granular flow. Phys. A Stat. Mech. Appl. **307**(3), 527–547 (2002)
3. Nagatani, T.: Self-organization and phase transition in traffic-flow model of a two-lane roadway. J. Phys. A Math. Gen. **26**(17), L781 (1993)
4. Lárraga, M.E., Río, J.A.D., Alvarez-Icaza, L.: Cellular automata for one-lane traffic flow modeling. Transport. Res. C Emerg. Technol. **13**(1), 63–74 (2005)
5. Esser, J., Schreckenberg, M.: Microscopic simulation of urban traffic based on cellular automata. Int. J. Modern Phys. C **8**(05), 1025–1036 (1997)
6. Maerivoet, S., De Moor, B.: Cellular automata models of road traffic. Phys. Rep. **419**(1), 1–64 (2005)
7. Wardrop, J.G.: Road Paper. Some Theoretical Aspects of Road Traffic Research. ICE Proceedings: Engineering Divisions, Thomas Telford (1952)

8. Nakayama, S., Kitamura, R., Fujii, S.: Drivers' route choice rules and network behavior: do drivers become rational and homogeneous through learning? Transport. Res. Record J. Transport. Res. Board **1752**(1), 62–68 (2001)
9. Ramming, M.S.: Network knowledge and route choice. (2001). Massachusetts Institute of Technology
10. Arslan, T., Khisty, J.: A rational approach to handling fuzzy perceptions in route choice. Eur. J. Oper. Res. **168**(2), 571–583 (2006)
11. Mahmassani, H.S., Liu, Y.-H.: Dynamics of commuting decision behaviour under advanced traveller information systems. Transport. Res. C Emerg. Technol. **7**(2), 91–107 (1999)
12. Abdel-Aty, M.A., Abdalla, M.F.: Examination of multiple mode/route-choice paradigms under ATIS. Intell. Transport. Syst. IEEE Trans. **7**(3), 332–348 (2006)
13. Adler, J.L.: Investigating the learning effects of route guidance and traffic advisories on route choice behavior. Transport. Res. C Emerg. Technol. **9**(1), 1–14 (2001)
14. Bogers, E.A., Viti, F., Hoogendoorn, S.P.: Joint modeling of advanced travel information service, habit, and learning impacts on route choice by laboratory simulator experiments. Transport. Res. Record J. Transport. Res. Board **1926**(1), 189–197 (2005)
15. Selten, R., Chmura, T., Pitz, T., Kube, S., Schreckenberg, M.: Commuters route choice behaviour. Games Econ. Behav. **58**(2), 394–406 (2007)
16. Zhang, L., Levinson, D.: Determinants of route choice and value of traveler information: a field experiment. Transport. Res. Record J. Transport. Res. Board **2086**(1), 81–92 (2008)
17. Dia, H.: An agent-based approach to modelling driver route choice behaviour under the influence of real-time information. Transport. Res. C Emerg. Technol. **10**(5–6), 331–349 (2002)
18. Adler, J.L., Satapathy, G., Manikonda, V., Bowles, B., Blue, V.J.: A multi-agent approach to cooperative traffic management and route guidance. Transport. Res. B Methodological **39**(4), 297–318 (2005)
19. Bazzan, A.L.C., Wahle, J., Klügl, F.: Agents in traffic modelling—from reactive to social behaviour. KI-99: Advances in Artificial Intelligence, pp. 303–306. Springer (1999)
20. Klügl, F., Bazzan, A.L.: Route decision behaviour in a commuting scenario: simple heuristics adaptation and effect of traffic forecast. J. Artif. Soc. Soc. Simulat. **7**(1) (2004)
21. Grimm, V., Berger, U., Bastiansen, F., Eliassen, S., Ginot, V., Giske, J., Goss-Custard, J., Grand, T., Heinz, S.K., Huse, G.: A standard protocol for describing individual-based and agent-based models. Ecological modelling **198**, 115–126 (2006)
22. Grimm, V., Berger, U., DeAngelis, D.L., Polhill, J.G., Giske, J., Railsback, S.F.: The ODD protocol: a review and first update. Ecological modelling **221**, 2760–2768 (2010)
23. Skiena, S.: Dijkstra's algorithm. Implementing Discrete Mathematics: Combinatorics and Graph Theory with Mathematica, pp. 225–227 Addison-Wesley, Reading, MA (1990)
24. Cappiello, A., Chabini, I., Nam, E.K., Lue, A., Abou Zeid, M.: A statistical model of vehicle emissions and fuel consumption. Intelligent Transportation Systems, 2002. Proceedings. The IEEE 5th International Conference on, IEEE (2002)

Agent-Based Modelling of Military Communications on the Roman Frontier

Nicholas M. Gotts

Abstract This paper describes a simulation of Roman military frontier communication systems. The aim is to elucidate how signalling installations, hypothesised on archaeological and textual evidence, could have been used, together with other methods of distance communication, to enhance the Roman army's performance in responding to actual or threatened cross-border attacks or raids.

Keywords Historical communication systems • Collective cognition • Embedded cognition

1 Introduction

The work reported is inspired by David Woolliscroft's work on Roman military signalling [1], which advances evidence that many siting decisions made in constructing and modifying Roman frontier defences were motivated by the need for rapid communication between fortifications as much as several kilometres apart. For example, some fortifications on Hadrian's Wall, which runs across northern England, have a restricted view to the north (from which any trouble would probably come), but a clear line of sight to one of the forts some kilometres south of the wall. What signalling systems were used, however, is not known. The ancient literature suggests possibilities using beacons or torches, but is short on clear accounts of what signals were used, and how they were encoded. The ROFS (ROman Frontier Signalling) model aims to elucidate how a range of signalling capabilities and systems could have helped the army secure the frontier. Development of an initial model (ROFS-1) has shown the need to include other ways in which the Roman army would have disseminated information: systems of regular patrols, and the dispatch of messengers, with reports, orders, and requests. The physical movements of troops and the command structures of the army also need to be modelled.

Studies of distributed intelligence under modern conditions generally assume that information-rich messages can pass rapidly between distant agents. Yet for most of

N.M. Gotts (✉)
24a George Street, Dunfermline KY11 4TQ, UK
e-mail: ngotts@gn.apc.org

© Springer International Publishing AG 2017   143
W. Jager et al. (eds.), *Advances in Social Simulation 2015*, Advances in Intelligent
Systems and Computing 528, DOI 10.1007/978-3-319-47253-9_12

human history, for most messages, the speed of human travel has been the limit. Apart from the use of carrier pigeons, almost all known long-distance pre-industrial signalling systems used light, usually in the form of fire, which can be visible from kilometres away day or night. Systems such as semaphore have advantages over fire-based systems (no need to keep the fire burning, or danger of it burning out of control), but while modern semaphore systems are used over considerable distances, they are coupled with telescopes, lacking in non-industrial cultures other than early modern Europe.

2 The Functions of Roman Frontier Fortifications

It might seem obvious that the function of a frontier wall or barrier, such as Hadrian's Wall and the Roman *Limes* on their German frontier, is to keep enemies out, but in the 1980s there were assertions that the purpose of Hadrian's Wall was "more to control local movement than to be a defensible frontier" [2]; gateways to the north from the small "milecastles" on the wall, and later from full-scale forts integrated into it, suggest a customs-post role. But [3] argues for a strong defensive role, based on new findings of additional defensive measures along both Hadrian's Wall, and the more northerly Antonine Wall (the frontier for a while in the later second and early third century); and some explicit statements in inscriptions from the Danube that forts were being constructed to prevent the "secret crossing of robbers." The sheer scale of Hadrian's Wall at least (stone-built, 117.5 km long, at least 3.5 m in height, wide enough for a walkway along the top, fronted and backed by defensive ditches) surely indicates a "national security" function. Several milecastles on Hadrian's Wall were sacked during the late second century, so clearly there was a significant military threat at a fairly large scale; and smaller-scale attacks—ambushes on patrols, raids on civilian settlements near the wall—may have been considerably more common. Moreover, for much of its period of operation forts existed a few kilometres north of the wall, so the wall itself probably functioned as a supply and reinforcement base for forward forces. It is assumed here that whatever their other functions, Roman frontier defences were primarily military in purpose; but [1], and the evidence of a walkway on top of Hadrian's Wall, suggest a need to view frontier defences as socio-technical systems, supporting their garrison in more active ways than simply as a barrier to movement, by facilitating and channelling the collection and dissemination of military information.

3 The ROFS Model

Woolliscroft's work [1] makes it highly likely frontier garrisons did employ some form of visual signalling system. However, most messages would have been carried by human messengers, mounted if the matter was urgent, and conveying the message either in speech or writing. While there is no specific evidence of them, the Romans certainly could have used carrier pigeons if they knew the practice (which is uncertain). Given what [1] indicates about the importance of good communication lines, it is also likely that redundancy would be built in, to compensate for the disadvantages of each means of communication: fire signals are less use in the day than at night, and much less so in thick fog or heavy rain or snow (all frequent occurrences in the areas concerned in winter). Hence we need to consider how the communication systems would have interacted, with each other and with the army's normal routines.

ROFS-1 is written in NetLogo 5.1.0 [4]. Its main purposes, as a first agent based simulation model in a relatively unexplored modelling domain, are to determine whether a model able to shed light on current historical issues concerning Roman frontier defences is feasible; and to raise and if possible resolve the main issues relevant to the design of such a model. These are briefly considered in turn:

1. Physical scale and level of detail. Modelling a strip along the frontier several kilometres deep, and at least as far along the frontier, is necessary to capture the dynamics of the wall garrison's activities: troops and messages would routinely have moved over such distances. Fine spatial detail is less important: NetLogo "patches" in ROFS-1 represent 25 m^2. The current environment is a generalised and simplified representation of Hadrian's Wall in the second century—it does not represent a specific location. The possibility of moving between any two patches, on foot or horseback, and the time taken to move if it is possible, varies with the type of patch, currently "pasture," "path," "road," "wall" or "settlement."
2. Social scale and level of detail. In ROFS-1, the agents are military formations (from a fort garrison of several hundred men to a patrol of two) rather than individuals; switching to individual representation would currently impose considerable computational cost for little representational benefit.
3. Activities of the agents. Each formation has a list of current activities, a set of "command" rules to prioritise and integrate them, and an episodic memory recording recent events. In each model time-step (representing 10 s), each formation continues all its current activities in parallel; some (such as moving from patch to patch) take multiple time-steps. Garrison formations may detach sub-formations, sending them on specific missions (routine patrols, message-carrying, or responses to reports of "barbarian" presence), and receive reports from incoming patrols and messengers. Actual fighting is not represented; the Romans react to signs of barbarian actions.

4 Results

Currently, evidence of a barbarian attack (on a patrol or a settlement) can be detected by a patrol, reported to a garrison (the patrol may turn back, or take short-cuts,

Table 1 Seconds mean delay between first attack and last alert, last dispatch of reinforcements

Scenario type	Ambush		Pillage		Pillage-Ambush	
Signalling	No	Yes	No	Yes	No	Yes
Mean delay to last alert	6025	2070	9427.5	5612.5	7870	4045
Mean delay to last dispatch of reinforcements	9832.5	8502.5	N/A	N/A	13,300	11,215

to report as soon as possible), and prompt the sending of a response force, which will mark the area of attack so that further alerts concerning it are not generated by later patrols. Messages can be sent between garrisons—by foot or horseback depending on the route, relaying news of the attack location, and if necessary, requesting reinforcements. Messages can be passed on to further garrisons which are not known to have received them (each message includes a list of garrisons to which it has already been sent). The model can be run with or without signalling capabilities: in the "signalling" condition, forts and milecastles, but not turrets or patrols, can send simple signals, notifying formations within range of an emergency, and optionally requesting reinforcements.

So far, only scenarios involving a single barbarian attack on a patrol or settlement, or one attack on a patrol and another on a settlement, have been tested. Twelve pairs of scenarios have been tested, four pairs each of three scenario-types: "ambush," involving a single attack on a patrol along the wall, "pillage," involving a single attack on a settlement behind the wall, and pillage-ambush, involving one attack of each kind. Within a scenario-type, the pairs differ only in the timing of the attack or attacks; within a pair, the difference is whether long-range signalling is available. Table 1 shows the number of seconds' delay from occurrence of the first attack to the alerting of the last garrison, and if reinforcements are requested, to the dispatch of the last reinforcements.

As expected, the availability of long-range signalling considerably shortens the delay in alerting garrisons, and to a lesser extent, in the dispatch of reinforcements (the latter difference is less because reinforcements are requested from nearby garrisons, while all garrisons require to be alerted). It should be noted that most of the delay in the signalling condition consists of the time between the attack, and the notification of the nearest garrison that it has taken place, which requires a patrol to discover it, and reach the garrison's base. Moreover, the integration of message-carrying and signalling is not perfect in the current version of the model: some unnecessary messages appear to be sent. Nevertheless, the results indicate that the model can be used to investigate the advantages of long-distance signalling, and its integration with message-carrying.

5 Discussion and Future Work

The next stage in this work is to test a range of possible signalling systems, varying in the distance over which signals can be sent, the resources in time and troops required to send them, and what can be conveyed. Reference [1] provides experimental evidence concerning likely range and resources for signalling systems described in ancient sources; and the texts cited and quoted provide some information about the complexity of messages that could be sent, although it is often unclear whether the systems described were actually used. Investigations are also planned into the advantages of signalling in terms of risk (a messenger can be intercepted, and this is most likely precisely when urgent messages need to be sent), and how robust the system would be to poor visibility conditions, and to signals being missed or misread through human error. Initially, to save time in model development, a range of model environments not corresponding to real places will be used; but work will then proceed to using model environments based on specific places, along Hadrian's Wall and perhaps the *Limes*.

The role of agent-based modelling in historical research, particularly in areas such as Roman military signalling where evidence is fragmentary, is as much in constructing possible scenarios and formulating hypotheses, as in hypothesis testing. However, ROFS-1 can at least test simple hypotheses: that even a rudimentary signalling capability could have reduced the time taken to disseminate information, the resources needed, and the risks involved in doing so. Investigations will involve varying the parameters of signalling referred to above (and those of message carrying by foot or horseback, although here there are fewer parameters to vary), and the types and intensities of barbarian activity; and looking at the resulting payoffs in terms of speed of response, and security against further barbarian attacks. The scenarios constructed should then assist in formulating and testing:

- Hypotheses about the siting of bases containing signalling systems: forts, milecastles and turrets in the case of Hadrian's Wall. For a range of signalling systems, and parameters concerning other means of communication, how much difference does it make to the functioning of the system to move a signalling site in a way that reduces inter-visibility, but does not increase journey times? A reduction in performance in some plausible cases would be evidence that the siting was indeed intended to facilitate signalling; no such reductions would tell against the hypothesis.
- Hypotheses about the signalling system itself. If siting makes a big difference in some cases, and little or none in others, then that would be evidence that the signalling system used was similar to those where it makes most difference.
- Linked to the above, hypotheses about training. There is no textual evidence that anything like a "signal corps" existed in the Roman army; but if advantages to specific siting decisions only appear with complex signalling, that would be (weak) evidence that some specialist training was involved.

Given its multiple information collection and communication roles (the likely use of the walkway for both routine observational patrols and special messengers, signalling along the wall and to forts on either side, and its role as a "one-way window" between Roman Britain and barbarian lands), we could consider Hadrian's Wall as the Roman army's information superhighway—but one that specifically facilitated the rapid concentration of lethal force. ROFS, in conjunction with existing evidence, should be able to elucidate its design and performance characteristics.

References

1. Woolliscroft, D.J.: Roman military signalling. Tempus Publishing (2001), reprinted by The History Press, Stroud, UK (2010)
2. Starr, C.G.: The Roman Empire 27 B.C.–A.D. 476: A Study in Survival. Oxford University Press, Oxford, UK (1982)
3. Hanson, W.S.: The nature and function of Roman frontiers revisited. In: Collins, R., McIntosh, F. (eds.) Life in the Limes: Studies of the People and Objects of the Roman Frontiers. Oxbow Press, Oxford, UK (2011)
4. Wilensky, U.: NetLogo. http://ccl.northwestern.edu/netlogo/. Center for Connected Learning and Computer-Based Modeling, Northwestern University, Evanston, IL (1999)

The Leviathan Model Without Gossips and Vanity: The Richness of Influence Based on Perceived Hierarchy

Sylvie Huet and Guillaume Deffuant

Abstract This chapter studies a model of agents having an opinion about each other agent and about themselves. During dyadic meetings, the agents change their opinion about themselves and about their interlocutor in the direction of a noisy perception of the opinions of their interlocutor. Moreover highly valued agents are more influential. This model produces several patterns depending on the parameter values. In particular, in some cases several highly influential agents (called leaders) emerge and sometimes the leaders have a low opinion of each other.

Keywords Opinion dynamics • Power structure • Leadership

The recently proposed Leviathan model [1] considers a population of agents, each characterised by its opinion (a continuous number between -1 and $+1$) about each of the agents (including itself), and a dynamics of these opinions through processes of opinion propagation and vanity, taking place during random dyadic encounters. This model is very rich in terms of emerging behaviors, and [1] focuses only on a limited set of parameter values. To go further, we propose here a more complete study of one of the model's basic processes which is the direct opinion propagation (without gossiping).

The particularity of this opinion propagation is that unlike many opinion models, in which the influence increases with the similarity between the agents (homophily [7–11]), the influence increases with the superiority of the speaker as perceived by the listener (difference between listener's opinion about the speaker and listener's self-opinion). More precisely, the influence function is a sigmoid of the difference between the listener self-opinion and its opinion about the speaker. This function is classically used as a smooth threshold function [5, 6]. This modelling choice can be grounded in the research in social psychology. Indeed, complementary to the classical approaches based on homophily, a large body of work considers the influence in terms of credibility of the source [2–4].

S. Huet (✉) • G. Deffuant (✉)
Irstea, UR LISC Laboratoire d'ingénierie des systèmes complexes, 9 Avenue Blaise Pascal, Aubière 63178, France
e-mail: sylvie.huet@irstea.fr; guillaume.deffuant@irstea.fr

© Springer International Publishing AG 2017
W. Jager et al. (eds.), *Advances in Social Simulation 2015*, Advances in Intelligent Systems and Computing 528, DOI 10.1007/978-3-319-47253-9_13

Surprisingly, and even in the absence of vanity and gossip, the model generates a rich variety of patterns, and this chapter aims at identifying and characterizing them. It is organised as follows. We start by a description of the dynamics. Then, we describe the observed patterns and offer some generalised definitions of their characteristics as well as how they appear during trajectories of the model. We especially point out the link between the influence function parameters, the level of agreement of agents about their respective order and the average sign of the opinion of the population. Finally, we discuss our results and identify some complementary studies to carry out.

1 Opinion Propagation Depending on Perceived Hierarchy

We consider a set of N agents, each agent i is characterised by its list of opinions about the other agents and about itself: $(a_{i,j})_{1 \leq i,j \leq N}$. We assume $a_{i,j}$ lies between -1 and $+1$, or it is undefined (equal to nil) if the agent i never met j. At initialisation, we suppose that the agents never met, therefore all their opinions are 0. When opinions change, we always keep them between -1 and $+1$, by truncating them to -1 if their value is below -1 after the interaction, or to $+1$ if their value is above $+1$. The individuals interact in uniformly and randomly drawn pairs (i, j) and at each encounter they try to influence each other on their respective values. We define one iteration, i.e. one time step $t \to t + 1$, as $N/2$ random pair interactions (each individual interacts one time on average during one iteration). To be more precise, one iteration involves the following steps:

Repeat $N/2$ Times
   Choose randomly a couple (i,j)
   Influence (i,j)
   Influence (j,i)

The influence (i,j) process is:

Influence (i,j)
   if $a_{ii} = nil$, $a_{ii} \leftarrow 0$
   if $a_{ij} = nil$, $a_{ij} \leftarrow 0$
   $a_{ii} \leftarrow a_{ii} + \rho p_{ij} \left(a_{ji} - a_{ii} + \text{Random} \left(-\delta, +\delta \right) \right)$
   $a_{ij} \leftarrow a_{ij} + \rho p_{ij} \left(a_{jj} - a_{ij} + \text{Random} \left(-\delta, +\delta \right) \right)$

We recognise the equations of opinion influence (or propagation) in which opinions attract each other, but with two differences. The first difference is that the strength of the propagation of opinion is ruled by a parameter ρ multiplied by a function $p_{i,j}$. Function $p_{i,j}$ implements the hypothesis that the more i perceives j as superior to itself, then the more j is influential on i. It is a logistic function (with parameter σ) of the difference between the opinion of i about j ($a_{i,j}$) and the opinion i about itself ($a_{i,i}$):

$$p_{ij} = \frac{1}{1 + \exp\left(-\left((a_{ij} - a_{ii})/\sigma\right)\right)}$$

$p_{i,j}$ tends to 1 when $a_{i,j} - a_{i,i}$ is close to 2 (i values j higher than itself), and tends to 0 when it is close to -2 (i values j lower than itself). When σ is small, p_{ij} rapidly changes from 0 to 1. When σ is large, this change is progressive.

At the first meeting, we suppose that the a priori opinion about j is neutral and we set $a_{i,j} \leftarrow 0$. Let us also observe that, at the initialisation, an agent has no opinion about itself thus we also set $a_{i,i} \leftarrow 0$ at the first discussion.

The second difference with simple attraction dynamics is the introduction of variable δ. This variable models the idea that an agent i has no direct access to the opinions of another one (j) and can misunderstand it. To take into account this difficulty, we consider the perception of the agent i as the value a_{jz} plus a uniform noise drawn between $-\delta$ and $+\delta$ (δ is a model parameter). This random number corresponds to a systematic error that the agents make about the others' opinions.

Note that the update for a pair meeting is synchronous: all opinion changes occurring during a meeting are computed on the same value of opinions taken at the beginning of a pair meeting and are applied at the same time at the end of the meeting.

Finally, the model has four parameters:

- N, the number of individuals;
- σ, the reverse of the sigmoidal slope of the propagation coefficient;
- δ, maximum intensity of the noise when someone is alluded to;
- ρ, the parameter controlling the intensity of the coefficient of the influence.

2 Exploring the Emerging Patterns in the Parameter Space

This section begins by presenting the emerging patterns that we can observe when the parameters vary. Then, their properties are studied further using an experimental design showing that the patterns appear in regular successions that depend on the parameter values. A last part points out the relation between parameters of the influence function, the level of agreement of agents on how they order each other and the sign of the average opinion in the population.

2.1 The Emerging Patterns

In the following figures the opinions of each agent is represented as the row of an $N \times N$ square matrix. The element $a_{i,j}$ from line i and column j is the opinion of agent i about agent j. We use colours to code for the opinions: blue for negative and red for positive opinions with lighter colours for intermediate values. In the tests, we consider $N=20$ agents and the parameter of propagation intensity $\rho=0.8$. We distinguish patterns with the following criteria:

- The presence of leaders: they are agents about which all agents have a very positive (close to +1) opinion; In our matrix representation, they appear as a red column;
- The level of agreement about the agents: agents agree more or less about their peers (the average value of opinions about an agent is called its reputation for sake of simplicity). In our matrix representation, when all the values in a column are close to each other, the level of agreement about the agent of this column is high. The distribution of reputations represents a more or less skewed hierarchy.

Moreover, we have characterised patterns using the sign of the average opinion of the population as well as the level of agreement on how they perceive each other (do they perceive each other in the same hierarchy), especially when they are leaders. In this latter case, we talk about disagreeing leaders or agreeing leaders.

Using these criteria, we have observed three main groups of patterns: dominance and crisis; disorder; and hierarchy. They are presented in the following.

Dominance and crisis. Dominance is characterised by the presence of one or two agents with a reputation close to +1, that we call leaders (it can very temporarily increase up to 6 but it does not last) while every other agent has a significantly lower reputation. Figure 1 shows the various types of dominance we observed plus the crisis in which there is no leader. Crisis (Fig. 1a) never appears alone; it alternates with the dominance (Fig. 1b). Moreover, various types of dominance can be observed, from negative (Fig. 1 b, c) to positive (Fig. 1d) as well as some in which leaders agree on their respective positions (Fig. 1b, d) or disagree (Fig. 1c). Figure 1c illustrates the disagreement between leaders: one can see the most positive leader (most intense red squares) in the center has a neutral opinion (white squares) about the second leader on the right; the second leader has also an opinion about the first leader which is close to 0 but it has a positive self-opinion.

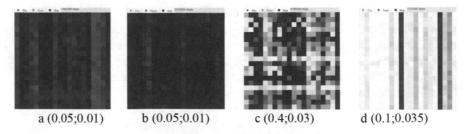

a (0.05;0.01)   b (0.05;0.01)   c (0.4;0.03)   d (0.1;0.035)

Fig. 1 *Dominance.* For various $(\delta;\sigma)$ noise bound and sigmoid parameter given below the graph and from left to right: Crisis in which no agent has any positive opinion; Negative dominance in which leaders agree; Negative dominance with disagreeing leaders about their respective leadership; Positive dominance with agreeing leaders. For low values of δ and σ, periods of crisis alternate with periods of dominance, even if less frequent. For larger values of values of δ and σ, crisis does not appear anymore and the level of agreement is lower due to the noise. $N=20$, $\rho=0.8$

Fig. 2 Disorder: negative on
the left; positive on the right.
They correspond to $(\delta;\sigma)$
couples of values given below
the graphs. $N=20$, $\rho=0.8$

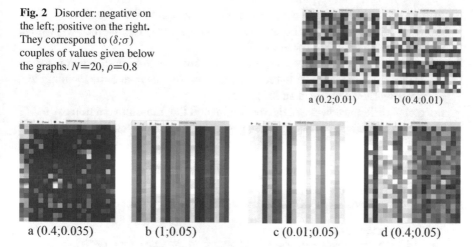

a (0.2;0.01)   b (0.4.0.01)

a (0.4;0.035)   b (1;0.05)   c (0.01;0.05)   d (0.4;0.05)

Fig. 3 *Hierarchy* patterns from left to right: (**a**) with disagreeing leaders and a majorly negative
population; (**b**) a majorly negative population with agreeing leaders; (**c**) a positive hierarchy with
agreeing leaders; (**d**) a majorly positive hierarchy with agreeing leaders. They correspond to $(\delta;\sigma)$
couples of value given below the graphs. $N=20$, $\rho=0.8$

Disorder. patterns are characterised by a very weak agreement on each other
values of opinion. Figure 2 shows the two cases that we observe: (a) the negative
disorder; and (b) the positive disorder. The positive disorder seems to exhibit a larger
number of positive self-opinion (see the red square on the diagonal from the top left
to the bottom right).

Hierarchy. in which the distribution of reputations is less bimodal than in
dominance patterns and the agreement on reputations is higher than in the disorder.
However, as shown in Fig. 3, different types can be identified: it can be more
negative than positive as in (a) and (b), or more positive than negative as in (c) and
(d); leaders can agree as in (b), (c) and (d), or totally disagree and have a negative
opinion of each other as in (a). This latter case is not so easily differentiable from
dominance with disagreeing leaders.

Starting from the various characteristics and types of pattern we observe,
we develop a procedure for automatically detecting the patterns and a larger
experimental design aiming at exploring the parameter space.

2.2 Systematic Exploration of the Parameter Space

An Algorithm Detecting the Pattern Types

We build a pattern-type detection algorithm based on a minimum number of
indicators:

- the average opinion for the definition of the main sign of the pattern;
- the bimodal properties of the distribution of opinions; the number of positive reputation to diagnose the absence of leader;
- the average distance between the maximum and the minimum opinion for an agent that we compare to 2δ in order to diagnose the disorder corresponding to a disagreement level higher than the noise δ;
- the global difference between the maximum and the minimum opinion to qualify the length of the opinion distribution, particularly to distinguish dominance from hierarchy.

The pattern identification algorithm is finally the following:

```
if (average opinion < 0) {
  if (average max-min distance > 2δ) pattern = disorder
  else {
    if (average opinion >= -0.5) {
      if (nb of positive reputation = 0) pattern = crisis
      else pattern = dominance}}
else {
  if (average max-min distance > 2δ) pattern = disorder
  else {
    if ((max opinion - min opinion)> 1) pattern hierarchy
    else dominance}}
```

Then we compute an experimental design running this diagnosis. The model includes four parameters. We fix N, the number of agents to 40, and ρ ruling the intensity of the opinion propagation coefficient to 0.8. This is in order to make tractable results of our study. We vary the other parameters as follows:

- δ, the intensity of noise disturbing the evaluation of other's opinions takes two different values: 0.01, 0.03, 0.05, 0.1, 0.2, 0.3, 0.4;
- σ, ruling the slope of the logistic function determining the propagation coefficients takes the values 0.01, 0.03, 0.05, 0.07, 0.1, 0.2, 0.3;

For each set of parameter values, we run the model for 200,000,000 iterations (one iteration corresponding to $N/2$ random pair interactions), and we repeat this for 10 replicas. We measure every 100,000 iterations a group of values allowing us to make conclusions about the properties of our patterns. The measured values over 10 replicas are averaged into indicator that is used in the next subsection to characterise our patterns.

The Properties of Patterns

Since patterns have been diagnosed using the algorithm presented in Sect. 1.2.1, we can compute their emerging properties which have not been used to diagnose them. For this purpose, complementary indicators inspired from what we have observed in Sect. 1.1 are collected during simulation time: the type of relations between leaders; the average agreement rate; the number of positive reputations;

the number of agents thinking they are the best over the whole population. The type of relation between leaders is defined by a minimum disagreement about the leadership between the leaders. The disagreement occurs when they disagree on their respective order (who is the first, who is the second; in most cases the two think they are the first). If more than 10 % of the leaders disagree, we say this is a disagreement (D) type of leadership; on the contrary it is called (A) agreement. The average agreement rate on order corresponds to the average percentage of dyadic meetings over all the meetings during which the two agents disagree about how to order them. Table 1 presents the results we obtained over our experimental design. We globally identified again what we have observed in Figs. 1, 2, and 3. A few configurations appear in only one among the 10 replicas of the 49 parameter sets. We decided to omit them in the table for sake of clarity.

The Table 1 confirms that the pattern dominance has one or two leaders even if a negative Hierarchy with disagreeing leaders also appears with 1.0 leader but it is probably a false diagnosis and it appears rarely (0.05 %).

The disorder in its positive form is very specific in terms of agreement on hierarchy but also in terms of number of agents thinking that they are the best; the two subcategories "Agreeing" and "Disagreeing" leaders are not really relevant since there is no leaders; they do not vary from each other and the "A" modality is very rare, probably due to the threshold chosen for the agreement. The disorder in its negative form is softer; it differs less from the other negative patterns.

Crisis always corresponds to disagreeing leaders (even if they are not positive in this case as for the other patterns).

Hierarchy, whatever its signs is characterised by a higher level of agreement rate. It is more characterised by disagreement between its leaders when the average opinion is positive while it is the contrary when the average opinion is negative. Overall, and except the case cited earlier, it shows more leaders than the dominance.

The "disagreement" between leaders corresponds to a smaller number of leaders in case of dominance and negative hierarchy. The negative patterns, crisis, dominance and disorder, whatever the type of relation between leaders, show a lower average agreement rate (between 0.6 and 0.7) than the positive ones.

From our results we also confirm that some patterns are only observable coupled to others in trajectories. Table 2 breaks down these observations.

We need 4.9 % of the runs to represent a total parameter set. Let consider the pattern appearing more than 4.9 % and see the average parameter values required to reach them. We distinguish clearly in Table 2 three areas: "dominance-disorder-crisis," "dominance-hierarchy" and hierarchy corresponding to the most frequent trajectories.

Trajectories in the Parameter Space

In order to understand the richness of the model, we locate these three trajectories into the parameter space in Table 3 with colour codes.

Table 1 The colours of the patterns are defined by the sign of the average opinion of the population (blue for negative; red for positive)

Pattern	Relation between leaders	Average number of positive reputations	Average agreement rate on hierarchy	Average number of agents thinking being the best	Frequency
Crisis	-	0,0	0,62	1,3	1,77%
Dominance	A	2,2	0,63	1,3	10,73%
	D	1,4	0,60	1,3	4,48%
Hierarchy	A	8,6	0,85	1,2	5,37%
	D	1,0	0,84	1,1	0,05%
Disorder	A	2,5	0,72	1,2	2,40%
	D	7,9	0,66	6,6	7,65%
Dominance	A	28,8	0,93	1,0	14,23%
Hierarchy	D	33,0	0,93	1,1	51,78%
Disorder	A	24,2	**0,47**	**23,2**	0,03%
	D	20,8	**0,55**	**20,4**	1,52%

Quantities are averages over all times and replicas and parameter values for which the diagnosis of pattern and the related relation type between leaders has been identified as described in the table

Table 2 Trajectories (set of patterns appearing during one execution of the model)

Set of patterns appearing in one trajectory	Trajectory sign	Relation between leaders	% of runs
Disorder	Negative		3,88%
	pAndN* and positive		2,24%
dominance disorder crisis	**Negative**	**Agree and disagree**	**12,04%**
dominance disorder	Negative	Agree and disagree	1,84%
dominance disorder crisis hierarchy	Negative	Agree and disagree	3,06%
dominance crisis hierarchy	Negative	Agree and disagree	0,82%
dominance disorder hierarchy	pAndN and negative	Agree and disagree	2,24%
Dominance	Positive	Agree	3,88%
dominance hierarchy	Negative	Agree and disagree	0,41%
	pAndN	**Agree**	**7,55%**
	Positive	**Agree**	**37,14%**
dominance hierarchy disorder	Positive	Agree	0,82%
Hierarchy	**positive or pAndN**	**Agree**	**19,80%**

Regarding the trajectory sign, we put altogether "positive" and "positive and negative" (*pAndN) when they represent less than 5 % of the runs

From the Fig. 2, we know that the empty bottom left cases of the Table 3 corresponds to Disorder (negative for σ 0.01 and δ 0.2; positive for σ 0.01 and δ 0.4). Disorder has not been identified as a main trajectory in the previous phase due to its low frequency of appearance. However, it is probably due to the diagnosis which is very demanding on the level of disorder: that is probable this level changes a lot during a simulation and only a part of measures satisfied the threshold of 2δ as the minimum average difference of opinion about one agent.

2.3 The Link Between Agreement and Sign

We observed in Table 1 that the average rate of agreement is a good indicator to distinguish negative from positive patterns. Figure 4 shows this indicator in the parameter space represented on abscissa for the positive (red) versus the negative diagnostics (blue). We observe that a strong change in the value of the indicator between positive and negative diagnosis is located in the space value of parameters as the frontier between the trajectory Dominance-Disorder-Crisis shown in Table 3. We observe again the various areas already noticed in the Table 3, especially regarding the sign (always positive or negative, or positive as well as negative and the level of agreement, weaker for low values of σ if δ is large enough). This is then relevant to understand how two agents start disagreeing on their perceived hierarchy.

That can be understood from the shape the influence function p_{ij}. Figure 5 reminds this shape for various values of σ. We observe for low values of σ (0.01, 0.05) that the weight becomes very quickly (ie for a small distance) 0 or 1. Then, when δ is close or higher than the distance making the influence close to or equal to 0

Table 3 Main trajectories in the parameter space: light orange is dominance-hierarchy positive; blue is dominance-disorder-crisis; brown is dominance-hierarchy positive and "positive and negative"; rose is hierarchy and dominance-hierarchy positive; red is hierarchy

δ / σ	0,01	0,03	0,05	0,07	0,1	0,2	0,3
0, 01	DH+	DH+	DH+/DH+-	DH+/DH+-	DH+/DH+-	DH+/DH+-	DH+/DH+-
0,03	DH+	DH+	DH+	DH+	DH+/DH+-	DH+/DH+-/DH+	DH+/DH+-
0,05	DDC	DH+	DH+	DH+	DH+	DH+/DH+-	DH+/DH+-
0,1	DDC		DH+	DH+	H/DH+	H/DH+	DH+
0,2		DDC		H	H	H/DH+	H
0,3		DDC	DDC	DDC	H	H	H
0,4			DDC	DDC	DDC	H	H

The white bottom left area corresponds to the pattern disorder in its negative shape then in its positive shape when δ increases

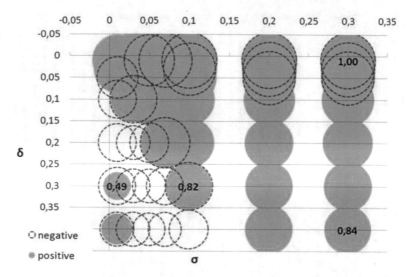

Fig. 4 Value of the average agreement rate order (size of the circle) for various δ (ordinate) and σ (abscissa) and for globally negative patterns (*blue*) or positive patterns (*pink*)

or 1, this means that an agent can consider its peer from influential to not influential (or almost) in just one meeting and just due to the noise varying from $-\delta$ to $+\delta$. When agent i becomes not influential (or almost) for agent j, i cannot change j's opinion about i if i gains a very high reputation, higher than the one of j. In such a situation, both agents think they are better than the other. We call the agent which is not anymore influenced by a peer an agent blind to this peer.

We now consider the smallest difference between the self-opinion and its opinion for one of its peer $(a_{ii}-a_{ij})$ for which the corresponding influence is a value ε which is small enough for neglecting the possible opinion changes. Derived from the influence function, the equation is:

$$a_{ii} - a_{ij} = \ln\left(\frac{1}{\varepsilon} - 1\right)\sigma \approx \sigma\ln\frac{1}{\varepsilon}$$

We observe comparing the Fig. 6 to the Table 3 that the result is close to define the frontier in the parameter space between the trajectory Dominance-Disorder-Crisis and the other trajectories, if we consider that δ is the typical distance between a_{ii} and the highest a_{ij}. This gives us some information about the frontier between negative and positive patterns.

For the parameters implying numerous disagreements on perceived hierarchies, the population tends to become negative and to show a dominance pattern. In the worst case, when σ is very small for large values of δ, a significant number of agents

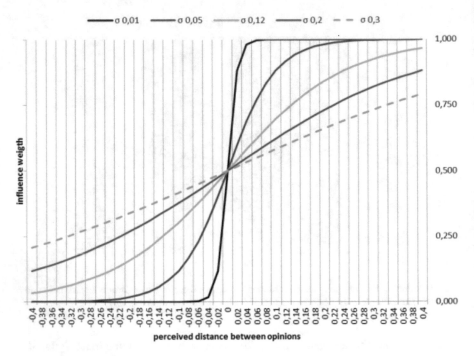

Fig. 5 Computation of the influence weight by the influence function for various values of σ and different distance between opinions (the peer one minus the self-opinion) for the listener

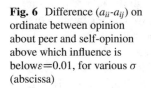

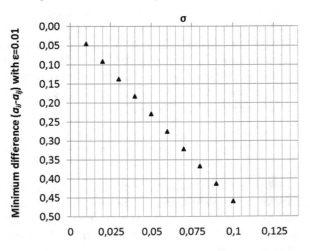

Fig. 6 Difference $(a_{ii}-a_{ij})$ on ordinate between opinion about peer and self-opinion above which influence is below $\varepsilon=0.01$, for various σ (abscissa)

consider themselves as superior to all the others: that is the case in the disorder pattern. In case of low disagreement, the observed patterns become positive and have the characteristics of the hierarchy.

3 Discussion: Conclusion

In this chapter, we study a simplified version of the Leviathan model, without the processes of vanity and gossiping, keeping only the direct opinion propagation. Like in the complete Leviathan model, we observe patterns but some of them are different from the ones identified in [1]. In particular, we identified two different types of leadership depending if leaders agree on their respective position on not. Overall we diagnosed several patterns organised in trajectories: dominance in which one or two leaders have a low opinion about all the other agents which have negative opinion about each other except about the leaders; it alternates during a simulation with a crisis pattern in which every agent has a negative opinion about all others, including itself. Disorder is a particular pattern in which the agents have different perceived hierarchies. During simulations several regular sequences of patterns appear. The main ones are: dominance-disorder-crisis, dominance-hierarchy, and hierarchy. Moreover these patterns/ sequences exhibit different average sign of opinions: most of them, except crisis, can be of both signs but while the hierarchy is more often positive, dominance is more often negative.

This work is a base for developing a new insight onto some behaviors of the Leviathan model by comparing the patterns emerging or not from the processes taking place in the model (i.e. opinion influence, vanity and gossip).

Moreover, the patterns of the model can be discussed in light of some sociopsychological researches.

The charismatic leaders at the beginning in the model are the individuals who are viewed as positive by every other. Similarly, the charismatic leadership [12] in the socio-psychological literature is defined as the one who benefits from a high esteem of the larger number of the others. For these same authors, the leadership derives its effectiveness from its influence on follower self-concept which is in turn a moderator of the leadership effectiveness. Identically, for [13] and [14], "Leadership is a relational term. It identifies a relationship in which some people are able to persuade others to adopt new values, attitudes and goals . . . ." We have seen in the model that overall leaders have a greater influence and tend to determine the reputation of everyone. However the criteria to define a leader, as well as the process of their emergence, are not so clear. Our modelling study has shown that in some cases leaders have the same perceived hierarchies whereas in other cases their perceived hierarchies are different. This leads to a question about what is more important: the popularity and the capacity to be open to others; or on the contrary, the stubbornness and strong resistance to influence? Our results can feed the debate in "leadership" research.

The agents tend to have positive opinions on average when the influence function is smooth (σ large), meaning that the differences of influence according to the perceived hierarchy are small, for a not too large noise (δ). For a sharp influence function and a large noise, agents tend to put themselves high in their perceived hierarchy, and for extreme values, the highest. This is due to the asymmetrical property of the influence function. This echoes the self-enhancement biases identified long ago in social psychology [15] particularly stresses the "illusory of superiority": when subjects estimate their relative position on a number of attributes, they typically report that they possess positive characteristics to a higher, and negative characteristics to a lower, degree than average or most others. This bias is often presented as an innate characteristic of the individual. The model suggests that it can be socially built from noisy interactions between agents with strong perceived hierarchies. This explanation would require some specific experiments for assessing its relevance.

References

1. Deffuant, G., Carletti, T., Huet, S.: The Leviathan model: absolute dominance, generalised distrust and other patterns emerging from combining vanity with opinion propagation. J. Artif. Soc. Soc. Simulat. **16**, 23 (2013)
2. Crano, W.D., Cooper, R.E.: A preliminary investigation of a hyperbolic model of attitude change. Midwest Psychological Association. Michigan State University, Cleveland, OH, May 4–6 (1972)
3. Pornpitakpan, C.: The persuasiveness of source credibility: a critical review of five decades' evidence. J. Appl. Soc. Psychol. **34**, 243–281 (2004)
4. Tormala, Z.L., Rucker, D.D.: Attitude certainty: a review of past findings and emerging perspectives. Soc. Personal. Psychol. Compass **1**, 469–492 (2007)
5. Peyton Young, H.: The diffusion of innovations in social networks. In: Durlauf, L.E.B.A.S.N. (ed.) The Economy as a Complex Evolving System, vol. III, pp. 20. Oxford University Press (2003)
6. Huet, S., Edwards, M., Deffuant, G.: Taking into account the variations of social network in the mean-field approximation of the threshold behaviour diffusion model. J. Artif. Soc. Soc. Simulat. **10**, (2007)
7. Castellano, C., Fortunato, S., Loreto, V.: Statistical physics of social dynamics. Rev. Mod. Phys. **81**, 591–646 (2009)
8. Byrne, D.: An overview (and underview) of research and theory within the attraction paradigm. J. Soc. Pers. Relat. **14**, 417–431 (1997)
9. Mark, N.P.: Culture and competition: homophily and distancing explanations for cultural niches. Am. Sociol. Rev. **68**, 319–345 (2003)
10. Takács, K., Flache, A., Mäs, M.: Is there negative social influence? Disentangling effects of dissimilarity and disliking on opinion shifts (June 3, 2014). SSRN: http://ssrn.com/abstract=2445649 or http://dx.doi.org/10.2139/ssrn.2445649 (2014)
11. Axelrod, R.: The dissemination of culture a model with local convergence and global polarization. J. Conflict. Resolut. **41**, 203–226 (1997)
12. van Knippenberg, D., van Knippenberg, B., De Cremer, D., Hogg, M.A.: Leadership, self, and identity: a review and research agenda. Leader. Q. **15**, 825–856 (2004)
13. Uhl-Bien, M.: Relational leadership theory: exploring the social processes of leadership and organizing. Leader. Q. **17**, 654–676 (2006)
14. Hogg, M.A.: A social identity theory of leadership. Pers. Soc. Psychol. Rev. **5**, 184–200 (2001)
15. Hoorens, V.: Self-enhancement and superiority biases in social comparison. Eur. Rev. Soc. Psychol. **4**, 113–139 (1993)

A Calibration to Properly Design a Model Integrating Residential Mobility and Migration in a Rural Area

Sylvie Huet, Nicolas Dumoulin, and Guillaume Deffuant

Abstract We propose a hybrid microsimulation and agent-based model of mobility integrating migration and residential mobility. We tested it on the evolution of the population of the Cantal, a French "département" with 150,000 inhabitants. We calibrated it using various data sources from 1990 to 2006, and tested its predictions on other data of the same period and on the period 2007–2012. The spatial heterogeneity of the evolution is well reproduced and the model makes surprisingly correct predictions despite numerous simplifying assumptions. From this calibration we learnt more about how to model residential mobility and migration considering an agent-based model approach.

Keywords Residential mobility • Migration • Calibration • Agent-based model • Microsimulation

In the literature [1], most dynamic microsimulation models of population evolution do not include spatial dynamics like the residential mobility. For example the microsimulation model of the French population called DESTINIE only includes the migration process, implemented by adding (or removing) individuals corresponding to the migratory balance. On the contrary, agent-based models and cellular automata easily integrate spatial dynamical processes. However, agent-based models which take into account all main demographic processes have not yet been developed [1] even if they have been involved in particular dynamics where they appear as more convenient, especially partnership formation and spatial mobility. More recently hybrid models combining advantages of the two modelling approaches are the subject of a growing interest [2] review the existing hybrid approaches as for example SVERIDGE [3] which integrates inter and intra migration [4] is a very interesting review on residential choice and household behaviour; it outlines several shortcomings of the current approaches:

S. Huet (✉) • N. Dumoulin • G. Deffuant
Irstea, UR LISC Laboratoire d'ingénierie des systèmes complexes, 9 Avenue Blaise Pascal, Aubière 63178, France
e-mail: sylvie.huet@irstea.fr; guillaume.deffuant@irstea.fr

© Springer International Publishing AG 2017   163
W. Jager et al. (eds.), *Advances in Social Simulation 2015*, Advances in Intelligent Systems and Computing 528, DOI 10.1007/978-3-319-47253-9_14

- The difficulty to connect the decision to move with the residential choice per se, and more generally the lack of retroaction between the demographic and the mobility processes;
- The decision to move is undoubtedly the most neglected aspect in the residential process, most models putting much more emphasis on the location choice;
- The location is generally decided using a discrete choice model which includes housing prices and housing and neighbourhood characteristics, then it is unclear whether this is a direct or indirect utility function;
- Regarding the location, the subset of alternatives is randomly and uniformly drawn from the whole set of vacant housings, disregarding any strategic consideration in the search process of the household;
- Migration and residential mobility are considered apart from each other, as independent decisions while migration and residential mobility cannot be distinguished for people living close to the border.

Thus, to overcome these limitations we propose a model coupling agent-based to microsimulation approaches and integrating demographic, migration, and residential mobility processes. The main features of the processes are derived from the French literature on residential mobility and migration. From [5, 6], we know that in France the main reasons to move are firstly related to family events [7, 8]. Established from the analysis of various French surveys, creating or splitting up a couple are such family events explaining most of the residential mobility. The second set of reasons is professional [8]. Notice that moving decreases with age and point out that the short distance mobility is rather linked to the modification of the family structure while the long distance mobility is more often associated to professional changes. The third type of reasons concerns the change in the tenure (mainly between renters and owners) [9]. From this literature review, we retained that a decision to move is due to: (1) the formation of a new couple; (2) the split of a couple; (3) a too long commuting time (higher than a threshold called *proximity* parameter) after a change of job, a new partnership, or a move; (4) a decrease in the housing satisfaction level due to a family event: we decide to capture this change through the compatibility of the housing size with the family size. As in [10], these events are taken into account in the decision function to move that we propose, while a second function is dedicated to the location choice of the new dwelling. Most of the demographic attributes and dynamics are parameterized from data extracted from various surveys, mainly the national Census and the Labour Force Survey. But some of the model parameters, especially those in relation with the mobility, have no documented value in the literature; we thus need to calibrate them. We argue that this calibration provides some insights on the relevance of our integrated model for mobility. Moreover, our model of residential mobility aims at reproducing the evolution of the population of the Cantal, a French "département" from 1990 to 2006. This problem shows some specific difficulties: (1) the population globally decreases but locally increases and shows a strong spatial heterogeneity; (2) the size of the basic spatial element, the municipality, is less than 1000 agents for 75 % of its 260 municipalities.

The next section presents the model, shortly for most of the dynamics for which the parameter values are directly derived from data, with more details for dynamics related to residential mobility which have to be calibrated. Then, the following section describes the calibration process and the final section proposes a discussion.

1   The Model

We have adopted a hybrid micro-modelling and agent approach. The purpose of the model is to determine how different dynamics at household level determine the population evolution in a network of rural municipalities. We assume that this evolution depends, on the one hand, on commuting flows and service, and on the other hand, on the number of jobs in various activity sectors (supposed exogenously defined by scenarios). Existing literature [11–14] stresses the importance of the different types of mobility between municipalities, commuting, residential mobility (short range distance), migration (long range distance) [4] and the local employment offer generated by the presence of the local population. Moreover, obviously, it appears also essential to include the demographic evolution of the municipality considering the strands explaining the local natural balance.

1.1   Main Entities, State Variables, and Scales

The model represents a network of municipalities and their population. The distances between municipalities are used to determine the flows of commuting individuals but also the flows of people making residential mobility as well as those looking for a partner. Each municipality comprises a list of households, each one defined as a list of individuals. The municipalities also include the offers of jobs, of residences and their spatial coordinates. Here is the exhaustive list of the main model entities with their main attributes and dynamics.

MunicipalitySet and Municipality. The set of municipalities can be of various sizes. It can represent a region of type NUTS 2 or NUTS 3,[1] or more LAU or intermediate sets of municipalities such as "communauté de communes" in France. Municipality corresponds to LAU2.[2] The municipality is the main focus of the model. It includes: A set of households living in the municipality; the set of jobs

---

[1]Eurostat defines the NUTS (Nomenclature of Territorial Units for Statistics) classification as a hierarchical system for dividing up the EU territory: NUTS 1 for the major socio-economic regions; NUTS 2 for the basic regions for the application of regional policies; NUTS 3 as small regions for specific diagnoses; LAU (Local Administrative Units 1 and 2) has been added more recently to allow local-level statistics.

[2]Consists of municipalities or equivalent units.

existing on the municipality (i.e. without those occupied by people living outside the modelling municipality set); the distribution of residences, or housings, on the municipality.

Each municipality has rings of "nearby" municipalities (practically every 3 km of Euclidian distance) with a maximum Euclidian distance of 51 km where individuals can find out jobs and partners while households can find lodgings. It is computed from the spatial coordinates of the municipalities. A threshold distance called "proximity" is the distance beyond which the municipalities are considered too far from each other for commuting between them.

There is a particular municipality, called "Outside," which includes available jobs accessible from municipalities of the set, but which are located outside this set. The job offer of Outside is infinite and the occupation is defined by the process ruling the probability of individuals to commute outside the municipality network.

The job and the residence. A job has two attributes, a profession and an activity sector in which this profession can be practiced. It is available in a municipality and can be occupied by an individual. The profession is an attribute of the individual at the same time it defines a job. In France, it takes six values. There are four activity sectors: Agriculture, Forestry and Fishing; Industry; Building; Services and Commerce. Overall, considering the six professions for four activity sectors, we obtain 24 jobs to describe the whole diversity of jobs in the Cantal "département." The residence has a type which is classically its size expressed in number of rooms. A residence is available in a municipality and can be occupied by 0, one or more households. Indeed several households can live in one residence for instance when a couple splits up and one of the partners remains in the common residence for a while. It is also the case in some European countries where it is customary for several generations to live under the same roof. The job and residence offers are initialised from Census data (see [15] for details).

Household. It corresponds to a nuclear family[3] and includes a list of individuals who have an occupation inside or outside the municipality. The first households located in each studied municipalities are built using algorithm described and studied in [16, 17]. At the initialisation time, households are associated randomly with residences. Then, before running, a first phase iterates moves in the same municipality in order to find a good fit between residences and households.

Individual. A new individual is instantiated from one of the adults of a household having the "couple" status in the birth method, or directly from the initialisation of the population, or by immigration. Her age of death, of entrance into the labour market, and of retirement are attributed when she is created. The values of these ages are drawn from distributions derived from French Census data. The age at which the agent becomes an adult and creates her own household is when she finds her first job; or she is chosen by a single adult as a partner; or she remains the only child in a household after her parents left or died while her age is higher than

---

[3] A nuclear family corresponds to the parents and the children; that is a reductive definition of the family corresponding to the most common one in Europe nowadays.

parameter firstAgeToBeAnAdult (15 years old for the French or other European National Statistical Offices). This change can lead her to move from the parental residence because of a low housing satisfaction level, but it's not always the case. The activity status of the individual defines her situation regarding employment, especially if she is looking for a job. The individual can quit a job, search for and change jobs ... Her profession is an attribute indicating at the same time her skills, level of education and the occupation she can look for.

1.2 Process Overview and Scheduling

The main loop calls processes ruling the demographic evolution, the migrations, the job changes, and their impact on some endogenously created services and/or jobs. First, the scenarios (i.e. external drivers) are applied to the municipalities. Then, demographic changes are applied to the list of households. Time is discrete with a time step of 1 year. The households are updated in a random order. The following gives more details about the dynamics.

Dynamics of offer for jobs, services and housing. In the municipality objects, changes in housing and job offers are specified in scenarios which are parameterised from data of the French censuses. Various dwelling sizes are considered in order to match the needs of households.

Dynamics of labour status and job changes. A newly born individual is initialised with a student status that she keeps until she enters the labour market with a first profession. Then, she becomes unemployed or employed with the possibility to look for a job. She may also become inactive for a while. When she gets older, she becomes a retiree. We describe rapidly these dynamics to situate them in the global picture of the model. They have been parameterised from data. The model and its parameterisation are described with more details in [10, 15, 18].

Demographic dynamics. A household can be created when an individual becomes an adult or when a new household arrives in the set of municipality (i.e. in-migration). The main reasons for household elimination are out-migration and death. Three main dynamics change the household type (single, couple, with or without children and complex[4]): makeCouple; splitCouple and givingBirth. Moreover, households change locations and that is the particular challenge of our modelling approach presented in this paper. The implementation of the "change location" related dynamics for Cantal is presented in the following (see also [10, 15]).

Household migration and mobility. In the process of changing residence, we include both residential migration and mobility without differentiating them on the distance of move, as it is often the case [4] in the literature. The submodel we

---

[4]A complex household is a household which is not a single, a couple with or without children.

propose directly manages both types of moving. However, it turned out easier for us to distinguish two categories of migration: the migration from outside to the set; the migration from inside the set.

The immigration into the set is an external forcing. Each year, a number of potential immigrants from outside the set are added to the municipalities of the set. These potential immigrants can really become inhabitants of the set if they find a residence by themselves or by being chosen as a partner by someone already living in the set in case they are single. That is the only actions they can do. Until becoming real immigrants, they are temporarily located into a municipality with a probability depending on its population size and its distance to the frontier of the set. A particular attraction of young agents for larger municipalities is also taken into account.

The mobility of agents already living inside the set of municipalities is mainly endogenous. Such a mobility can lead the household simply to change residence, municipality or to quit the set of studied municipalities. Overall, a household decides to look for a new residence when:

- a couple splits: one of the partners has to find out another residence even if she remains for a while in the same residence (creating her own household);
- one of its adults finds a job away from the current place of residence (beyond the proximity parameter which has to be calibrated) and she is the one deciding for the household (the leader is chosen at random every time a decision should be taken);
- it is not anymore satisfied by the residence. This satisfaction is assessed through a function possibly considering two dimensions identified in the literature (see our introduction): the difference of size between the occupied size and an ideal size for this household; the average age of the household. The calibration procedure has to determine if both dimensions should be taken into account to reproduce available data about residential mobility and migration.
- a student or a retiree decides to move outside the set (parameterised by data);

These are the events leading to a decision for changing residence. This is important to notice that other events possibly imply these events and lead to a decision for a residential mobility. This is the case of most of the family events with have an impact on the distance to commuting, the size of the household . . .

When a household has decided to move, the principle for the search of a new residence is the same as the one for searching a place to work. The leader of the household (chosen at random among the adults each time a decision has to be taken for the household) looks from the closest to the furthest job offer, considering successive rings of range of distance[5] of size 3 km for France (i.e. the same basic distance used to search for a job). She starts from her place of work (or residence if she does not work), meaning at a distance at most 3 because the first range is 0–3 km (i.e. first ring for search). If she can't find a satisfying place to live in, she

---

[5]The distance definition depends on the parameterization of the model.

continues looking from 3 to 6 km. She iterates the procedure, from 6 to 9 and so on, until finding a not empty list of possible residence offers or reaching a ring which is too far and the research stops. Before accepting to consider an offer, she checks the residence offer is not too far from the place of work of her partner (if she has one). She can also move outside. The decision moving outside depends on the parameterisation. The searching procedure finished, if the agent had not found out a residence in her current municipality and if she has found possibilities elsewhere, she decides to move outside using the probability to move outside knowing her municipality of living. Finally, if she does not move outside the set of municipalities and has found a residence, she chooses at random a municipality of residence in the list of collected residence offers. The probability to move out of the set of municipalities varies with the age of the individual. What is an acceptable residence offer to collect during the search procedure depends on the size of the housing. The level of acceptation of a possible size is 0 if the size does not respect the fact that the household wants to increase or decrease the current size of its house. If the offered size respects this tendency its probability to be collected for the list of choices decreases with the difference of size between the offered and the ideal size.

The way a household decides if a residence offer is satisfying has to be parameterised too. The level of satisfaction of a possible place of residence depends on two dimensions: the municipality where the proposal is; the size of the proposed housing. We have noticed from the literature [9] that most young couples want to become owners and quit renting. In rural areas, they often prefer a house instead an apartment. Such housing is much more rare and expensive in town than in the countryside. Thus, a municipality is examined as a place to reside if it satisfies the need for a house (and not for an apartment) for the household of size higher than one. Practically, a municipality has a ratio of house offer over its residence offer and it is considered in the research procedure with a probability equal to this ratio for every household larger than one member. Only the three main towns of the Cantal have a ratio possibly lower than 1. This ratio should be parameterised by the calibration process. The result of the calibration indicates how useful is considering a more detailed model of the offer. The second dimension influencing the place of residence is the size of the residence offer: we simply state it has at least to respect the need for increasing size or decreasing size of the household.

Overall, the parameters for immigration are: yearly migration rate; number of out of the set migrants in year $t^0 - 1$; probabilities for characteristics of the immigrants (size of the households, age of individuals . . .); distance to the frontier of the region of each municipality. Regarding the endogenous residential mobility, we have to parameterise or calibrate: a function defining the satisfaction of the size of housing and the satisfaction of job and workplace; migration rates of different agent categories extracted from data (as for example in France, for students and retirees).

Most of the events related to couples (formation, splitting, birth) can impact on residential mobility. Moreover they are poorly known from a modelling point of view and should be calibrated considering possibly different dimensions. They are presented in the following.

MakeCouple. The couples are made of individuals both existing in the population (either already living in dwelling of the municipality or a potential migrant). Therefore, two "compatible" individuals have to be identified for creating a couple. The usually procedure presented in the literature [1, 19–29] appears very heavy to compute for our case, especially because we have to take some spatial constraints into account while they don't. Moreover, it generally forbids that the individual lives other events during the year she tries forming a couple, such as moving for example. Thus we opt for a close procedure, significantly lighter computationally. We decompose the couple creation dynamics in two subdynamics: the decision searching for a partner; the search for the partner. The implementation for Cantal is:

- at each time step, each single individual searches for a partner following a probability; we do not consider a pool of single who are looking for another one as it is classically done in the literature (simply, a searching single can meet another single who is not necessarily searching a partner);
- when she searches, she tries a given number of times in her place of residence before trying in every municipality close to her own and her place of work to find someone who is also single and whose age is not too different (given from the average difference of ages in couples and its standard deviation from INSEE[6]). She begins searching very close and goes to search further until reaching a distance equal to the threshold parameter "proximity" (in the same way she looks for a residence or a job). She can search among the inhabitants or the potential immigrants.

Overall the probability to search and the maximum number of trials have to be calibrated since they cannot be derived from available data. The calibration should indicate if these two parameters are really useful (see discussion). Notice that when a couple is formed, the new household chooses the larger residence (the immigrating households always go into residences of their new partners) and this move can lead one member to commute very far (at a distance higher than the MunicipalitySet parameter proximity). This situation can change only when she is becoming the leader triggered by the job search method and implying that the household will aim to move closer to her job location, or if she changes of job.

SplitCouple. Even if the union duration is often used to model the union dissolution [1, 19, 20, 23–25], it appears that observed splitting probability is more strongly correlated with the date of union creation than with the duration of the union [30]. However, we do not have in our data when the couples existing in 1990 were formed. Thus, to limit the number of parameters and the complexity of the model, we use a constant probability to split that we will calibrate. When the split takes place, the partner who works further from the residence leaves the household

---

[6]French Institute of Statistics and Economical Studies

and creates a new household, which implies that she searches for a new residence. When there are children, they are dispatched among the two new households at random.

Giving birth. Following the reviews of the literature [1, 19, 20], the most common variables used for the fertility are the age, the marital status, and the number of children already born. However, in practice the total number of variables used to define the probability of birth is often very high. The same type of approach is developed in [23]. We have not enough data related to the Cantal to envisage such a data-based approach. In our model, only "couple" households can have children, and one of the adults should be in age to procreate. A couple has an annual probability to have a child computed from an "average number of children by individual." This "average number of children" has to be calibrated because it cannot be directly derived from existing data such as the fertility rate.[7] The calibration should indicate us if this "average number of children" is a constant or increase over time. The other parameters for "giving birth" are the age bounds to give birth: they are 18 and 45 for Cantal.

2 The Calibration of the Residential Mobility Dynamics

2.1 The Method: The Particle Swarm Optimization (PSO)

The particle swarm optimisation (PSO) [31] is a heuristic inspired from a social metaphor of individuals using their knowledge as well as the information given by their neighbours in order to find out a good solution. PSO considers a swarm of particles moving in the space of the parameter values. Each particle knows its position and the associated value of the function to minimise, the best position found by one of its neighbours, and its best position since the beginning. At each time step, every particle moves in the space with a velocity which is a compromise between its previous velocity, the knowledge about its best position and the best position in her neighbourhood. It has been shown to successfully optimise a wide range of continuous functions [32]. Some theoretical results about its convergence and good practices for choosing its parameters are available [33]. We parameterised as follows the algorithm:

---

[7]The total fertility rate (TFR), sometimes also called the fertility rate of a population is the average number of children that would be born to a woman over her lifetime if: (1) she were to experience the exact current age-specific fertility rates through her lifetime, and; (2) she were to survive from birth through the end of her reproductive life. It is obtained by summing the single-year age-specific rates at a given time. This computation is equivalent to give an identical weight to each age range, whatever their real weight in the population. It suppresses the structural effect linked to the distribution by age ranges of women in age to procreate.

- 15 particles: it has been shown few particles are sufficient—the choice is empirical depending on the perceived difficulty of the problem (values between 10 and 60 are used in practice). Their initial positions are randomly chosen in the space of research where particles are confined during the iterations.
- A "ring" neighbourhood: each particle has itself and the two particles located before and after it on the ring for neighbours.
- 10 replicas for 700 iterations have been executed to define our best result.

2.2 The Criteria for Calibration

The function to minimise is the sum of the differences between indicators computed from simulation results and the data. The 518 indicators we used are described in the following as well as the results of the calibration process (8 at the regional level; 54 at the "canton" level; 456 at the municipality level).

2.3 The Results of the Calibration Process

The exact values of the nine calibrated parameters are given and discussed in the following concluding section. The present subsection shows how 50 replicates of these 9 calibrated parameters fit the indicators composing the calibration objective function. Figure 1 show the fit level of the first set of 8 indicators. The value is given 1 year less for the simulated results compared to the target value of the INSEE in order to make them comparable because the simulated results are measured at the

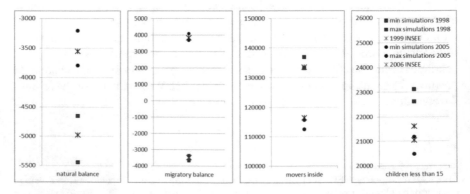

Fig. 1 Comparison between the simulated values and the target value. Target values from INSEE are represented by a cross while minimum and maximum values obtained over 50 replicas of the best parameter set are represented by *red squares* in 1998 and by *black circles* in 2005. From *left* to *right*: natural balance; migratory balance; number of movers (i.e. people changing residence) inside the Cantal; number of children less than 15

end of the year while the INSEE gives its value on the 1st January of a year. We can see on the figure that: the natural balances from 1990 to 1999 and from 2000 to 2006; the migratory balances from 1990 to 1999 and from 2000 to 2006; the number of individuals changing residences inside the Cantal from 1990 to 1999 and from 2000 to 2006; the number of individuals less than 15 years old in the Cantal in 1999 and 2006, are well fitted by the model since the targeted values are embedded in the simulation result ranges over 50 replicas. Except the number of individuals being less than 15 years old in the Cantal in 1999 and 2006 the values from data are within the interval of variations of the 50 replicas. It might not be relevant to try to obtain a better fit because the data giving the target value are not so stable or reliable:

- the number of children results partly from the immigrating children who are potential immigrants really come in the set of municipalities; these potential immigrants are built regarding their age on reference data measured only on 25 % of the population. Then it is possible the difference between the target and the model comes from this approximation;
- the targeted number of movers inside does not correspond, as measured in the model, to the number of agents moving from one residence to another. Indeed, in the census, this number corresponds to the number of inhabitants having a different address since the last census, but we don't know if they moved several times.
- the initial population in the simulation model has been built directly from the detailed data of the 1990 census. However, the next target in 1999 and 2006 has been built on the following census which has changed the counting method. Then, instead the 159,245 counted by the 1990 census and implemented in the simulation model, the next census which are corrected "a posteriori" assesses the population in 1990 at 158,723 (522 less).

The other 4 basic indicators are averaged errors over spatial objects. They relate to: (1) Population of each of the 27 "Cantons" of the Cantal (error based on the average of the 27 differences between the simulated value and the targeted one) in 1999 and 2006 (forming overall 54 indicators since two dates are considered); (2) Population of each of the 228 municipalities of at most 1000 inhabitants in 1990 (error based on the mean of 228 differences between the simulated value and the targeted one over the target value) in 1999 and 2006 (forming overall 456 indicators since two dates are considered).

The two first dedicated to the 27 "Cantons" of the Cantal in 1999 and 2006 are shown in Fig. 2. Only 48 % in 1998 and 44 % in 2006 of the "cantons" are given by the simulations without error (target comprised in the interval of replicas' results). For the others "cantons," the errors given by the average value over the replicas and expressed relatively to the target values are below 0.2.

Despite these errors the global evolutions of the "cantons" in terms of decrease or increase are well respected (see Fig. 3). Indeed, only one "canton" over 27 increases instead of decreasing but the error is not so large since in fact the population of the canton 1523 [Vic-sur-Cère] (on the right side of Aurillac) increases for an average

174   S. Huet et al.

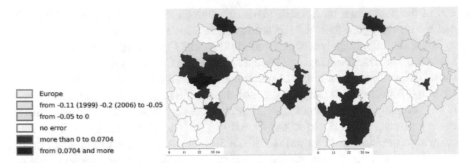

Europe
from -0.11 (1999) -0.2 (2006) to -0.05
from -0.05 to 0
no error
more than 0 to 0.0704
from 0.0704 and more

Fig. 2 "Canton" maps. Average relative errors over the targeted population of each "canton" for 50 replicas (considering the bounds defined by the minimum and maximum results over 50 replicas, the error is computed as the (larger distance between one bounds and the target)/target) in 1999 (on the *left*) and 2006 (on the *right*). Then, the value no error means that the target value is comprised in the interval defined by the minimum and the maximum results over 50 replicas. A negative value means model gives a larger population than the target while a positive value means the contrary

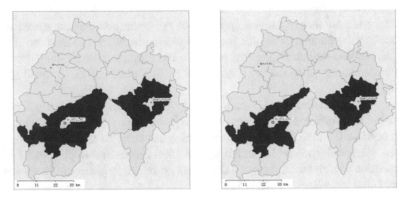

Fig. 3 "Canton" maps. *Left*: average absolute model evolution 2006-1990 over 50 replicas, in terms of decrease in *yellow* (−1043.5 to 0), and increase in *red* (0 and more)—*Right*: INSEE evolution 2006-1990, *yellow* −1296 to 0; *red* 0 and more

of 133.26 agents while the INSEE predicts a decrease of 295 individuals. Then the error is about 430 agents while the whole population of the Cantal is about 150,000.

Concerning the two last indicators dedicated to the population of each of the 228 municipalities of at most 1000 inhabitants in 1990 in 1999 and 2006: 68 % of these municipalities in 1998 (154 municipalities over 228) and 69 % in 2005 (157 municipalities over 228) have populations for INSEE (i.e. the reference) comprised in the minimum–maximum range of the model's results given by the 50 replicas of the best set parameter values.

3 Conclusion: Discussion

We propose a model of mobility integrating commuting, family events, satisfaction of housing, job availability, and migration. We assume that the decision to move depends on: some family events as union dissolution and couple formation; the distance of commuting; the adequacy of the size of the housing expressed in number of rooms to the size of the household. Nine parameters of these dynamics cannot be derived directly from available data and we chose to calibrate them. The calibration provides values for which the results from simulations are close to the reference data. The following presents these parameters and their calibrations values for our implementation of the whole population of the Cantal region (starting with one agent "individual" for each "real" individual of the region, with virtual "individuals" composing "households" located in municipalities formed of the same number of households as "real" municipalities). They are discussed in order to define what we have learnt regarding how to model mobility.

Two parameters relate to the formation of couples, the calibration yields a probability to search of 0.22 and a number of search trials of 5. This means only one over 4 or 5 singles decide 1 year to look for a partner (not everyone), knowing that she is susceptible to meet any single agent since she has a congruent age to her and using the location rule described in this paper. This result indicates us the utility of the two parameters: indeed, considering everyone search for a partner and just limiting the number of trials is not a sufficient model to fit data of reference. A two-step process is required based on a decision phase and a search phase.

One parameter drives the decision to split for a couple for which we made the simplest hypothesis considering the average duration of a couple is constant. Our calibration yields a splitting probability of 0.107. This means on average couples have duration of about 9–10 years. Even if we know from literature a cohort approach is more relevant and then that we overestimate the dissolutions of older people, the choice for a constant probability to split is sufficient for calibrating the model properly. This is very useful since in our case, as almost always, we have not enough data to implement a cohort approach.

Regarding the birth rate, the question was to know if it is necessary to consider an increasing probability to give birth for each couple to obtain the right number of births (remaining almost constant over the time while the fertility increases during the considered periods in Cantal). Our calibration confirms that the model has to consider an increasing individual probability to give birth. Indeed, the slope of our linear function giving the average number of children of a woman a given year (and then the searched probability) is 0.063 and the intercept is 2.38.

Except for splitting of a couple or the migration of students and old people, the decision to move is associated to the satisfaction level about the current dwelling. This satisfaction depends on the size of the dwelling compared to the size of the household, and possibly from the average age of the household. How high the satisfaction increases with age whatever the dwelling size is, controlled by the parameter *pres*. It has been parameterised to 0.045. This value confirms us older people does not want to move even if their dwelling is too large: if it were not the case, it would be 0.

We also assume a searching process for a housing, for a job or for a partner (as finding a partner implies changing residence for at least one of the partner) starting from the best location and enlarging the search progressively if the object of search (house or partner) is not found. For an occupied active individual, the best location of housing is the municipality of work. For the other individuals, the best location is the current municipality of residence. The issue of the calibration process is to define if the individual searches until she finds or if there is a maximum distance where she stops searching. The calibration suggests that the search should be bounded by 18 km (Euclidian distance). The quality of this result can be evaluated through the capacity of the model to give values describing the commuting distance and the residential mobility distance distributions close to the references. Figure 4 shows the two commuting and residential mobility distributions given by the model and the INSEE. They are close enough even if the model tends to underestimate the probability to move in the same municipality (3 km or less). The result shows a reasonable fit given the hidden complexity of the dynamics behind such a distribution.

Only for the search process for dwelling, the calibration has to indicate if a specific mechanism should be added for avoiding sometimes the three largest municipalities of Cantal as residence locations: we should know if bigger municipalities are less preferred by families for their residence. The parameter value for "dispo threshold" means they are maintained in the searching zone only for 32 % of the searches susceptible to contain them. Overall, we conclude that both mechanisms, one pushing agents to live outside the main municipalities, the other bounding the area where people search for new housing or partner, are necessary to reproduce the spatial evolution of the population. It is probable, even if the parameters change depending on case study these dynamics should always be both present to model the spatial evolution.

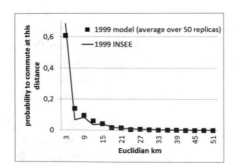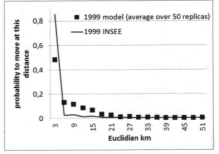

Fig. 4 Commuting (on *left*) and residential mobility (on *right*) distance distributions for INSEE (*line*) and the model based on results from calibration (*squares*) averaged over 50 replicas

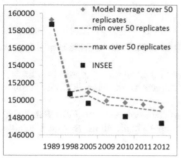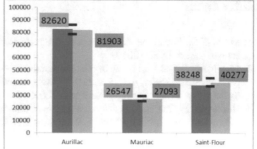

Fig. 5 Population in 2012: (1) on the *left* at the Cantal level; (2) for the three various "arrondisse-ments" composing the Cantal "département" given by INSEE (in *blue*) and the model (on average over 50 replicates in rose); The red lines correspond to the minimum and the maximum results over the 50 replicates

The last challenge we had to deal with was about the small size of our municipalities (on average 440 inhabitants without the two largest). Despite this difficulty impacting the statistical robustness, the results at a global level seem reasonably well reproduced. At a lower level, the canton is a more reliable level to look at the results since it is composed on average of 4500 inhabitants (without the two largest). Overall the model correctly reproduces the spatial heterogeneity of the evolution sign of the municipalities. Moreover, the model seems to keep some relevance to predict the "future." Indeed, while it has not been calibrated from 2006 to 2012, the total 2012 population predicted by the model exhibits 1 % error. On Fig. 5 (1), the reference of INSEE is just below the curves given by the model and we have to remind the change made by INSEE in the counting method making afterwards the starting population 522 less than the one we built from the 1990 Census. Figure 5 (2) shows the population of the three subspaces corresponding to "arrondissement" level. The error for the "arrondissements" are 1, 2 and 5 % on average but the target defined by the INSEE reference is between the minimum and the maximum results over 50 replicates.

Finally our model has coped with many difficulties related to the classical independent views of the mobility. However, its many integrated processes makes necessary a sufficient quality of related process involving a decision to move and a location choice. Reaching this quality is a challenge for each of them. Several processes, for instance the search for a job, remain to validate and detailed data are not always available, especially in rural areas.

Acknowledgements This work has been funded by the PRIMA (Prototypical policy impacts on multifunctional activities in rural municipalities) collaborative project, EU 7th Framework Programme (ENV 2007-1), contract no. 212345. We also thank for data and tools: (1) Maurice Halbwachs (www.cmh.ens.fr/); (2) Irstea, DTM, Grenoble (siddt.grenoble.cemagref.fr/).

References

1. Morand, E., Toulemon, L., Pennec, S., Baggio, R., Billari, F.: Demographic modelling: the state of the art. INED (2010)
2. Birkin, M., Wu, B.: A review of microsimulation and hybrid agent-based approaches. In: Heppenstall, A.J., Crooks, A.T., See, L.M., Batty, M. (eds.) Agent-Based Models of Geographical Systems, pp. 51–68. Springer, New York (2012)
3. Holme, E., Holme, K., Makila, L., Mattson-Kauppi, M., Mortvik, G.: The SVERIGE Spatial Microsimulation Model: Content, Validation and Example Applications. Spatial Modelling Centre, Sweden (2004)
4. Coulombel, N.: Residential choice and household behavior: state of the art. Ecole Normale Supérieure de Cachan (2010)
5. Minodier, C.: Changer de logement dans le même environnement. Données sociales. La Société Française 515–523 (2006)
6. Gobillon, L.: Emploi, logement et mobilité résidentielle. Economie et Statistique 349–350, 77–98 (2001)
7. Debrand, T., Taffin, C.: Les facteurs structurels et conjoncturels de la mobilité résidentielle depuis 20 ans. Economie et Statistique 381–382, 125–146 (2005)
8. Debrand, T., Taffin, C.: Les changements de résidence: entre contraintes familiales et professionnelles. Données sociales. La Société Française 505–513 (2006)
9. Djefal, S.S.E.: Etre propriétaire de sa maison, un rêve largement partagé, quelques risques ressentis. CREDOC 177, 4 (2004)
10. Huet, S.: Individual based models of social systems: data driven hybrid micro-models of rural development and collective dynamics of filtering or rejecting messages. Computer Science, vol. Doctorate, p. 252. Université Blaise Pascal—Clermont II, Ecole Doctorale Sciences pour l'Ingénieur de Clermont-Ferrand, Irstea (2013)
11. Davezies, L.: L'économie locale "résidentielle". Géographie, économie, société 11, 47–53 (2009)
12. Blanc, M., Schmitt, B.: Orientation économique et croissance locale de l'emploi dans les bassins de vie des bourgs et petites villes. Economie et Statistique 402, 57–74 (2007)
13. Perrier-Cornet, P.: La dynamique des espaces ruraux dans la société française: un cadre d'analyse. Territoires 2020(3), 61–74 (2001)
14. Dubuc, S.: Dynamisme rural: l'effet des petites villes. L'Espace Géographique 1, 69–85 (2004)
15. Huet, S., Lenormand, M., Deffuant, G., Gargiulo, F.: Parameterisation of individual working dynamics. In: Smajgl, A., Barreteau, O. (eds.) Empirical Agent-Based Modeling: Parameterization Techniques in Social Simulations, vol. 1, The Characterisation and Parameterisation of Empirical Agent-Based Models, p. 22, 133–169. Springer, Heidelberg (2013)
16. Gargiulo, F., Ternès, S., Huet, S., Deffuant, G.: An Iterative Approach for Generating Statistically Realistic Populations of Households. PLoS One 5, 9 (2010)
17. Lenormand, M., Deffuant, G.: Generating a synthetic population of individuals in households: Sample-free versus sample-based methods. Journal of Artificial Societies and Social Simulation 16(4), 15 (2012)
18. Huet, S., Lenormand, M., Deffuant, G., Gargiulo, F.: Parameterisation of individual working dynamics. In: Smajgl, A., Barreteau, O. (eds.) Empirical Agent-Based Modeling: Parameterization Techniques in Social Simulations, p. 22. Springer, New York (2012)
19. Bacon, B., Pennec, S.: APPSIM—Modelling Family Formation and Dissolution. University of Canberra (2007)
20. O'Donoghue, C.: Dynamic microsimulation: a methodological survey. Braz. Electron. J. Econ. 4, 77 (2001)
21. Li, J., O'Donoghue, C.: A Methodological Survey of Dynamic Microsimulation Models. Maastricht University, Maastricht Graduate School of Governance, The Netherlands (2012)
22. Abelson, R.P., Miller, J.C.: Negative persuasion via personal insult. J. Exp. Soc. Psychol. 3, 321–333 (1967)

23. Duée, M.: La modélisation des comportements démographiques dans le modèle de microsimulation Destinie. INSEE (2005)
24. INSEE, D.R.E.P.S.: Le modèle de simulation dynamique DESTINIE. Série des documents de travail de la Direction des Etudes et Synthèses Economiques **124** (1999)
25. Robert-Bobée, I.: Modelling demographic behaviours in the French microsimulation model Destinie: an analysis of future change in completed fertility. INSEE (2001)
26. INSEE: Enquête sur l'histoire familiale. Data distributed by the Centre Maurice Halbwachs, 48 bd Jourdan, Paris (1999)
27. Blanchet, D., Buffeteau, S., Crenner, E., Le Minez, S.: Le modèle de microsimulation Destinie 2: principales caractéristiques et premeirs résultats. Economie et Statistique 441–442, 101–122 (2011)
28. Blanchet, D., Chanut, J.M.: Projeter les retraites à long terme. Résultats d'un modèle de microsimulation. Economie et Statistique, **315** (1998)
29. Poubelle, V.: Prisme, le modèle de la Cnav. Retraite et Société **48**, 202–215 (2006)
30. Vanderschelden, M.: Les ruptures d'unions: plus fréquentes mais pas plus précoces. Insee Première novembre, **4** (2006)
31. Kennedy, J., Eberhart, R.: Particle swarm optimization. IEEE 0-7803-2768-3/95, 1942–1948 (1995)
32. Poli, R., Kennedy, J., Blackwell, T.: Particle swarm optimization. An overview. Swarm Intell. 33–57 (2007)
33. Clerc, M., Kennedy, J.: The particle swarm—explosion, stability, and convergence in a multidimensional complex space. IEEE Trans. Evol. Comput. **6**, 59–73 (2002)

Modeling Contagion of Behavior in Friendship Networks as Coordination Games

Tobias Jordan, Philippe de Wilde, and Fernando Buarque de Lima-Neto

Abstract It has been shown that humans are heavily influenced by peers when it comes to choice of behavior, norms, or opinions. In order to better understand and help to predict society's behavior it is therefore desirable to design social simulations that incorporate representations of those network aspects. We address this topic, by investigating the performance of a coordination game mechanism in representing the diffusion of behavior for distinct data sets and diverse behaviors in children and adolescent social networks. We introduce a set of quality measurements in order to assess the adequacy of our simulations and find evidence that a coordination game environment could underlie some of the diffusion processes, while other processes may not be modeled coherently as a coordination game.

Keywords Social simulation • Social networks • Coordination games • Threshold models • Agent-based computational economics

1 Introduction

Christakis and Fowler [1] suggest that our social contacts heavily influence our decisions, opinions, and behavior. By proposing new approaches to tackle existing problems in crowd behavior, these findings may be helpful for people that aim to understand, guide, or control the behavior of societies. For example, the knowledge that the political opinions of spouses strongly affect each other seems to be of high value to politicians during election periods. Physicians and other health care professionals could use the findings that adolescents affect each other's smoking, drinking, and sexual-interaction behavior. Thus, the existing research on contagion

T. Jordan (✉) • P. de Wilde
University of Kent, Canterbury, UK
e-mail: tj202@kent.ac.uk; P.Dewilde@kent.ac.uk

F.B. de Lima-Neto
University of Pernambuco, Recife, Brazil

University of Münster (ERCIS), Münster, Germany
e-mail: fbln@ecomp.poli.br

© Springer International Publishing AG 2017
W. Jager et al. (eds.), *Advances in Social Simulation 2015*, Advances in Intelligent Systems and Computing 528, DOI 10.1007/978-3-319-47253-9_15

of behavior and spread of information within social networks can give good hints for campaigns and political action in order to achieve a desirable behavior. Current research proves that such influence exists and is also able to point out situations where the effects are stronger or weaker. However, to adequately model the effects of contagion and information spreading for simulation and prediction models, a coherent representation is required. This is the motivation of the work presented below. Our research implements the contagion of behavior within friendship networks as a coordination game [2, 3] where single individuals within the observed systems coordinate their actions with their neighbors. We assume that individuals benefit from compliant behavior. We adapt the coordination game mechanism to two different data sets, each containing information about friendship ties, as well as time-dependent information about specific behaviors or behavioral outcomes. To evaluate the performance of the implementation of the coordination game, we simulate the coordination dynamics on the given friendship network starting with the real initial situation and subsequently compare the state of the system after the simulation with the state observable in reality after one or more time steps.

Synopsis We review the state of relevant research regarding diffusion of behavior and information in social networks and present the data sets used within this work in Sect. 2. Section 3 presents the implementation of the coordination game and its adaption to the different data sets while Sect. 4 contains experimental setup and results. We discuss the experimental results in Sect. 5 and conclude with Sect. 6.

2 Background

2.1 Diffusion Processes in Social Networks

Existing research indicates that human decisions, opinions, norms, and behavior are influenced by the social environment [4]. Social influence and contagion as well as spread of behavior and information through social networks has been documented in a wide range of cases [1]. For instance, diffusion of voting behavior [1] and obesity [5] has been proven statistically. Moreover, cooperative behavior has been shown to be contagious, though depending on tie structure and dynamics [6] and recent studies revealed contagiousness of emotions [7]. Other behaviors do not spread like sexual orientation [8]. This indicates the existence of those effects on other individual behaviors of children and adolescents such as "commitment to school education," substance use, or sport. Marques [9] reveals the huge differences between social networks of the poor and those of more wealthy people. Considering the above, this further encourages the modeling and simulation of social network effects in order to understand social phenomena and to guide political decision making. Related research has also been conducted in children and adolescent networks. For instance, roles of nodes within a network of school children have already been identified [10] and diffusion of social norms and harassment behavior in adolescents school networks have been empirically studied and evidenced [11].

It then has been reasonably shown that behavior, norms information, and opinions flow within social networks of adults and children. Approaches to model this diffusion come, for example, from the field of Social Psychology, like the concept of Social Influence Network Theory [12] from Friedkin. Another approach is the modeling as a coordination game [2, 3]. Those models may be considered as advanced threshold models [13–15] that incorporate social network structure instead of simple crowd behavior. The coordination game as implemented in [3] is characterized by the assumption that individuals benefit when their behavior matches the behavior of their neighbors in the network. Hereby a node within a network can adopt one of two behaviors A or B. The node receives pay-off a when equaling its behavior with a neighbor that adopts *behaviorA*. b, respectively, denotes the pay-off a node receives when both, her and her neighbor adopt *behaviorB*. When choosing different behaviors, nodes receive a pay-off of 0 (other implementations may introduce negative pay-offs for noncompliance). The total pay-off for each node can accordingly be calculated as presented in (1) and (2). Here P_i^a denotes the total pay-off for node i from choosing behavior A (respectively, Behavior B for P_i^b), d_i denotes the degree of node i and n_i^a (same for n_i^b) denotes the number of neighbors of node i adopting behavior A (respectively, Behavior B).

$$P_i^a = an_i^a \tag{1}$$
$$P_i^b = b(d_i - n_i^b) \tag{2}$$

This determines that the best strategy for node i is to choose behavior A if $n_i^a \geq bd_i$ and behavior B otherwise. Rearranging the inequality in (3), we get:

$$r \geq T \text{ with } T = \frac{b}{a+b} \text{ and } r = \frac{n_i^a}{d_i} \tag{3}$$

In the absence of knowledge of the individual pay-offs a and b, a global threshold T may be found experimentally, as shown in the remainder of this paper.

2.2   Data

We perform our experiments on two different data sets, both contain information about adolescent friendship ties, as well as about different types of behavior.

(i) The first data set stems from the study *"Determinantes do desempenho escolar na rede de ensino fundamental do Recife"* [16]. The survey was conducted by Fundação Joaquim Nabuco (FUNDAJ) in 2013, gathering data from more than 4000 pupils in public schools in the North-Eastern Brazilian city Recife. Those data contain among others the social network of the pupils and their performance in the subject maths at the beginning and at the end of the year. Children were asked to nominate their five best friends. In this way, a network containing 4191 students was generated. However, 573 students that did not nominate any friend within their class were removed from the data set, leading to a total number of 3618 vertices.

(ii) The second data set is a selection of 50 girls from the social network data collected in the Teenage Friends and Lifestyle Study [17]. Here the friendship network as well as behavior in sports and substance use of students from a school in Scotland was surveyed. The survey started in 1995 and continued for 3 years until 1997. Students were 13 years old when the study started. The study counted 160 participants of whom 129 participated during the whole study. The friendship networks were surveyed asking the pupils to name up to twelve friends. Pupils were also asked to report their behavior related to sports, smoking as well as alcohol and cannabis consumption.

3 Diffusion of Behavior Modeled as a Coordination Game

We model the imitation of behavior of neighbors within the friendship network according to the coordination game as presented in Sect. 2. As indicated in Sect. 2, we posses no information about possible pay-offs a and b or eventual costs of transition and hence aim to find the threshold GT experimentally. This means that a vertex within the network changes its state over time depending on the state of its neighbors. For simplicity, the vertices may adopt one of two different states according to the investigated behavior. Hereby one state indicates that the vertex adapted behavior A, the other possible state indicates the adaption of behavior B. For each iteration, the current ratio r_i is being calculated. Here a_i denotes the number of neighbors of node i that adapt behavior A and n_i denotes the total number of neighbors of node i.

$$r_i = \frac{a_i}{n_i} \tag{4}$$

If the perceived ratio r_i is higher than the global threshold GT and the state of node i is B , the node changes its behavior towards Behavior A. Conversely, if r_i is below GT and Node i's behavior is A, it changes its behavior towards B. Due to the differences in data representation, the coordination game had to be implemented slightly differently for the two settings, as follows.

3.1 FUNDAJ

The only information available for more than one moment in time of the FUNDAJ survey is the mark of the pupils in the subject maths for the beginning and the end of the year. Although marks are not a behavior in themselves, they stem among others from individual behavior such as doing homework, paying attention, and studying frequently. Marks are therefore considered a good indicator for the behavior *engagement at school*. They are represented as numeric values between 0 and 100. In order to differentiate between two behaviors, students are classified as *good students* or *bad students* according to their mark. Students whose mark lies below the threshold *tm* are thereby classified as bad students and vice versa. The setting of *tm* defines hereby the number of *good students* (positives) and *bad students*(negatives) and hence affects heavily if nodes are predominantly connected to positives or negatives. High values for *tm* generate large numbers of *bad students* and smaller numbers of *good students* and vice versa. The ratio r_i from Eq. (4) is being calculated for each student at each iteration of the simulation. If required, the mark for the next time step m_{i+1} is being multiplied by the factor $1 + f$ in order to alternate the state of the node:

$$m_{i+1} = m_i * (1 + f) \tag{5}$$

Parameter *GT* sets the affinity of the nodes to change behavior. Thus, depending on the proportions of positives and negatives, it either yields a volatile or a stable system. Adaption parameter f also influences the stability of the system, where volatility increases with increasing values of f.

3.2 Scottish Dataset

The Scottish data set contains information about four different behaviors, which are practicing sports, drug (cannabis) use, alcohol-use, and smoking behavior. Characteristic values differ slightly for the distinct behaviors, as there are, for example, two increments representing the intensity of sports but four increments for drug use intensity. Thus, we classified the characteristic values in order to obtain a simplified two status situation. Table 1 presents the characteristic values and their classification as *Behavior A*, all other values are accordingly classified as *Behavior B*. In contrast to the FUNDAJ data, the representation of behavior by discrete values

Table 1 Classification of characteristic values for behavior

Behavior	Characteristic values	Class as behavior A if:
Sports	1 (non regular); 2 (regular)	≥ 2
Drugs	1 (non), 2 (tried once), 3 (occasional), and 4 (regular)	≥ 2
Alcohol	1 (non), 2 (once or twice a year), 3 (once a month), 4 (once a week), and 5 (more than once a week)	≥ 2
Smoke	1 (non), 2 (occasional), and 3 (more than once a week)	≥ 2

required a slightly different imitation process. Hence, for the Scottish data set, if a vertex changes its state, it, respectively, raises the behavior value by 1 if it aims to adopt behavior A or, decreases the behavior value by 1 if it aims to adopt behavior B. Information is available for three consecutive years. Hence the starting value for each vertex in the coordination game is its behavior in year one. The quality of the simulation is measured comparing the state of the simulation after a certain number of iterations with the state of the real system after 2 years, here referred to as *benchmark t+1* or after 3 years, denominated as *benchmark t+2*. Moreover, the friendship network of the girls in the study has been surveyed for each of the 3 years, the study lasted. This yields the three slightly different networks g_1 at the first survey, g_2 after 1 year, and g_3 after 2 years. This implicated for the simulation that the neighbors that a vertex considers for the calculation of its state vary for each year t according to the network g_t. In order to incorporate those network dynamics into the simulation, we changed the network used to define the adjacent vertices of a node after completing 50 % of iterations. Experiments indicated that network combination of g_1 as representation for the friendship network in period between year 1 and year 2, and g_2 representing the friendship network in the period from year 2 to year 3 outperformed the results for network combination (g_2, g_3). Hence, we assume that the more appropriate network combination is the former. Therefore experiments and results presented in the remainder of this paper refer to network combination (g_1, g_2).

4 Experiments

4.1 Experimental Setup

Experiments were run for the two coordination game settings with varying parameters in order to find a parameter setting that leads to plausible results. As for the simulation with FUNDAJ data, the simulation was conducted with all combinations of the parameters GT (global threshold) and f (adaption parameter) for $GT, f \in [0, 0.2, 0.4, 0.6, 0.8, 1]$ and tm (classification of marks) with $tm \in [20, 40, 60, 80]$. For simulations with the Scottish data set the parameter GT was set to values $GT \in [0.0, 0.05, 0.1, \ldots, 1.0]$.

4.2 Quality Measurement

In order to assess the quality of the respective simulation, four distinct quality measures were applied: (1) match quality, (2) ROC-curves (3) graph-based quality measures, and (4) average estimation error.

(i) The most intuitive measure for the simulation quality is to compare the state of each vertex v_s after a certain number of simulation iterations with its state in reality v_r in *benchmark t+1* or *benchmark t+2*. Hereby, we denote the case when $v_s = v_r$ as *match* and accordingly the case $v_s \neq v_r$ as *miss-match*. This quality measure is named *match-quality* and denoted as q for the rest of this work. The match quality q of the simulation can then be assessed as in (6), where n denotes the total number of vertices:

$$q = \frac{\sum_{i=1}^{n} match_i}{n} \qquad (6)$$

However, for skewed attribute distributions, this measure favors estimates with high numbers of positive or, respectively, negative estimates and hence fails to mirror the quality of the simulation when the distribution of attributes is skewed.

(ii) The ROC-metric [18] sets the number of true positives (*Recall*) in relation to the number of false positives (*Fallout*). *Recall* is the ratio of correctly estimated positives values, the *true-positives*, and the total number of positive values n^p. *Fallout* denotes the ratio between wrongly estimated positive values *false-positives* and the total number of negative values n^n.

$$Recall = \frac{true - positives}{n_p} \qquad (7)$$

$$Fallout = \frac{false - positives}{n_n} \qquad (8)$$

The ROC-curve displays, respectively, *Recall* values for each simulation on the ordinate and *Fallout* values on the abscissa. Values above the diagonal of the graph indicate the existence of a signal and values below the diagonal may be interpreted as noise. Thus, this metric provides a clearer picture of simulation quality. Best estimates can be found mathematically maximizing the *Youden-Index* [19] y as presented in (9).

$$y = Recall - Fallout \qquad (9)$$

(iii) For global analysis it might not be necessary to simulate the state of each vertex correctly, as long as the system state can be predicted adequately. Thus, as third quality measure, *behavior distribution in friendship-patterns* was implemented. Hereby we define friendship-patterns in the network using a modified version of NEGOPY [20]. According to NEGOPY, we define vertex types as isolate, dyad, liaison, and group member. As we deal with undirected networks, we do not classify tree-nodes. According to Richards, an isolate is an individual with at maximum one friend. Two persons connected only to each other are denoted as dyad. Liaisons are individuals with more than 50 % connections to members of different groups. Liaisons can also be nodes that are mostly connected to other liaisons and with less than 50 % links to group members. A composition of minimum three individuals is referred to as group, if the individuals share more than 50 % of their linkage, build a connected component and stay connected if up to 10 % of the group members are removed. For measuring the quality, the number of positive vertices n_k^p in each friendship-pattern class k is calculated after each iteration of the simulation. Subsequently, the error e_k is calculated as difference between n_k^p of simulated and real values. The *average-error e* denotes the weighted average error of the simulation and n_k the number of vertices in friendship-pattern k :

$$e = \frac{\sum_{k=1}^{n} e_k * n_k}{\sum_{k=1}^{n} n_k} \tag{10}$$

(iv) The average estimation error ϵ assesses the average difference between simulated values for behavior and real behavioral outcomes. Here n denotes the total number of nodes in the simulation, while the difference between simulation and reality for node i is represented by ϵ_i.

$$\epsilon = \frac{\sum_{i=1}^{n} \epsilon_i}{n} \tag{11}$$

4.3   Results

This subsection presents the results from the experiments presented earlier. We first present results for the experiments with FUNDAJ data and subsequently report results for experiments with the Scottish data set.

Results: FUNDAJ

Figure 1 presents the results for simulations with FUNDAJ data for 15 iterations. Figure 1a contains ROC-curves for the experimental results with varying settings of tm, GT, and f. For each investigated value of mark threshold tm, the figure illustrates an individual ROC-curve. The dashed lines indicate the ROC-level of the respective setting for tm before starting the simulation. Thus only parameter settings leading to ROC-values situated above the respective dashed line can be considered as settings that improve the quality of the simulation. The colored lines in Fig. 1b represent the development of quality indicators q and e for distinct parameter settings and also indicate the average estimation error ϵ during the run time of the simulation.

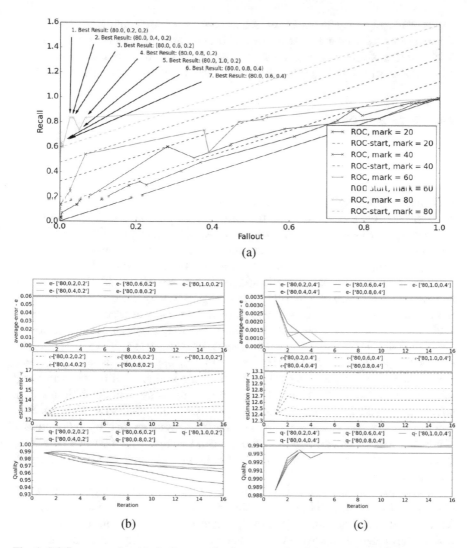

Fig. 1 ROC-curve and analysis for coordination game simulations with FUNDAJ data—15 iterations; for varying threshold marks *tm*, classifying the pupils as *good students*, if their mark is greater than *tm* or *bad students* if their performance is below *tm* and for varying settings of *GT* and *f*, as pointed out in parentheses (*tm,GT,f*). $Recall = \frac{true-positives}{n^p}$; $Fallout = \frac{false-positives}{n^n}$. (**a**) ROC-curve FUNDAJ. (**b**) Analysis FUNDAJ-*f* = 0.2. (**c**) Analysis FUNDAJ-*f* = 0.4

The results with the highest *Youden-Index* in simulations with FUNDAJ data set are indicated by arrows pointing from the respective parameter settings for mark threshold *tm*, global threshold *GT*, and adaption parameter *f* in parentheses as (*tm*, *GT*, *f*) in Fig. 1a. The results for *q*, *e*, and *ε* of those most promising parameter settings are presented in Fig. 1b and c. The more detailed analysis of the five parameter settings that were performing best in ROC-curve analysis in Fig. 1b yields increasing e and increasing estimation error ϵ while q continuously decreases. However, as presented in Fig. 1c the second best performing parameter settings from ROC-curve analysis lead in general to a decay of e and significant growth of q whereas at least one setting (80, 0.2, 0.4) also decreases estimation error ϵ slightly.

Results: Scottish Data

Figure 2 illustrates the results for experiments with the Scottish data set for 50 iterations for each of the investigated behaviors. The solid lines in Fig. 2a illustrate the *Recall–Fallout* relation for varying parameter settings and for different behaviors. The black diagonal line in this graph indicates *Recall–Fallout* ratios that represent random processes, while the dashed lines indicate the ROC-level of the start situation. Since experiments with Scottish data were run with two different networks as explained in Sect. 3, analysis of q, e, and ϵ in Fig. 2b–e contains blue lines, indicating the values calculated in relation to *benchmark t+1* and red lines, representing the results calculated in relation to *benchmark t+2*. ROC-curve for the simulation of diffusion of behavior *sport* in Fig. 2a is very close to the diagonal of the graph, indicating that the simulation is rather a random process. Furthermore ROC-values cannot reach the ROC-level before starting the simulation indicated by the dashed line. However, there are two values for t that yield ROC-values above the diagonal of which $t = 0.55$ generates the most promising results. Hence, q, e, and ϵ development are analyzed over the whole run time in Fig. 2b. It is observable that e in $t + 1$ indicated by the blue line decreases significantly until the 25th iteration, which is when the network g_t is replaced by network g_{t+1}. After the 25th iteration, e in $t + 2$ decreases heavily. ϵ decreases slightly for benchmark $t + 2$ but increases if compared to benchmark $t + 1$. Although decreasing for the first five iterations, q remains stable during the following 20 iterations and slightly improves after 25 iterations.

ROC-curve for smoking behavior in Fig. 2a yields positive results for t 0.35, 0.4, and 0.45, significantly outperforming the initial ROC-value indicated by the dashed line. A deeper examination of q, e, and ϵ development during run time in Fig. 2c shows that as compared with benchmark $t + 1$ neither q, nor e or ϵ develop positively. Though, compared with benchmark $t + 2$ a strong improvement of q as well as a significant decrease of e and a slight decrease of ϵ is observable.

The t values indicated by the ROC-curve for alcohol-use in Fig. 2a do not reach initial ROC-level and yield decreasing q and increasing e until the underlying network is changed after 25 iterations, initiating a slight improvement of those values for both benchmark values as presented in Fig. 2d. Nevertheless, q never reaches a value higher than the start value, also e does not drop under its start value and ϵ remains on an equal level. ROC-curve for drug-use in Fig. 2a yields positive results for t 0.35, 0.4, and 0.45, slightly exceeding the initial ROC-value. Figure 2e presents decreasing e and ϵ, as well as increasing q over the run time for benchmark value $t + 2$, while all quality measures develop negatively for benchmark $t + 1$.

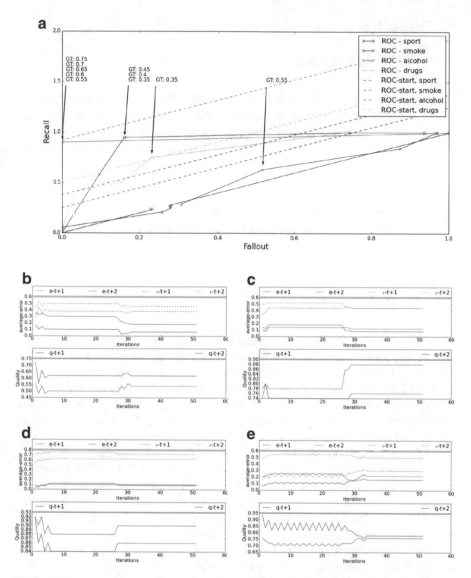

Fig. 2 ROC-curves and analysis for coordination game simulations with Scottish data set—50 iterations. For varying behaviors and for varying settings of GT. $Recall = \frac{true-positives}{n^p}$; $Fallout = \frac{false-positives}{n^n}$. (**a**) ROC-curves Scottish data. (**b**) Analysis sport—$GT = 0.55$. (**c**) Analysis smoking—$GT = 0.45$. (**d**) Analysis alcohol—$GT = 0.65$. (**e**) Analysis drugs—$GT = 0.35$

5   Discussion

5.1   FUNDAJ Data Set

As pointed out in Sect. 4, the parameter setting (80, 0.2, 0.4) performs best as under this setting average-error e is being more than halved (approximately 75 %). For this setting also match quality q increases slightly, ϵ shows a small decay, and *Youden-Index* improves. This indicates that the setting reasonably approximates the real system state. However, simulation is not very adequate in estimating the individual behavior. Thus we argue that diffusion of marks can be reasonably modeled as a coordination game if the researcher is willing to disregard individual states and is interested in the global state of the network instead. Results further indicate that 15 iterations under the given parameter setting are good for approximating one school year.

5.2   Scottish Data Set

Simulating the coordination game spread for behavior *sport* with $GT = 0.55$ yields a relatively small *Youden-Index* and cannot improve the ROC-level of the initial situation. However, the development of *average-error* for benchmark $t + 1$ and $t+2$ yields improvement of the overall state of the network, while decreasing *match-quality* q and increasing ϵ for both benchmarks. Although improving the estimation of the general network state, setting $GT = 0.55$ cannot improve the estimation quality and can therefore not be considered a good setting for GT.

As for simulating the spread of behavior *smoke* throughout the given network, we found strong evidence for the suitability of parameter $GT = 0.35, 0.4$, and 0.45 in the ROC-curve. Run time analysis of e, ϵ, and q indicates that the parameter setting when run on network g_1 cannot reproduce the spreading during the first year, since the former grows while the latter declines for the first 25 iterations. However considering benchmarks $t + 2$ and network g_2, all three quality indicators support the hypothesis that spreading occurs as a coordination game with $t = 0.35$, 0.4, or 0.45. Recall that children were around age 13 when the study started, this discrepancy may be explained by the nature of the behavior smoking, which probably has a higher attraction to children aged 14–15 than to children aged 12–13. Similar but not as striking evidence can be found when examining behavior *drug-use*. As *drug-use* has been explicitly surveyed as the use of cannabis, this seems coherent, since tobacco use does commonly precede cannabis use. Conversely, for the behavior *alcohol-use*, results are not clear. ROC-curves indicate that parameter settings yielding reasonable estimates of the real situation exist. Yet, run time analysis of those cases show that those promising parameter settings do not lead to an improvement of the estimation. Hence, we argue that for alcohol-use we cannot find evidence that spreading of behavior can be modeled as a coordination game within the given data set. This might also be related to the age of the students, since parents influence might be stronger during this period. Additionally due to the restriction of available data to female students the lack of spreading could be gender related.

6   Conclusion

In this paper we adopted the coordination game mechanism for simulating the spreading process of behavior throughout social networks. We ran the simulation on two different data sets, the FUNDAJ study with school children from metropolitan area of Recite and the study from Scottish female pupils. We investigated the spread of behavior "commitment to school education" represented by the marks of the pupils in the FUNDAJ study, as well as the behaviors "Substance use" for tobacco, drugs, and alcohol and the behavior "practicing sports" as surveyed in the Scottish data set. We found good indications that a coordination game mechanism underlies the spread of behavior "commitment to school education" as well as "smoking" and "drug-use" but could not find comparable evidence for behavior "Alcohol-use." Results for behavior "practicing sports" were not clear. We argue that the missing evidence for behavior "Alcohol-use" may stem from the nature of the data set, since surveyed individuals were below 16 years of age until the end of the survey. Moreover only female pupils participated. Since male adolescents are more susceptible to early alcohol-use, this could be an explanation for the lack of evidence, for that particular aspect.

This work serves as a first step in simulating the spread of behavior throughout social networks, since it provides evidence that (1) there is an underlying game-environment for the agents within the social system (2) that it can be modeled as a coordination game. However, the players of this game, the bounded rational agents [21] might be equipped with decision finding mechanisms that better approximate human decision making. Though driving the social systems from a real start situation towards the state in reality after one or, respectively, two years, the investigated deterministic mechanism still lead to a considerable difference between the real and the simulated system. Hence, we argue that the deterministic mechanism is not fully capable to simulate human bounded rationality and the lack of information humans face within their decision process. Besides this, eventual noise within the data and external influences may not be represented by a deterministic mechanism. Future work should therefore deal with the creation of a heuristic decision mechanism for the individual agents, that better represents human decision making within a coordination game setting. In addition, a binary behavioral variable is an extreme simplification for the on continuous scales measured nuances of human behavior such as sports activities, drug and alcohol consumption, or school performance. It is therefore desirable to investigate how more complex scaling systems influence the outcomes of this research. Furthermore, inter-temporal components shall be introduced, representing an "aging" of relations and behaviors, modifying the influence of neighbors according to the "age" of the friendship, as well as according to the past behavior of the neighbor. In the same sense, friendship weights may be modified according to the position of friends within the individual networks since people may tend to follow "role models." Finally, the presented mechanism must be applied to different data sets in order to empirically verify the results.

Acknowledgements This research was partially funded by Andrea von Braun Stiftung. The authors thank Fundação Joaquim Nabuco for providing data and domain expertise.

References

1. Christakis, N.A., Fowler, J.H.: Connected: The Surprising Power of Our Social Networks and How They Shape Our Lives. Little, Brown, New York (2009)
2. Easley, D., Kleinberg, J.: Networks, Crowds, and Markets: Reasoning About a Highly Connected World. Cambridge University Press, Cambridge (2010)
3. Lelarge, M.: Diffusion and cascading behavior in random networks. Games Econ. Beh. **75**(2), 752–775 (2012)
4. Latane, B.: The psychology of social impact. Am. Psychol. **36**(4), 343 (1981)
5. Christakis, N.A., Fowler, J.H.: The spread of obesity in a large social network over 32 years. N. Engl. J. Med. **357**(4), 370–379 (2007)
6. Jordan, J.J., Rand, D.G., Arbesman, S., Fowler, J.H., Christakis, N.A.: Contagion of cooperation in static and fluid social networks. PloS One **8**(6), e66199 (2013)
7. Kramer, A.D., Guillory, J.E., Hancock, J.T.: Experimental evidence of massive-scale emotional contagion through social networks. Proc. Natl. Acad. Sci. **111**(24), 8788–8790 (2014)
8. Brakefield, T.A., Mednick, S.C., Wilson, H.W., De Neve, J.-E., Christakis, N.A., Fowler, J.H.: Same-sex sexual attraction does not spread in adolescent social networks. Arch. Sex. Behav. **43**(2), 335–344 (2014)
9. Marques, E.: Redes sociais, segregação e pobreza. Editora Unesp (2010)
10. Kratzer, J., Lettl, C.: Distinctive roles of lead users and opinion leaders in the social networks of school children. J. Consum. Res. **36**(4), 646–659 (2009)
11. Paluck, E.L., Shepherd, H.: The salience of social referents: a field experiment on collective norms and harassment behavior in a school social network. J. Pers. Soc. Psychol. **103**(6), 899 (2012)
12. Friedkin, N.E.: A Structural Theory of Social Influence, vol. 13. Cambridge University Press, Cambridge (2006)
13. Granovetter, M.: Threshold models of collective behavior. Am. J. Sociol. **83**, 1420–1443 (1978)
14. Granovetter, M., Soong, R.: Threshold models of interpersonal effects in consumer demand. J. Econ. Behav. Organ. **7**(1), 83–99 (1986)
15. Schelling, T.C.: Dynamic models of segregation. J. Math. Sociol. **1**(2), 143–186 (1971)
16. Coordenação de Estudos Econômicos e Populacionais, Fundação Joaquim Nabuco Fundaj. Determinantes do desempenho escolar na rede de ensino fundamental do Recife (2013)
17. Michell, L., Amos, A.: Teenage friends and lifestyle study dataset (1997)
18. Fawcett, T.: An introduction to ROC analysis. Pattern Recogn. Lett. **27**(8), 861–874 (2006)
19. Youden, W.J.: Index for rating diagnostic tests. Cancer **3**(1), 32–35 (1950)
20. Richards, W.D., Rice, R.E.: The NEGOPY network analysis program. Soc. Networks **3**(3), 215–223 (1981)
21. Todd, P.M., Gigerenzer, G.: Bounding rationality to the world. J. Econ. Psychol. **24**(2), 143–165 (2003)

A Model of Social and Economic Capital in Social Networks

Bogumił Kamiński, Jakub Growiec, and Katarzyna Growiec

Abstract We propose an agent based model allowing to simulate social and economic capital stock on aggregate and individual level in the economy. It is shown that different topologies of the social network lead to varying levels of both types of capital in the economy and that there is a trade-off between social and economic utility which they ultimately convey. Therefore, under mild conditions "small world" type networks should provide an optimal balance between both types of capital. We also find that individuals who form "bridges" in the social network can benefit from their importance in the network both in terms of their social and economic utility.

Keywords Bonding social capital • Bridging social capital • Economic capital • Social network structure

1 Introduction

The objective of this work is to identify the impact of selected structural characteristics of individuals' social networks on their willingness to cooperate, social trust, and—in consequence—on their labor productivity and earnings. As these relationships are studied both at the individual and the aggregate level, we shall thus also address the question, to which extent population-wide social network structures (and the implied aggregate social capital stocks) affect the average willingness to cooperate and average labor productivity in an economy.

The direct reason for undertaking the current research is the fact that—although many of the theoretical definitions of social capital invoked in the literature relate directly to the structure of social networks [1, 2]—empirical studies typically rely on

B. Kamiński (✉) • J. Growiec
Warsaw School of Economics, Al. Niepodległości 162, 02-554 Warsaw, Poland
e-mail: bkamins@sgh.waw.pl; jakub.growiec@sgh.waw.pl

K. Growiec
Department of Psychology of Personality, Faculty of Psychology, University of Social Sciences and Humanities, ul. Chodakowska 19/31, 02-554 Warsaw, Poland
e-mail: katarzyna.growiec@swps.edu.pl

© Springer International Publishing AG 2017
W. Jager et al. (eds.), *Advances in Social Simulation 2015*, Advances in Intelligent Systems and Computing 528, DOI 10.1007/978-3-319-47253-9_16

heavily simplified operationalizations, largely due to the problems with availability of sufficiently detailed data. In particular, according to our best knowledge, the literature lacks a thorough empirical quantification of structural characteristics of social network such as their local density, heterophily of the social network, or the durability of social ties. Simultaneously, the available mathematical models of economies which attempt to capture the economic role of social capital, e.g., [3, 4], generally achieve analytical tractability at the cost of making strong prior assumptions about unrealistic network structures.

2   Model Description

The model is designed and analyzed according to the multi-agent approach. We assume a population of $n \in \mathbf{N}$ *agents* who are placed along a circle—thus agent 1 is the neighbor of agents 2 and n. Agents are linked to one another. By a link between agents i and j we mean their *social tie*, in line with the definition due to Bourdieru [5].[1]

The binary matrix R of dimension $n \times n$ captures the links between agents and is generated according to the Watts–Strogatz algorithm, cf. [6]. If $R_{i,j} = 1$, then we say that agent i is linked to agent j with a social tie. Otherwise $R_{i,j} = 0$. It is assumed that social ties are symmetric, so that $R_{i,j} = R_{j,i}$ for all pairs i, j. By $D_i = \sum_{j=1}^{n} R_{i,j}$ we denote the number of agent i's social ties (*degree* of the node).

The Watts–Strogatz algorithm has three parameters that influence the structure of generated network: population size n, *rewiring probability* μ, and *radius r*. In the initiation phase agent i is connected to agents that are within the radius r, i.e., its set of neighbors is $\{j : 0 < \min\{|i-j|, n-|i-j|\} \le r\}$. Next each connection is changed to a different random connection with probability μ. Hence, the "rewired" network is always between lattice network ($\mu = 0$) and random network ($\mu = 1$). For moderate values of μ we obtain the so-called *small world* networks.

We assume that the location of agents on the circle reflects their *similarity*, e.g., membership in the same social groups. Formally we define similarity of agents i and j as their distance along the circle normalized to the interval $]0, 1]$:

$$S_{i,j} = 1 - 2 \min\{|i-j|, n-|i-j|\}/n. \tag{1}$$

Observe that initially (before rewiring) Watts–Strogatz algorithm connects similar agents. This is a desired property from the perspective of the proposed model as it implies a situation where agents do not contact other agents that are significantly

---

[1]Bourdieu [5]: "Social capital is the aggregate of the actual or potential resources which are linked to possession of a durable network of more or less institutionalized relationships of mutual acquaintance and recognition—or in other words, to membership in a group—which provides each of its members with the backing of the collectivity-owned capital, a 'credential' which entitles them to credit, in the various senses of the word." (p. 128).

different than them (eg. belong to a different social group). After rewiring we can get the other extremum when $\mu = 1$ and the network is fully random: similarity has no influence on the structure of social ties of agents.

Putnam [1] defines bonding social capital measures linkages between similar agents. Therefore we define agent i's *bonding social capital stock* as:

$$A_i = \sum_{j=1}^{n} R_{i,j} S_{i,j}^{\alpha}, \qquad \alpha > 0. \tag{2}$$

Under this definition the social capital of an agent will be maximal for $\mu = 0$ and decrease as μ grows. We introduce control parameter α to be able to test how the speed of decay of value of similarity between agents influences the results.

The bonding social capital stock will also be called the *social utility* of an agent. The average level of bonding social capital in the society can be computed as:

$$\bar{A} = \sum_{i=1}^{n} A_i / n. \tag{3}$$

The heterophily coefficient of agent i's social network is defined as:

$$H_i = 1 - A_i / D_i. \tag{4}$$

This coefficient helps assess to which extent the average person with whom agent i maintains social ties differs from herself in terms of their social characteristics (i.e., along the *similarity* dimension $S_{i,j}$). We assume that the more diverse the set of agents known by the given agent is the more she will be willing to cooperate in economic interactions.

The graph R is then used to define the *social distance* between agents, $L_{i,j} \in \mathbb{N} \cup \{\infty\}$, equal to the length of the shortest path connecting agents i and j. The interpretation of $L_{i,j}$ is how many "hops" between people linked with social tie are needed to connect i and j. We assume that the higher the $L_{i,j}$ the lower the trust between the agents is when they interact.

Following Xianyu [7], we consider the case where agents interact economically with others by playing the "prisoner's dilemma" game with them. In the game if two agents are engaged in an economic interaction, then they both independently have to decide if they want to cooperate. We assume that if the agent decides to exploit (does not want to cooperate, is a "free rider") his partner, then she gets a better payoff than when she would make an effort to cooperate. However, if both parties play exploitation strategy they are worse off than when they would cooperate. This set of assumptions naturally leads to prisoner's dilemma setup.

We assume that the individual i's willingness to cooperate depends on the heterophily coefficient of her social ties (trust through "experience") as well as the social distance between i and j (trust through "acquaintance"). This probability equals:

$$P_{i,j} = (1 - \beta + \beta H_i)L_{i,j}^{-\gamma}, \tag{5}$$

where $\beta \in [0,1]$ and $\gamma > 0$. As for α the parameters β and γ can be varied to analyze how sensitive the results are to the relative importance of experience and acquaintance factors (we take that if $L_{i,j} = \infty$, then $L_{i,j}^{-\gamma} = 0$). In particular if $\beta = 0$, then H_i is not important at all and if $\beta = 1$, then it is very important.

Agent i's average willingness (probability) to cooperate in the prisoner's dilemma game is identified with agent i's *social trust* level:

$$T_i = \frac{1}{n-1} \sum_{j=1}^{n} P_{i,j}. \tag{6}$$

Thus we assume that social trust of agent i is proportional to (1) the average shortest path length between agent i and a randomly chosen other agent, and (2) the heterophily coefficient of agent i's social ties. This assumption is in line with the empirical literature which finds that social trust can be acquired via demonstration. The average level of social trust in the society is defined as:

$$\bar{T} = \frac{1}{n} \sum_{j=1}^{n} T_i. \tag{7}$$

The bridging social capital stock of agent i measures how important she is in providing connections in the social network and is defined as follows. First, we consider the expression:

$$\frac{1}{(n-1)(n-2)} \sum_{j \neq i, k \neq i, j \neq k} L_{j,k}^{-\gamma}. \tag{8}$$

It is the average path length between random two agents other than agent i under the transformation $L_{j,k}^{-\gamma}$ used in the formula (5). Then we consider a social network where the node i has been removed. We compute the updated path lengths $\widetilde{L_{j,k}}$ and evaluate:

$$\frac{1}{(n-1)(n-2)} \sum_{j \neq k} \widetilde{L_{j,k}}^{-\gamma}. \tag{9}$$

The difference between formulae (8) and (9) signifies the extent to which the presence of agent i in the network has been reducing the distances between any two other agents in terms of the measure $\bar{T}$. This difference will be denoted as B_i and called the bridging social capital stock of agent i. The average bridging social capital stock in the society is denoted as

$$\bar{B} = \frac{1}{n} \sum_{i=1}^{n} B_i. \tag{10}$$

The concept of *economic utility* of agent i is based upon the assumption that each agent i plays the prisoner's dilemma game will all other agents with whom she maintains social ties. Then, agent i's payoff matrix can be normalized to:

$$G = \begin{bmatrix} 1 & 0 \\ g_{dc} & g_{dd} \end{bmatrix}, \tag{11}$$

where agent i's strategies are indicated in rows, and the opponent's strategies—in columns. The first row/column denotes the decision to cooperate, and the second one—to defect. For the prisoner's dilemma game we have $1 < g_{dc} < 2$ and $0 \leq g_{dd} < 1$.

When agent i meets agent j, each of them independently decides whether to cooperate or not. Agent i's probability to adopt the cooperating strategy equals to $P_{i,j}$. Hence, the mixed strategy of player i against player j is: $[P_{i,j} \ 1 - P_{i,j}]$. For such a setup, agent i's expected payoff when playing against j equals:

$$E_{i,j} = [P_{i,j} \ 1 - P_{i,j}] G \begin{bmatrix} P_{j,i} \\ 1 - P_{j,i} \end{bmatrix}. \tag{12}$$

In conclusion, agent i's economic utility measure, assuming that she plays the prisoner's dilemma game with every other agent $j \neq i$, is equal to:

$$E_i = \frac{1}{n-1} \sum_{i \neq j} E_{i,j}. \tag{13}$$

The average economic utility in the society amounts to $\bar{E} = \sum_{i=1}^{n} E_i / n$.

In the ensuing analysis, we are going to assume that agents intend to maximize their total utility, which is positively related to both social utility A_i and economic utility E_i. Under such an assumption, our model allows for analyzing the theoretical relationships between social network structures and economic outcomes both at the aggregate and the individual level.

3 Preliminary Results

Simulation analysis of the proposed model shows that there is a general negative relationship between the aggregate social utility (A is maximized for regular graphs) and economic utility (E is maximized for random graphs), in line with the empirical findings. Therefore, under the assumption of a constant expected marginal rate of substitution between both sources of utility, "small world" networks should be socially optimal.

At the individual level, our results indicate that the agents' bridging social capital stocks (*B*) are positively correlated both with their social utility (*A*) and economic utility (*E*). This confirms that "bridges" in the social network can benefit from their importance [8] in both dimensions of their utility.

Acknowledgements The research was financed by the grant number 2013/11/B/HS4/01467 from National Science Centre, Poland.

References

1. Putnam, R.: Bowling Alone. Collapse and Revival of American Community. Simon & Schuster, New York (2000)
2. Lin, N.: Social Capital. Cambridge University Press, Cambridge (2001)
3. Durlauf, S.N., Fafchamps, M.: Social capital. In: Aghion, P., Durlauf, S.N. (eds.) Handbook of Economic Growth. Elsevier, Amsterdam (2005)
4. Growiec, K., Growiec, J.: Social capital, trust, and multiple equilibria in economic performance. Macroecon. Dyn. **18**, 282–315 (2014)
5. Bourdieu, P.: The forms of capital. In: Richardson, J.C. (ed.) Handbook of Theory and Research of Sociology of Education. Greenwood Press, New York (1986)
6. Watts, D., Strogatz, S.: Collective dynamics of "small-world" networks. Nature **393**, 440–442 (1998)
7. Xianyu, B.: Prisoner's dilemma game on complex networks with agents' adaptive expectations. J. Artif. Soc. Soc. Simul. **15**, 3 (1998)
8. Burt, R.S.: Decay functions. Soc. Netw. **12**, 1–28 (2000)

The Impact of Macro-scale Determinants on Individual Residential Mobility Behaviour

Andreas Koch

Abstract Models of urban residential mobility usually take a bottom-up perspective by emphasising individual (household) decisions on locational preferences. These decisions are, however, framed by social, economic, and political forces such as markets, legal rules, and norms. The chapter therefore presents a preliminary attempt to incorporate macro-scale determinants, such as housing market mechanisms and urban planning decisions, into an agent-based model (ABM) to simulate patterns of residential mobility more comprehensively. The case study refers to the city of Salzburg, Austria.

Keywords Human agency • Social forces • Urban segregation • Agent-based modelling

1 Introduction

Theories of urban residential mobility in general and of segregation processes such as gentrification in particular can be roughly distinguished into micro-level approaches of human agency and macro-level approaches of social forces. The first set of approaches tries to explain patterns of socio-spatial fragmentation with small-scale clusters of socially similar distributions of households embedded into a heterogeneous and diverse composition of communities at the urban scale by individual autonomous decision-making processes. The second set focuses predominantly on large-scale economic and cultural forces which act contemporarily under neoliberal regimes of profit making. The impact of bottom-up processes on the structuring of housing allocation can be properly analysed by theories of emergence and, methodologically, by the application of agent-based modelling [1–3]. Top-down influences or determinations, in turn, can be adequately described by downward causation in theory and, methodologically, once again by agent-based modelling or microsimulation approaches [4–6].

A. Koch (✉)
Department of Geography and Geology, University of Salzburg, Salzburg, Austria
e-mail: andreas.koch@sbg.ac.at

© Springer International Publishing AG 2017   201
W. Jager et al. (eds.), *Advances in Social Simulation 2015*, Advances in Intelligent
Systems and Computing 528, DOI 10.1007/978-3-319-47253-9_17

The chapter presents a preliminary attempt to incorporate macro-scale determinants, such as housing market mechanisms and urban planning decisions, into an agent-based model (ABM) to simulate patterns of residential mobility more comprehensively. In doing so, this model serves as an interim stage for the ultimate goal of designing a model that represents socio-spatial and economic-spatial micro-macro linkages of urban housing dynamics in a more sophisticated way. Theoretical references draw, among others, on the rent gap theory—the hypothesis of a "disparity between the potential rents that could be commanded by [. . .] properties and the actual rents they are commanding" ([7], p. 145)—and theories of housing segregation that consider the allocation of living space and city districts to households in explicit ways [8, 9].

The ABM used here builds upon an existing model of residential mobility which has been created for the city of Salzburg, Austria, to provide bottom-up simulations for patterns of spatial distribution of households with different socio-economic status and social preference [10]. In the remainder a brief description of the original ABM and its macro-structure extension is given, followed by some first and preliminary results.

2   The Model Design

The current version of the segregation model is a *conceptual* simulation model. Its primary aim is to verify the model's purpose, which is to achieve an understanding of intra-urban residential mobility in order to detect mechanisms of emergence. The newly built-in component aims to represent the targets of urban planning and real estate agencies that are practically realised through the implementation of large housing estates. The question raised here is how these large housing estates affect individual migration behaviour. We use two types of housing estates: the first one consists of more or less homogeneous units of social housing; the second is a mix of social housing and private properties.

The agent population is subdivided into four groups, representing households with different socio-economic and socio-demographic characteristics. These include social status, as well as different preferences towards their social neighbourhood environment. Agents are forced to move if they cannot afford living at the current location. In addition, they are able to move voluntarily if they are dissatisfied with the social environment in their neighbourhood. In contrast to other ABMs of segregation, however, not all dissatisfied agents actually do move, due to different reasons (e.g. relocation is expensive, existing social ties are threatened to become dissolved, potential new ties are unsure). On the other hand, a small proportion of households chooses to move, even though they are satisfied with their place of living (again, due to different reasons such as changes in household size, aspiration for a newly built home or changes in the services infrastructure). The resolution of the city is set to approximately 15,000 spatial entities inhabited by 150,000 citizens. A number of 4000 agents is selected as potential intra-urban movers, which is a conservative estimation of 2.7 % of the total population. Initially, agents of all four subgroups are randomly distributed over the urban space, according to the price per patch, i.e. initially, all agents can afford the dwelling they are living in (for further details, see [10]). With regard to the standard model (without implementation of large housing estates), two phenomena are interesting and noteworthy: first, compared to a Schelling-type model, segregation is no longer a common phenomenon that is evenly distributed over the urban space. Instead, segregation is spatially concentrated on some places of the city. Second, segregation takes place notwithstanding.

For the extended model two large housing estates have been inserted into two different city districts. The socially homogeneous housing estate with a high proportion of social housing has been realised in a less affluent district (Itzling) in order to support affordable living conditions for the local communities. This housing estate represents an actually existing urban planning realisation. The second estate, following a "social mix" housing concept, has been realised in a wealthy city district of Salzburg (Aigen) and represents an actual realisation of future urban planning. In order to analyse the question of how large housing estates influence individual migration decisions under the pre-conditions of individual affordability and preferences, the size of the two housing estates has been varied by three steps. In addition, migration dynamics have not only been investigated within the housing estates, but also in the immediate local surroundings. The ABM has been created with NetLogo 5.2.0 [11]. Eleven repetitions of simulation runs have been conducted for each initialisation of the variable set, and the mean values of these 11 runs have been used. The simulation time has been set to 150 steps.

3   Some Selected Model Results

A comparison of the allocation patterns of the four agent groups of the standard model with the extended model reveals, among others, one common result: the two large housing estates do not exert a significant influence on the large-scale level of the entire city. In comparison, the most affluent agent population, defined as "red agents" (with a strong affinity to a "social status" neighbourhood similarity and less pronounced "social attitude" neighbourhood similarity), appears to be easily able to create neighbourhoods of similar agents. This is, however, also true for the least affluent agent population, the "blue agents" (with opposite characteristics compared to those of the "red agent" population). The blue agents are, however, not as often able to do so, because they concentrate on places of less expensive districts which hardly exist. The middle-income agent populations, the "green agents" and the "yellow agents" (with more balanced preferences towards social status and social attitude), exhibit less pronounced segregation tendencies.

More interesting are the local dynamics within and around the newly created large housing estates. The case of Itzling (high proportion of social housing) illustrates that all four agent populations can achieve quite similar levels of

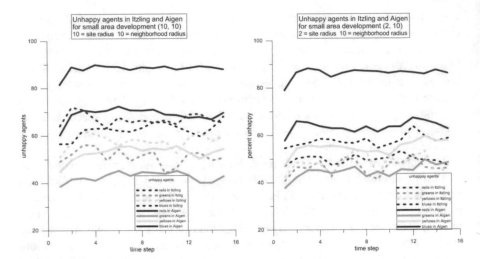

Fig. 1 Dissatisfied agents in Itzling and Aigen, compared across agent populations. *Left*: largest housing estate (radius of ten cells for site and for neighbourhood development); *right*: smallest housing estate (radius of two cells for site development)

dissatisfaction (see Fig. 1), which are relatively independent from the size of the estate. The larger the housing estate area is, the less easily are red agents capable of realising their preferences and needs (this is also true for green and yellow agents, but to a lesser degree). One explanation might be that they are not as easily able to achieve their goals in terms of status satisfaction, which is obvious from an empirical perspective, as social housing is dedicated to less affluent households. The case of Aigen (mix of social housing and private property) is somewhat different. The blue population has the highest percentage of dissatisfied agents, irrespective of the size of the housing estate; living in a high-price place is most difficult for them, although urban planning has developed strategies of inclusion. Surprisingly, unhappy red agents are ranked second, but maybe this is again because of the mixing strategy, which makes it more difficult to achieve the goal of homogeneous social neighbourhoods in sustainable ways.

A change of the basis of comparison from across-agent to within-agent population comparisons leads to another insight: the different dynamics of how agents struggle with their spatial and social environment. In this case the migration behaviour (measured again by levels of dissatisfaction) of the four agent populations over the two districts is compared. As Fig. 2 illustrates, this dynamic is less pronounced for the red and blue agents, as is the case for the green and yellow agents (see Fig. 3).

While red agents feel most comfortable in a small housing estate in Itzling and much less comfortable in larger units in both districts, the blue agent population exhibits a sharp distinction between Itzling (smaller degrees of unhappiness) and Aigen. The common competition for affordable living space gives the red agents an edge due to a higher degree of economic power, which in turn provides them with

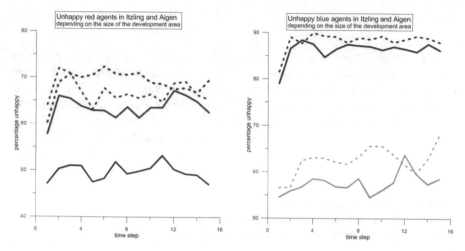

Fig. 2 Dissatisfied red agents (left; *purple* = Itzling, *red* = Aigen) and *blue agents* (right; *light blue* = Itzling, *dark blue* = Aigen), compared across districts. *Solid lines* represent small housing estates, *dotted lines* large housing estates

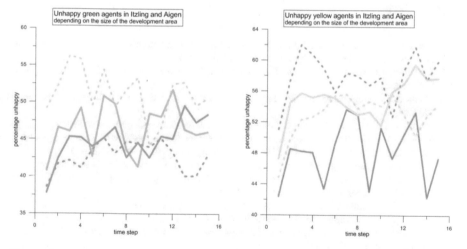

Fig. 3 Dissatisfied green agents (left; *light green* = Itzling, *dark green* = Aigen) and *yellow* agents (right; *light orange* = Itzling, *yellow* = Aigen), compared across districts. *Solid lines* represent small housing estates, *dotted lines* large housing estates

more opportunities to live in close proximity—even in areas that are spatially less attractive. The blue agents, on the other hand, are displaced to locations with a high degree of social housing, preventing social interactions with other social groups.

The migration behaviour of the green and yellow agent populations is much more volatile, even though they show lower levels of dissatisfaction. Furthermore, there is no clear segregation pattern, neither in terms of the size of the housing estate nor of the city district and thus the planning strategy (social housing or social mix).

4 Conclusion

The extension of the standard model towards an incorporation of macro-level determinants is necessary to better understand the mutual relationships between individual residential mobility behaviour and social impacts affecting it. Some results can claim epistemological plausibility, such as the evolution of dissatisfaction of the red and blue agent populations. However, other results need further investigation since they exhibit unexpected social and spatial effects of initial model settings. The results presented here are partly based on reliable variable settings (e.g. the relationship between income and rent price development), partly they lack justification (e.g. all four agent populations are of the same size). Therefore a more thorough analysis of value ranges of the variables in use is necessary. On the other hand, and due to the conceptual status of the current model, it is possible to generalise the results to some degree—based on the given model design and its purpose.

References

1. Ioannides, Y.: From Neighborhoods to Nations. The Economics of Social Interactions. Princeton University Press, Princeton and Oxford (2013)
2. Schelling, T.C.: Micromotives and Macrobehavior, Revised editionth edn. Norton & Co., New York (2006)
3. Zhang, J.: Tipping and residential segregation: a unified Schelling model. J. Reg. Sci. **51**, 167–193 (2009)
4. Birkin, M., Wu, B.: A review of microsimulation and hybrid agent-based approaches. In: Heppenstall, A.J., Crooks, A.T., See, L.M., Batty, M. (eds.) Agent-Based Models of Geographical Systems, pp. 51–68. Springer, Dordrecht, Heidelberg, London, New York (2012)
5. Sawyer, K.R.: Interpreting and understanding simulations: the philosophy of social simulation. In: Edmonds, B., Meyer, R. (eds.) Simulating Social Complexity, pp. 273–289. Springer, Dordrecht, Heidelberg, London, New York (2013)
6. Jordan, R., Birkin, M., Evans, A.: Agent-based modelling of residential mobility, housing choice and regeneration. In: Heppenstall, A.J., Crooks, A.T., See, L.M., Batty, M. (eds.) Agent-Based Models of Geographical Systems, pp. 511–524. Springer, Dordrecht, Heidelberg, London, New York (2012)
7. Knox, P., Pinch, S.: Urban Social Geography, 5th edn. Pearson Education Limited, Harlow (2006)
8. Dangschat, J.: Segregation. In: Häußermann, H. (ed.) Großstadt. Soziologische Stichworte, 3rd edn. VS Verlag, Wiesbaden (2007)
9. Dangschat, J.: Sag' mir wo du wohnst, und ich sag' dir wer Du bist! Zum aktuellen Stand der deutschen Segregationsforschung. PROKLA **27**, 619–647 (1997)
10. Koch, A.: The individual agent makes a difference in segregation simulation. Proceedings of the 10th Conference of the European Social Simulation Association, pp. 12. (2014)
11. Wilensky, U.: NetLogo. Center for Connected Learning and Computer-Based Modeling. Northwestern University, Evanston, IL (1999). http://ccl.northwestern.edu/netlogo/

Modelling the Energy Transition: Towards an Application of Agent Based Modelling to Integrated Assessment Modelling

Oscar Kraan, Gert Jan Kramer, Telli van der Lei, and Gjalt Huppes

Abstract To attain a better understanding of the energy transition we have applied Agent Based Modelling (ABM) to Integrated Assessment Modelling (IAM) in an abstract model with which we developed a proof of concept model of society's response to a changing climate and energy system. Although there is no doubt that large scale neoclassical IAMs have provided key insights for business decisions and policy makers, we argue that there is a need for an approach that focuses on the role of heterogeneous agents.

With our abstract ABM based on agents with heterogeneously spread discount rates we were able to give a new perspective on appropriate discount rates in the discussion between mitigation and adaption to climate change. We concluded that applying ABM to IAM yields good prospects to the further development of the implementation of society's response to a changing environment and we propose future additions of the model to include adaptive behaviour.

Keywords Integrated assessment modelling • Agent based modelling • Cost–benefit analysis • Mitigation • Adaptation • Climate change

1 Introduction

Since we only have one Earth and hence no possibilities to experiment, we use energy models and their resulting scenario's to understand the dynamics of the energy transition from a fossil fuels based to a zero-carbon emission energy system and quantify narratives about how this transition could evolve [1]. The response

O. Kraan (✉) • G.J. Kramer • G. Huppes
Institute for Environmental Sciences (CML), Leiden University, Leiden, The Netherlands
e-mail: o.d.e.kraan@cml.leidenuniv.nl; Kramer@cml.leidenuniv.nl; huppes@cml.leidenuniv.nl

T. van der Lei
Faculty of Technology, Policy and Management (TPM), Delft University of Technology, Delft, The Netherlands
e-mail: li.lei-van-der@dsm.com

© Springer International Publishing AG 2017
W. Jager et al. (eds.), *Advances in Social Simulation 2015*, Advances in Intelligent Systems and Computing 528, DOI 10.1007/978-3-319-47253-9_18

207

of society to a warming climate with its associated unclear consequences on the economy and biosphere are substantially uncertain because it is faced with difficult trade-offs that have different time horizons to address the problem [2].

With this in mind it is not surprising that there is a growing scientific recognition that there is a need to focus attention to model the economy, the energy system and its environment from the bottom up concept of complex adaptive systems (CASs) [3–5]. CASs are systems that are shaped by decision by heterogeneous adaptive agents on different levels such as countries, companies and individuals. We have applied this concept with the use of agent based modelling (ABM) with which we could focus our attention to the integration of society's response to a changing climate and energy system into integrated assessment models.

At the moment, most large scale top down models that combine climate and economy, the so-called integrated assessment models (IAM), rely on more or less elementary forms of the prevailing neoclassical theory of economic growth modelled with computer equilibrium models. Although there is no doubt these IAM and other neoclassical energy models (for an overviews look at [6]) have provided key insights for business decisions [7] and policy makers, "a basic problem is the underlying paradigm of an intrinsically stable economic system that follows an optimal growth path governed by the investment of perfectly informed rational actors maximizing a universal intertemporal utility function" [8]. Other researchers have distinguished the same problem [9, 10].

The development of behavioural economics in the field of economics [11] and the development of the field of complexity science in computer science has given rise to increased attention to the integration of society's response to the energy transition in IAM with ABM [3, 8, 9, 12]. ABM is used to simulate complex adaptive systems (CAS) such as the energy system [12] and is well suited to model adaptive heterogeneous agents that, based on their decisions, can be part of emergent system behaviour. Whereas previous studies such as [13] used ABM to show the role of adaptive change of agents this study focuses on the heterogeneity of agents.

To address the need for a better understanding of the energy transition and to quantify narratives of worldviews on how this transition can happen based on heterogeneous adaptive agent decisions, the conceptualization of an ABM should start with the simplification of the energy transition to its key characteristics and the assumption on how agents make decisions.

This study shows the results of a proof of concept agent based model of the energy transition within which heterogeneous agents apply a classical cost–benefit analysis (CBA) to the problem of mitigation versus adaptation. Future adaptive behaviour aspects to add to this model are proposed.

2 Background

2.1 CBA and Discounting

Research has put a lot of attention to the timing of mitigation versus adaptation [14] with the use of CBA [15]. CBA is an economic analysis to evaluate options generating costs and benefits on different time-scales [15]. These costs and benefits are evaluated on their present value by multiplying them with the discount factor which depends on the applied discount rate. The assumption that humans and animals discount the future has been proven by empirical studies in economics and behavioural ecology [16]. The large time lag between when society incurs the cost and reaps the benefits of decisions by agents to mitigate climate change makes a CBA sensitive to the discount rate.

Because of the non-linear characteristics of earth's climate the exact relationship between GHG emissions and a warming climate (the climate sensitivity) and the effect of a warming climate on our economy (the socio-economic sensitivity) [2] and biosphere are for a large part uncertain. This gives rise to different ethical worldviews about how to solve the problem of a warming climate. These different worldviews translate to some extent to the different discount rates researchers apply.

Researchers have questioned how these different worldviews should be incorporated in IAMs and what discount rates would be appropriate [16–18]. The spectrum is stretched by on the one hand Stern applying a near zero pure rate of social time preference, and Nordhaus applying a market conform discount rate. However there is general consensus among scholars that total climate change damages are larger with larger cumulative CO_2eq stabilization levels [19].

Our ABM can account for these different worldviews by modelling heterogeneous agents that apply different discount rates in their individually applied CBA. Other scholars have argued that CBA is of limited use to evaluate decisions to mitigate climate change because of deep uncertainty in the climate and socio-economic sensitivity [20]. By acknowledging their contribution to the discussion on CBA in IAM, we argue that by addressing the uncertainty with sensitivity analysis on key variables, the model is fit for purpose.

3 Conceptualisation of the Model

3.1 Purpose

The purpose of the model is to simulate the energy transition and quantify narratives about how such a transition can evolve by narrowing the system down to its main characteristics. More specific we try to give a new perspective on the appropriate discount rate in models that discuss mitigation and adaptation to climate change. The emergent system is described by the CO_2eq emission level and the system costs at the end of the runtime of the model. The model is written in the software environment of Netlogo. Its code and the exact equations for the variables can be made available by the author upon request. The model has been validated with recording and tracking behaviour, single-agent testing and multi-agent testing as proposed by Van Dam [21].

3.2 Agents and Their Environment

The system is composed by one type of agent, that represent members of society in all its forms, business decision makers, country representatives or individual consumers that use fossil fuels and can make a decision to invest in a GHG-mitigating technology based on individually performed CBA. The timescale is arbitrary but notionally equivalent to the year 2100.

Agents are assumed to emit a standard unit of CO_2eq emissions which over time results in cumulative stock of CO_2eq emissions. Agents have a binary choice to mitigate these emissions completely. This decisions results in a cumulative investment in mitigation technology. The investment costs of a mitigation technology are assumed to go down exponentially based on the learning curve of these technologies.

How the adaptation costs, in our model equivalent with the climate change damage costs, actually will evolve is faced with uncertainty. The model assumes a climate change damage function which represents the adaptation costs agents have to make. This damage function is a function of cumulative CO_2 emissions with a large parameter bandwidth reflecting the deep uncertainty on climate and socio-economic sensitivity. This parameter bandwidth is expressed with the curvature of the adaptation-cost function, as well as it's begin and end points. When referred to the "normal" adaptation-costs function, we refer to an exponential upward curve as other researchers have identified as most probable [16].

The worldview of an agent on how adaptation costs will evolve is expressed by the discount rate which translates in a discount factor that exponentially depends on time and the discount rate. The discount rate is randomly given to agents based on an exogenous discount rate distribution, is fixed to the agent and is uncorrelated to their.

The summation of the present value of benefits an agent will gain by mitigating now is the summation of the present value of avoided climate change damages over the years, reflecting the difference between climate change damages with or without the cumulative emissions an agent would have emitted when he would not have made the investment (business as usual (BAU)) (Fig. 1).

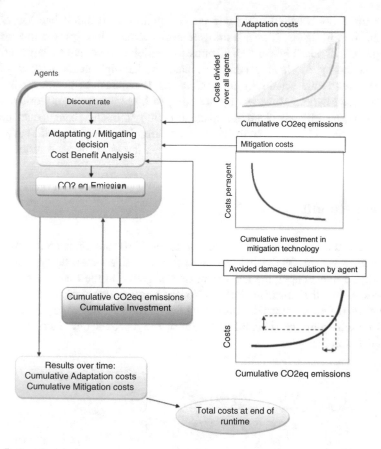

Fig. 1 System description

4 Narratives

The conceptualisation of the models finds it origin by agents in reality at different levels. Individual consumers will realistically not take into account the specific adaptation costs they prevent when they invest in solar panel because of the large sensitivity in the climate and socio-economic system. However, mitigating agents intuitively do understand that they make a small contribution to a better world.

If we look at country level we could argue that agents will take the benefits for avoided climate change adaptation more seriously as they can mitigate a larger percentage of the total cumulative GHG emissions in the BAU scenario by imposing policies and regulation on to their agents. However, due to the political lifetime in the different political systems they will have a longer or shorter foresight which they take into account.

The trade off between mitigating and adaptation is difficult because of their different time horizons. If agents mitigate now because they apply a low discount rate, agents will possibly avoid GHG emission which will not have a large influence on the economy for later generations. Due to the high uncertainty in climate and socio-economic sensitivity mitigation can been seen as insurance for more influential adaptation. However, by investing to late because of a worldview that supports a low climate change damage function, climate change damages can hardly be avoided because relatively small contribution mitigating will have on the cumulative CO_2eq emissions and of the large time lag in the climate system.

5 First Results

In Figs. 2 and 3 the first results of our model with 10 runs on each setting of 100 agents are presented. On the first row on the left hand side we see the typical analysis by Stern; all agents applying a discount rate of zero. On the first row on the right hand side we see the typical analysis by Nordhaus, all agents applying a relative high discount rate. Further down the rows, we have introduced heterogeneity among the agents by enlarging the standard deviation of the discount rate distribution which are depicted in Fig. 4.

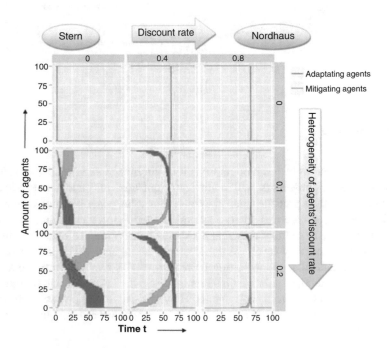

Fig. 2 Development of types of agents over time

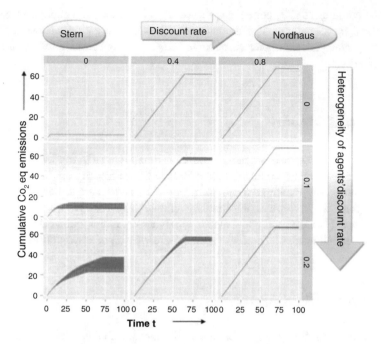

Fig. 3 Development of cumulative CO_2 eq emissions of all agents over time

Fig. 4 Distribution of
discount rates

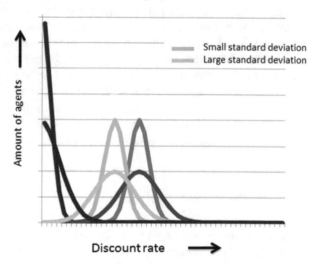

This resulted in a bandwidth of pathways as indicated by the coloured area in
Fig. 3. The actually figures are arbitrary but we can distinguish behaviour features
of our model. As expected, with no heterogeneity, all agents move together, the
higher discount rate they apply, the later they move from adaptation to mitigation.

In the cumulative CO_2 equivalent curves, Fig. 3, we see that under our assumptions about the climate change damage function and mitigation technology costs, mitigating later yields a higher cumulative CO_2 equivalent level stabilization level, as expected.

6 Future Additions

We have used heterogeneous agents within a neoclassical IAM model of the energy transition and with that we have made a start with introducing ABM within IAM. To get a more realistic simulation, we propose to add adaptive behaviour to our agents. Agents worldviews, represented by applied discount rates would not only be heterogeneously spread, but could also made dynamic, individual worldviews could change under influence of different factors. Here we propose two options.

6.1 Agents Within Agents

We can argue that the energy system and its decision makers are actually agents within agents. Ostrom supported a polycentric approach to battle climate change in which she argued that policies and regulations should be discusses at various levels, not only from top down [22]. Individual consumers make up the decision structure within in a country which is an agent at the negotiator table on international conferences. More concrete, if a large enough critical mass of agents supports mitigation, an agent at the second level will decide that all agents at the first level will have to mitigate. This structure of agents within agents can of course be stretched to include city councils, companies, provinces, and NGOs. but fact is that agents can be formed at different levels. In this way agents will influence each other and the evolution of institutions on various levels can be investigated.

6.2 Multi-criteria Analysis

Decisions between mitigating and adaptation are a combination of economic considerations and societal and political judgments. Therefore we propose another way to implement adaptive behaviour and bounded rationality in to agents with which we can let agents make decisions between mitigating and adaptation on more arguments than only our classic CBA. Ostrom and other scholars have tried to distinguish several design features of systems where this group rationality has the biggest chance to flourish [22].

The integration of different criteria on which agents make decisions can be done with the use of multi-criteria analysis. In multi-criteria analysis agents can give weight to different aspects which are scored from 1 to 10. The result of the overall score gives a measure for decision making.

Factors that could be included are the fact that agent are motivated by leadership, non-monetarily expresses ecosystem services, reciprocity of their network and reputation.

7 Conclusion

Like other researchers before us, we have argued that models that try to simulate the energy transition should use integrated assessment modelling combined with agent-based modelling to more realistically model society's response to climate change. Although the results can be discussed in view of the many uncertainties, simplifications and assumptions, we do feel that with the conceptualization of our model we have presented some basic aspects of behaviour of members of society which are present in the real world. We have done this by given a new dimensions to the ethical discussion on the use of appropriate discount rates to use in the light of possible consequences of climate change by distributing heterogeneous discount rates among the agents within the ABM.

Our model gives a first proof of principle of the use of agent based modelling within integrated assessment modelling with the aim to further develop models with more realistic agent behaviour.

We can conclude that applying ABM to IAM yields good prospects to the further development of the implementation of society's response to a changing climate and energy system.

References

1. Moss, R., Edmonds, J., Hibbard, K., Manning, M., Rose, S., Van Vuuren, D., Carter, T., Emori, S., Kainuma, M., Kram, T., Meehl, G., Mitchell, J., Nakicenovic, N., Riahi, K., Smith, S., Stouffer, R., Thomson, A., Weyant, J., Wilbanks, T.: The next generation of scenarios for climate change research and assessment. Nature. **463**, 747–756 (2010)
2. IPCC: Summary for policymakers. In: Climate Change 2014: Impacts, Adaptation, and Vulnerability. Part A: Global and Sectoral Aspects. Contribution of Working Group II to the Fifth Assessment Report of the Intergovernmental Panel on Climate Change, pp. 1–32. Cambridge University Press, Cambridge UK; New York, NY (2014)
3. De Vries, B.: Interacting with complex systems: models and games for a sustainable economy. Netherlands environmental assessment agency (PBL), (2010)
4. Morgan, M., Kandlikar, M., Risbey, J. Dowlatabadi, H.: Why conventional tools for policy analysis are often inadequate for problems of global change. Climatic Change. **41**, (1999)

5. Palmer, P., Smith, M.: Model human adaptation to climate change. Nature **512**, 365–366 (2014)
6. Martinot, E., Dienst, C., Weiliang, L., Qimin, C.: Renewable energy futures: targets, scenarios and pathways. Annu. Rev. Environ. Resour. **32**, (2007)
7. Royal Dutch Shell. 40 Years of Shell Scenarios. http://s03.static-shell.com/content/dam/shell/static/future-energy/downloads/shell-scenarios/shell-scenarios-40yearsbook061112.pdf. Accessed 01 August 2015
8. Giupponi, C., Borsuk, M., De Vries, B., Hasselman, K.: Innovative approaches to integrated global change modelling. Environ. Model Softw. **44**, 1–9 (2012)
9. DeCanio, S., Howarth, R., Sansted, A., Schneider, S., Thompson, S.: New directions in the economics and integrated assessment of global climate change. Prepared for the Pew Center on global climate change. (2000)
10. Ackerman, F., DeCanio, S., Howarth, R., Sheeran, K.: Limitations of integrated assessment models of climate change. Clim. Chang. **95**, 297–315 (2009)
11. North, D.: Economic performance trough time. Am. Econ. Rev. **84**(3), 359–368 (1994)
12. Gerst, M.D., Wang, P., Roventini, A., Fagiolo, G., Dosi, G., Howarth, R.B., Borsuk, M.E.: Agent-based modeling of climate policy: an introduction to the ENGAGE multi-level model framework. Environ. Model Softw. **44**, 62–75 (2012)
13. Janssen, M.A., De Vries, H.J.M.: The battle of perspectives: a multi-agent model with adaptive responses to climate change. Ecol. Econ. **26** (1998)
14. Bosello, F., Carraro, C., De Cian, E.: Climate policy and the optimal balance between mitigation, adaptation and unavoided damage. Fondazione Eni Enrico Mattei Working paper Series. **32**, (2010)
15. De Bruin, K., Dellink, R., Agrawala, S.: Economic aspects of adaptation to climate change: integrated assessment modeling of adaptation costs and benefits. No 6. OECD Environment Working Paper. **6** (2009)
16. Nordhaus, W.: The Climate Casino. Yale University Press, New Haven, USA (2013)
17. Stern, N.: The Econimcs of Climate Change: The Stern Review. Cambridge University Press, Cambridge, UK (2007)
18. Weitzman M.: On modeling and interpreting the economics of catastrophic climate change. Rev Econ. Stat. **91**(1), (2009)
19. Dietz, S.: A Long-run Target for Climate Policy: The Stern Review and Its Critics. Grantham Research Institute on Climate Change and the Environment/Department of Geography and Environment, London School of Economics and Political Science, London (2008)
20. Ackerman F.: Debating climate economics: the Stern review vs. its critics. Report to Friends of the Earth England, Wales and Northern Ireland. (2007)
21. Van Dam, K., Nikolic, I. Lukszo, Z.: Agent-based modeling of socio-technical systems, Vol. 9, Agent-based social systems. (2013)
22. Ostrom, E.: A polycentric approach for coping with climate change. Background paper to the 2010 World Development Report (2010)

A Spatially Explicit Agent-Based Model of the Diffusion of Green Electricity: Model Setup and Retrodictive Validation

Friedrich Krebs and Andreas Ernst

Abstract The purpose of this chapter is to propose and illustrate a validation procedure for a spatially explicit ABM of the diffusion of green electricity tariffs in the German electricity market. We focus on two notions of model validity: We report on structural validity by describing the model setup and its empirical and theoretical grounding. Then we challenge simulation results with a rich spatially explicit historical customer data set thus focusing on retrodictive validity. In particular the latter validation exercise can be prototypic for the class of spatially explicit diffusion ABMs in data-rich domains because it systematically scrutinises validity on different levels of agent aggregation.

Keywords Green electricity • Innovation diffusion • Spatial patterns • Validation

1 Introduction

The share of renewables among the total electricity production in 2014 in Germany has reached 26 % [1]. Most of it, however, is not being produced in Germany itself, but is imported mostly from Scandinavian hydroelectric power plants. In turn, grey energy from German nuclear or coal plants is sold back, leading to a mere relabelling of green energy. To counter this, a small number of "pure" green electricity providers have started to exclusively sell energy that is simultaneously produced by renewable resources like photovoltaics or wind power (i.e. no relabelling), and to invest continuously in extending their stock of renewable energy plants. By doing this, they ensure that their economic activity contributes to a durable transition towards renewable energies. Household electricity use constitutes a significant part of the consumption of energy and thus represents an important factor in the transformation of the energy market.

F. Krebs (✉) • A. Ernst
Center for Environmental Systems Research, Kassel, Germany
e-mail: krebs@usf.uni-kassel.de; ernst@usf.uni-kassel.de

© Springer International Publishing AG 2017
W. Jager et al. (eds.), *Advances in Social Simulation 2015*, Advances in Intelligent Systems and Computing 528, DOI 10.1007/978-3-319-47253-9_19

It is the goal of the agent-based model (ABM) developed in the course of the SPREAD project (Scenarios of Perception and Reaction to Adaptation) to provide a psychologically founded dynamical reconstruction of the adoption of pure green electricity by German households. The presented ABM of the demand side of the German energy market includes habitual behaviour and deliberative decisions, communication over personal networks, and sensitivity towards external events such as price changes and important events. In the model agents represent households and the behavioural target variable is the selected energy provider, i.e. either green energy or grey energy. Each agent is characterised by its position in geographical space, its affiliation to one of five lifestyles according to the Sinus-Milieus® for Germany [2] and its position in an artificial social network.

In the last few decades such ABMs of innovation diffusion have gained substantial maturity (see [3] for an up to date and comprehensive overview). ABMs of innovation diffusion are successful because the method bridges the gap between the empirically found heterogeneity of customers (for instance in terms of the classical adopter types introduced by [4]) and empirically observed macro-level patterns of a diffusion process. The model presented in this chapter belongs to the class of spatially explicit ABMs which have experienced growing attention over the last few years [5–9]. These studies are grounded in rich empirical data sets for very specific application domains and aim to contribute to policy analyses and decision support. In such application contexts, the common strong argument of the method of ABM concerns the validity of the model structure for instance in terms of the representation of micro-level individual decision processes or inter-individual social dynamics. On the other hand, the models deliver very rich and detailed simulation results of future scenarios which can for instance be presented in the form of maps of certain model indicators. In stakeholder dialogue, the credibility of the simulation assessments increases with the degree to which statements on model validity can be (and are) made.

The purpose of this chapter is to propose and illustrate a validation procedure for the ABM of the diffusion of green electricity tariffs in the German electricity market. We focus on two notions of model validity [10–13]: Section 2 reports on *structural validity* by describing the model setup and its empirical and theoretical grounding. Section 3 challenges simulation results with existing historical data thus focussing on *retrodictive validity*. In particular the latter validation exercise can be prototypic for the class of spatially explicit diffusion ABMs in data-rich domains because it systematically scrutinises validity on different levels of agent aggregation. Section 4 summarises and discusses.

2   Model Setup and Empirical Initialization

The agent-based model (ABM) presented in this chapter relies in large part on the LARA architecture [14]. While this behavioural architecture has been tested in previous projects and relies itself on a vast variety of empirical data, the green electricity adoption model is its first application in the energy domain. Thus, in order to endow the agents with as realistic characteristics as possible, a large set of empirical data was collected (from surveys, psychological experiments, and publicly available statistics) to feed the model's parameters with. Most important, one of the four large German pure green electricity providers, the Elektrizitätswerke Schönau (EWS) has made available data about their customers on a weekly basis and by postcode area [15]. A more detailed description of the ABM and its psychological underpinning is found elsewhere [16].

2.1 Agent Goals

In multi-attribute decision making, the goals as the criteria for weighting the options are of outmost importance. Based on earlier research on environmental behaviour (e.g. [17]) and in accordance with literature from resource use (e.g. [18]), we consider three following goals to be decisive in environmental behaviour in general: Ecological orientation, economic orientation (cost minimization), and social conformity. While the first two clearly are bound to specific outcomes (the greener the energy, the better for the ecology goal, the cheaper the energy, the better from the perspective of the economic goal), the latter orientation is a two-sided sword. A strong goal of social conformity will as long hinder the adoption of some innovation as long as the adopters in the personal network are in the minority. Only when a majority of friends has adopted the innovation, the agent will deliberate about adopting it, too. For the domain of green energy adoption, a fourth goal is introduced. The supply of electricity has to be reliable to an agent, and the perception of reliability of provision may differ between green and grey electricity.

2.2 Habitual and Deliberative Decisions

Agents decide either in deliberative or in habitual mode. These modes are activated depending on the current internal state of the agents together with the influence of their social, economic and media related environment. The default mode is the habitual decision. The action chosen in the last time step is simply updated with the current environmental constraints. In deliberative decision mode however, the agent weighs all known utilities of the behavioural options with respect to its various goals. This mode is triggered by internal or external events. An agent's deliberative decision mode is activated by a weighted sum of the contributions of a number of triggers:

$$s = \alpha + \beta_{pc}\,\delta_{pc} + \beta_{ps}\,\delta_{ps} + \beta_{cd}\,\left|w_{ecol} - q_{green}\right| + \beta_{nc}\,nc + \beta_{me}\,\delta_{me} \qquad (1)$$

where

- α is an additive constant;
- δ_{pc} is 1 if the agent perceived a change in the relative electricity prices, 0 otherwise; a change of electricity prices is perceived as such only if its relative amount exceeds a certain threshold which is a model parameter, and this change is in general not perceived in the current time step, but only with a latency of up to 12 months which is determined randomly;
- δ_{ps} is 1 if there is another agent initiating a personal communication with the focal agent in the current time step and 0 otherwise;
- w_{ecol} is the importance (weight) the focal agent attaches to ecological orientation; q_{green} is the descriptive social norm which is operationalized as the proportion of agents in the focal agent's social network who have chosen green electricity in the last time step; the absolute value of the difference of w_{ecol} and q_{green} serves as a measure for the cognitive dissonance between the agent's ecological orientation and descriptive social norm;
- n_c is the agent's need for cognition;
- δ_{me} is 1 if there is a media reported relevant event in the current time step and 0 otherwise;
- the β's are weights for price change, personal communication, cognitive dissonance, need for cognition, and media events, respectively.

The results of the above equation are bounded to values between 0 and 1 and give the agent's probability of entering the deliberative decision mode in the current time step.

Once the deliberative decision mode is triggered, the agent uses an algorithm that maximises its utility using a multi-attributive utility function including all its weighted agent goals (ecological orientation, economic orientation, social conformity, and reliability of provision). Formally, the utility U_i of a behavioural option i is given by

$$U_i = \sum_{j \in G} w_j u_{ij} \tag{2}$$

where

- w_j is the weight of goal j,
- G is the set of all considered goals and
- u_{ij} is the partial utility of behavioural option i with respect to goal j.

The goal weights are milieu group specific parameters empirically determined by the results of our own survey. The goal weights of each individual agent are randomly distributed in a uniform way around the empirical mean, with the maximum deviation being a goal specific model parameter that is the same for all milieu groups of agents.

Each of the agent goals defines a partial utility according to Eq. 2. For the goal of social conformity, that partial utility of a behavioural option i depends linearly on the share of adopters of green electricity among the friends in the agent's simulated social network:

$$u_{i, socConf} = 2\,q_i - 1 \qquad (3)$$

where q_i denotes the proportion of agents in the considered agent's social network who have chosen that behavioural option i in the last time step.

The goal of cost minimization uses the price of grey electricity as a reference point, and the economic utility of green electricity is calculated based on its relative price difference to grey electricity, which is set to 0 per definition:

$$u_{green,costMinim} = \frac{p_{grey} - p_{green}}{p_{grey}} \qquad (4)$$

where p_{grey} and p_{green} are the prices of grey and green electricity, respectively.

2.3 External Events

To reflect important external events that are transported in the media (like the Fukushima catastrophe the concern about nuclear power raised sharply), the model allows agents to switch from habitual to deliberative decision mode, and to increase the strengths of those goals that are addressed by the news. The size of the goal weight change is assumed to depend not only on the strength of the external event but also on the starting conditions of the agent, e.g. its goal strength before the event. Thus, the new weight of an affected goal is calculated from the old one according to the following transformation function:

$$w_{new} = w_{old} + c w_{old} \left(1 - \frac{w_{old}}{r}\right) \qquad (5)$$

where

- $c\,(-1 \leq c \leq 1)$ is the strength of the simulated event reported in the media and
- $r > 0$ is the maximum value the weight can take (its minimum value being 0).

2.4 Information Diffusion

The outlined cognitive processes of individual decision-making are supplemented by a (less cognitive) process of information diffusion representing the individual awareness of EWS as a green energy brand. An agent can become aware of EWS by word-of-mouth process of information exchange over the ties of its network: With a given probability an agent probes its direct network alteri and if at least one of them knows of EWS, the focal agent also becomes aware of EWS as a green energy provider. Once an agent decides to change from grey to green energy by deliberation it becomes an EWS customer if it is aware of EWS; otherwise the agent is counted as green energy customer of a different brand.

2.5 Initialization and Spatial Upscaling

Simulations start at January 2005 and each agent in the model represents 125 real-world actors that share the same characteristics, so that the model comprises of an artificial population of about 300,000 agents. The upscaling method described in [19] uses the lifestyle typology introduced before [20] to provide the agents, e.g. with heterogeneous goals or preferences on their personal network and communication. The milieus' spatial distribution stemming from marketing data [21] is used to allocate the agents according to their type in space, i.e. on a map representing Germany. The agents are embedded in a network of other agents they are connected with and with whom they communicate. That artificial social network is constructed based on theoretical assumptions and empirical evidence [22] and includes homophily of agents as well as their spatial proximity. Between agents of the same milieu, social connections are generated with a higher probability than between agents of different milieus. The probability of connections between two agents decreases sharply with increasing distance between the agents. In the simulation reported here, the network consists of more than 3.5 million connections, and conforms to small world [23] and scale free [24] characteristics. Agents' social perception in the simulated network relates to the choice of electricity provider of their friends, and communications pertains to informing their friends about their own choice or the stimulation of some other agent to reflect about changing its electricity provider by switching from habitual to deliberative decision mode for one time step.

The January 2005 time slice of the empirical EWS customer data is used to initialise agents having already adopted green electricity from EWS and agents initially aware of EWS. This initialization is done on a spatial resolution of postcode areas. Finally, we initialise agents as customers of different (pure) green electricity providers based on the December 2004 time slice of the yearly market shares of the three major providers of pure green energy in Germany from 2004 to 2013. Data stem from Greenpeace Energy ([25], p. 25–25), Lichtblick [26], and Naturstrom [27]. Initialization of green electricity customer agents excluding EWS is done randomly, i.e. each agent (not already initialised as an EWS customer) becomes an green electricity customers with a probability of the total market share of the three other providers of pure green energy.

3   Results and Validation

One of the goals of the ABM outlined in the previous section is to provide a psychologically grounded dynamical reconstruction of the EWS customer development. Therefore, the assessment of the model involves a retrodictive validation which relates the spatio-temporal patterns found in historical market data to the patterns generated by simulations of the respective time range. To do so, we compare empirically observed indicators of the diffusion process to the respective indicators calculated from two typical simulation runs which differ in their parameter settings. The rationale of the validation exercise is to start from most aggregated indicators and step-wise increase the temporal and spatial resolution

We begin with a subsection describing the empirical data sets used for validation. Then, three proceeding subsections report on simulation and validation results at different levels of aggregation.

3.1   Empirical Data and Their Mapping to Model State Variables

The presented validation exercise focusses on two agent-level state variables of the outlined ABM. The first variable is the electricity provider selected by the respective agent in the model. This variable can take three states: Green electricity from EWS, pure green electricity from a provider other than EWS, or grey electricity. The second model variable investigated in the validation exercise can take two states and represents whether an agent is aware of EWS as an energy brand. The two model variables are indicative of the outcome of two different (interacting) agent-level processes in the model namely the outcomes of deliberation or habitual decision-making, and information diffusion through social networks. During model initialisation, both agent state variables are set up according to the December 2004 time slice of empirical time series data (see Sect. 2.5).

The purpose of a retrodictive validation is to make statements on the relation of model-generated time series starting from January 2005 to the respective observed historical time series. For the market share of the different electricity providers, validation data sets are given by the spatially explicit EWS customer data and by the spatially aggregated customer numbers of the three major providers of pure green energy in Germany from 2004 to 2013. For the market awareness of EWS no direct empirical validation data are available. Therefore, brand awareness of EWS was extrapolated depending on the brand's market share. To do so, a regression analysis of existing survey data on market share and customer awareness of various electricity brands was performed. Then, EWS brand awareness was estimated by the regression equation based on the described spatially and temporally explicit EWS customer data [28].

3.2   Temporally and Spatially Aggregated Validation

We begin model assessment by determining three aggregated indicators for the German green electricity market from January 2005 to December 2011 for each of the two simulation runs and the historical data respectively. To do so, we aggregate the December time slices of the EWS customer data spatially for Germany to obtain yearly totals of EWS customers. This step is performed for the historical data and for the simulation runs. Then we calculate the mean of the aggregated customer numbers of EWS and of the other green energy providers over the considered time span. Accordingly, we can determine spatially and temporally aggregated data on the energy providers selected by the German households. The respective indicator values are displayed in Table 1.

The first two rows of Table 1 indicate that both simulations appear to get close to the total market share of green energy and to the market share of EWS. However, simulation 1093 misses in particular the share of EWS within the green energy market (third row). This is consistent when comparing to simulation 1089.

3.3   Spatially Aggregated/Temporally Disaggregated Validation

This section scrutinises the temporal dimension of the diffusion process. The introduced model indicators are spatially aggregated for Germany and have a monthly resolution. Historical data of the EWS customer development are likewise spatially aggregated and shown at monthly time steps. The historical development of the overall green energy market is shown on a yearly resolution. Figure 1 illustrates the respective time series.

Diagram (a) of Fig. 1 shows the temporal development of the overall green energy market. Qualitatively both simulations reproduce the increase of the green energy market share after the Fukushima nuclear hazard (see vertical reference line). However, in accordance with the assessments shown in Table 1 the two simulations differ drastically in magnitude and we see that simulation 1089 much

Table 1 Comparison of spatio-temporally aggregated indicators of the German green electricity market from January 2005 to December 2011

	Historical data (%)	Simulation 1093 (%)	Simulation 1089 (%)
Average market share of green electricity from one of the three major providers or from EWS	1.45	0.51	0.99
Average market share of EWS	0.17	0.17	0.15
Average market share of EWS within green energy market	11.7	34.11	14.61

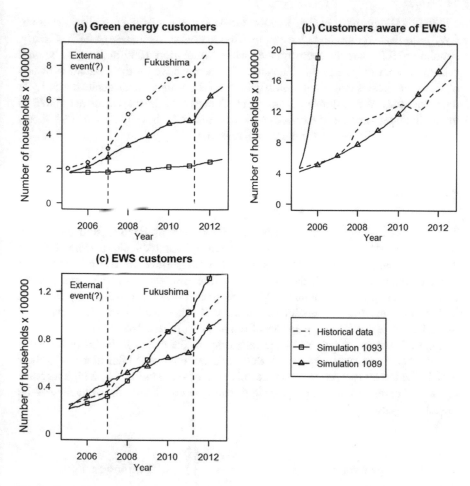

Fig. 1 Spatially aggregated development of the German green electricity market from January 2005 to December 2011. Diagrams show historical data and two simulation runs. Diagram (**a**) displays the total number of households buying green electricity from one of the three major providers or from EWS. Diagram (**b**) displays the degree of awareness of EWS in terms of the number of households "knowing" EWS. The third diagram (**c**) shows the number of EWS customers

better reproduces the historical data also in terms of temporal development. Diagram (c) of Fig. 1 displays the temporal development subset of green energy customers buying from EWS. Here, both simulation curves are similarly close to the validation data. Still, the decrease of the EWS customer numbers from 2010 to 2011 is not covered by the model. Likewise, the second bend in the empirical customer curves of diagrams (a) and (c) at year 2007 (see reference line) is not visible in the simulation results. These observations indicate the presence of some additional external event not represented in the model structure (see Sect. 2.3).

Finally, Diagram (b) of Fig. 1 shows the development of the brand awareness of EWS. The dashed line shows the extrapolated historical awareness of EWS. Clearly, simulation 1093 largely overestimates while simulation 1089 appears to be well calibrated with respect to this indicator. This perspective on the simulation results obviously indicates that simulation 1093 can safely be discarded despite its ability to reproduce the EWS customer development. Considering the market share of EWS as the main target variable of the model, the conclusion is that simulation 1093 shows (numerically) valid results for the wrong reasons!

3.4 Spatially Disaggregated/Temporally Aggregated Validation

This section investigates the spatio-temporal spread of EWS customers. To do so, we generate maps of the average increase in market share from 2005 to 2011 at a spatial resolution of 50 km for Germany (total of 169 grid cells). To calculate the respective indicator for a cell on the grid, we divide the mean number of agents/households becoming EWS customers during 2005–2011 by the total number of agents/households located in the cell. Then we classify the increase of market share of EWS into categories of range 0.025 for the 169 cells.

Figure 2 shows histograms of the categories of the cells for the historical EWS customer data and simulation 1089. Both distributions are similar in terms of their peak in the lowest categories and in the existence of some outliers with high increase in market share. Still, the simulation significantly overestimates the frequency of the lowest category.

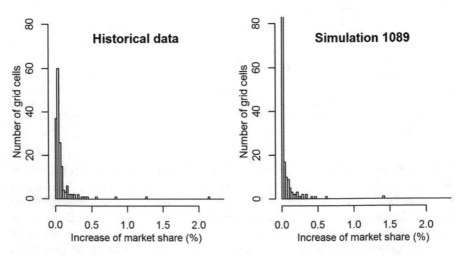

Fig. 2 Average increase in market share from 2005 to 2011 spatially aggregated on a grid of 50 km for Germany (169 grid cells) in categories of range 0.025. In *right histogram* the first category has a count of 108

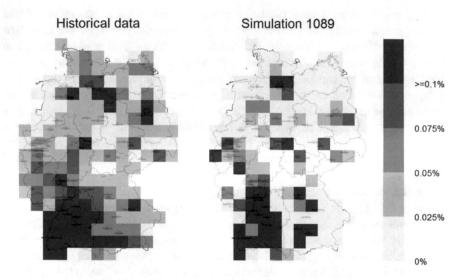

Fig. 3 Spatial patterns in the average increase in market share from 2005 to 2011. Data is spatially aggregated on a grid of 50 km for Germany

The spatial dimension of the diffusion is illustrated in Fig. 3. Here, the respective categories of the cells are shown in grey shades where cells with an increase of at least 0.1 % are shown in black. Apparently, the simulation roughly captures some of the spatial patterns in the historical data: a relatively high increase in market share located in a contagious area in the southwest of Germany and around three larger cities starting from the North Hamburg, Berlin, and Dresden. In contrast to this characteristic pattern of high diffusion speed, regions with low increase of market share are not well captured by the simulation. This is indicated by the comparatively large white areas on the simulation map.

4 Discussion

Validation of ABMs is a recurring and extensively discussed issue. The common strong argument of the method of ABM is its capability for structural validity, i.e. the potential of the models to capture structural properties of a target system in order to propose explanations of collective phenomena. However, once detailed empirical datasets and real-world geography is included in the simulations, the credibility of ABMs is significantly increased if the (degree of) fit of simulation results and empirical data can be stated, in particular when offering decision support in policy contexts. The latter (retrodictive) validation exercise comes to be increasingly challenging as more and more rich social science data sets become available.

This chapter reports on a validation procedure conducted in an ongoing ABM project of the demand side of the German green electricity market. We show that the presented ABM obtains structural validity from its grounding in psychological theory and survey data. Furthermore, the model is spatially explicit in terms of the locations of the modelled household agent. A particular feature of the project setup is that detailed customer data of a green electricity provider on spatial and temporal resolution are available. In principle, the format of this historical data matches the main agent behavioural variable both in time and space. The rationale of the validation procedure was to start from most aggregated indicators derived from historical data and simulation results, and stepwise increase the temporal and spatial resolution.

In the multilevel retrodictive validation, each level of aggregation offered different insights: On a spatially and temporally aggregated level, we showed that the magnitude of the overall market share of green electricity as well as the market share of EWS can be well reproduced by the simulations. Investigating the temporal dynamics of spatially aggregated indicators revealed that the modelled temporal development of the EWS customer numbers gets reasonably close to the historical data. Furthermore, the customer increase due to the Fukushima accident is well represented in the simulations. However, a second bend in the empirical customer curve around 2007 is apparently not represented in the model. This hints at a reconsideration of the external events represented in the model structure. Finally, we disaggregated data spatially and investigated the spatial distribution of the increase of EWS market share. By visual comparison of frequency distributions and spatial patterns, similarity of the spread in historical and simulated data could be assessed. Future work should objectify this notion of mere face validity and introduce numerical measures, for instance statistical signatures of the distributions or measures of spatial fit [29].

Acknowledgements The research presented in this chapter was partly funded by the German Ministry for Education and Research (BMBF) under contract no 01UV1003A, "Scenarios of Perception and Reaction to Adaptation (SPREAD)".

Special thanks go to Ramón Briegel for model design and implementation, and Angelika Gellrich and Sascha Holzhauer for carrying through parts of the empirical research reported here.

References

1. Strom-Report.de.: Stromerzeugung 2014 nach Energieträgern. (2015). http://strom-report.de/strom-vergleich/#stromerzeugung. Accessed 23 April 2015
2. Sinus Sociovision GmbH.: Die Sinus-Milieus® in Deutschland 2007. (2007), http://www.sinus-sociovision.de/2/2-3-1-1.htm. Accessed 9 April 2009
3. Kiesling, E., Günther, M., Stummer, C., Wakolbinger, L.: Agent-based simulation of innovation diffusion: a review. Cent Eur J Oper Res **20**(2), 183–230 (2012). http://dx.doi.org/10.1007/s10100-011-0210-y
4. Rogers, E.M.: Diffusion of Innovations. Free Press, New York, NY (1962)

5. Dunn, A.G., Gallego, B.: Diffusion of competing innovations: the effects of network structure on the provision of healthcare. J. Artif. Soc. Soc. Simul. **13**(4), 8 (2010). http://jasss.soc.surrey.ac.uk/13/4/8.html
6. Krebs, F., Holzhauer, S., Ernst, A.: Modelling the role of neighbourhood support in regional climate change adaptation. Appl Spat Anal Policy **6**, 305–331 (2013). http://dx.doi.org/10.1007/s12061-013-9085-8
7. Schwarz, N., Ernst, A.: Agent-based modeling of the diffusion of environmental innovations – An empirical approach. Technol Forecast Soc **76**(4 SI), 497–511 (2009)
8. Sopha, B.M., Klöckner, C.A., Hertwich, E.G.: Adoption and diffusion of heating systems in Norway: Coupling agent-based modeling with empirical research. Environ Innovat Soc Transit **8**, 42–61 (2013). http://www.sciencedirect.com/science/article/pii/S2210422413000427
9. Sorda, G., Sunak, Y., Madlener, R.: An agent-based spatial simulation to evaluate the promotion of electricity from agricultural biogas plants in Germany. Ecol. Econ. **89**, 43–60 (2013)
10. Amblard, F., Bommel, P., Rouchier, J.: Assessment and validation of multi-agent models. In: Phan, D. (ed.) GEMAS studies in social analysis. Agent-based modelling and simulation in the social and human sciences, pp. 93–114. Bardwell Press, Oxford (2007)
11. Ormerod, P., Rosewell, B.: Validation and Verification of Agent-Based Models in the Social Sciences. In: Squazzoni, F. (ed.) Lecture Notes in Computer Science. Epistemological Aspects of Computer Simulation in the Social Sciences, pp. 130–140. Springer, Berlin, Heidelberg (2009)
12. Troitzsch, K.G.: Validating simulation models. In: Horton, G. (ed.) Networked simulations and simulated networks. 18th European Simulation Multiconference, June 13th–16th, 2004, pp. 265–270. SCS Publication House, Erlangen (2004)
13. Windrum, P., Fagiolo, G., Moneta, A.: Empirical validation of agent-based models: alternatives and prospects. J. Artif. Soc. Soc. Simul. **10**(2) (2007). http://jasss.soc.surrey.ac.uk/10/2/8.html
14. Briegel, R., Ernst, A., Holzhauer, S., Klemm, D., Krebs, F., & Martínez Piñánez, A.: Social-ecological modelling with LARA: A psychologically well-founded lightweight agent architecture. In: Seppelt, R., Voinov, A.A., Lange, S.S., Bankamp, D. (Eds.), International Congress on Environmental Modelling and Software. Managing Resources of a Limited Planet. Sixth Biennial Meeting. Leipzig, Germany. (2012)
15. Elektrizitätswerke Schönau (EWS). Anonymized customer data time series. (2013)
16. Ernst, A., Briegel, R.: A dynamic and spatially explicit psychological model of the diffusion of green electricity across Germany. Submitted to the Journal of Environmental Psychology. (2015)
17. Ernst, A., Spada, H.: Modeling actors in a resource dilemma: a computerized social learning environment. In: Towne, D., de Jong, T., Spada, H. (eds.) NATO ASI Series. Simulation-Based Experiential Learning, pp. 105–120. Springer, Berlin Heidelberg (1993)
18. Messick, D.M., Brewer, M.B.: Solving social dilemmas: a review. In: Wheeler, L., Shaver, P. (eds.) Personality and social psychology review, pp. 11–44. Sage, Beverly Hills, CA (1983)
19. Ernst, A.: Using spatially explicit marketing data to build social simulations. In: Smajgl, A., Barreteau, O. (eds.) Empirical Agent-Based Modelling – Challenges and Solutions, pp. 85–103. Springer New York, New York, NY (2014)
20. Sinus Sociovision GmbH: Die Sinus-Milieus in Deutschland 2005: Informationen zum Forschungsansatz und zu den Milieu-Zielgruppen. Sinus Sociovision GmbH, Heidelberg (2005)
21. Microm.: Microm Consumer Marketing. (2015), Retrieved March 10, 2015 http://www.microm-online.de/zielgruppe/strategische-zielgruppen/microm-geo-milieusr/
22. Holzhauer, S., Krebs, F., Ernst, A.: Considering baseline homophily when generating spatial social networks for agent-based modelling. Comput. Math. Org. Theor. **19**(2), 128–150 (2013). http://dx.doi.org/10.1007/s10588-012-9145-7
23. Watts, D.J., Strogatz, S.H.: Collective dynamics of 'small-world' networks. Nature **393**(6684), 440–442 (1998)

24. Albert, R., Barabási, A.-L.: Statistical mechanics of complex networks. Rev. Mod. Phys. **74**(1), 47–97 (2002)
25. Greenpeace Energy eG.: *Geschäftsbericht 2012*. from http://www.greenpeace-energy.de/fileadmin/docs/geschaeftsberichte/Greenpeace_Energy_Geschaeftsbericht2012.pdf (2013)
26. LichtBlick, S.E.: Yearly customer numbers of Lichtblick for Germany from 2004–2013 (Email) (2014, May 23)
27. Naturstrom, A.G. Yearly customer numbers of Naturstrom for Germany from 2004–2013 (Email) (2014, April 15)
28. Heister, J.: Agentenbasierte Analyse der Informationsdiffusion in empirisch fundierten sozialen Netzwerken. Unpublished Bachelor Thesis. Kassel: Center for Environmental Systems Research, University of Kassel (2014)
29. Krebs, F.: Towards an empirical validation of spatial patterns of simulated innovation diffusion. Paper to be presented at the 11th Artificial Economics Conference, Porto, Portugal (2015)

A Network Analytic Approach to Investigating a Land-Use Change Agent-Based Model

Ju-Sung Lee and Tatiana Filatova

Abstract Precise analysis of agent-based model (ABM) outputs can be a challenging and even onerous endeavor. Multiple runs or Monte Carlo sampling of one's model (for the purposes of calibration, sensitivity, or parameter-outcome analysis) often yields a large set of trajectories or state transitions which may, under certain measurements, characterize the model's behavior. These temporal state transitions can be represented as a directed graph (or network) which is then amenable to network analytic and graph theoretic measurements. Building on strategies of aggregating model outputs from multiple runs into graphs, we devise a temporally constrained graph aggregating state changes from runs and examine its properties in order to characterize the behavior of a land-use change ABM, the RHEA model. Features of these graphs are transformed into measures of complexity which in turn vary with different parameter or experimental conditions. This approach provides insights into the model behavior beyond traditional statistical analysis. We find that increasing the complexity in our experimental conditions can ironically decrease the complexity in the model behavior.

Keywords Agent-based model analysis • Graph representation • Network analysis • Complexity metrics • Land-use change

Paper presentation for the Social Simulation Conference 2015 (SSC, ESSA), Groningen, The Netherlands, Sept. 16–18, 2015.

J.-S. Lee (✉)
University of Twente, Enschede, The Netherlands
Erasmus University Rotterdam, Rotterdam, The Netherlands
e-mail: lee@eshcc.eur.nl

T. Filatova
University of Twente, Enschede, The Netherlands
e-mail: t.filatova@utwente.nl

W. Jager et al. (eds.), *Advances in Social Simulation 2015*, Advances in Intelligent Systems and Computing 528, DOI 10.1007/978-3-319-47253-9_20

231

1 Introduction

Agent-based models (ABMs) are capable of producing a plethora of complex output which require sophisticated techniques for analysis and visualization. An overview of current ABM output analysis and visualization techniques [8] highlights the breadth of techniques necessary for ABM output analysis. Still, there remain many advanced approaches that have yet to be commonly employed by ABM researchers. In this paper, we explore an network analytic approach for investigating ABM behavior and demonstrate the utility of this approach by applying it to an agent-based land-use change (LUC) model called RHEA (**R**isks and **H**edonics in an **E**mpirical **A**gent-based land market) [3].

1.1 Networked ABM Output

While the interaction of agents in ABMs often constitute networks of varying modalities (i.e., multiple classes of entities) and topologies (i.e., the shape of the overall structure), the model's state transitions from one discrete time period to the next can also be represented as network, specifically a *weighted directed graph*, and then subjected to network analytic methods particularly drawn from graph theory and *social network analysis* (SNA). The weights of the edges in such graphs would represent the number of times the two connected states were traversed in one or more runs of the simulation. Crouser et al. [2] propose the use of *aggregated temporal graphs* (ATGs), the vertices of which encapsulate a unique state configuration of the model. Multiple model runs are then aggregated into a single graph.

Examination of a smaller, more focused set of unique states would yield fully connected graphs (or cliques) in which only the edge weights vary, and the aggregation would lose much of the model's complexity. For example, a model that oscillates between two states in a staggered pattern would appear as a small graph with only two vertices and a set of bidirectional edges while the pattern of transitions would be lost. To address this limitation, we apply a variant of this approach in which time is disaggregated in the graph portrayal; we call this variant *temporally constrained aggregated graphs* (TCAGs). Each vertex under this approach is identified uniquely by both its state and time signature. The multiple edges (or edge weight) between two vertices would then indicate the number of simulation runs that traverse the states represented by those two vertices at that specific time interval. While a plot of lines may also be used to portray these state changes, its visual limitations make it unsuited for our analyses.

While the two aforementioned network analytic approaches bear some resemblance to *time aggregated graphs* [6, 9], their techniques are quite different from the latter, which entails an aggregation of the edges for a fixed vertex set (of dynamic social and transportation networks) rather than an aggregation of vertices for graphs of varying vertex sets.

1.2   The RHEA Model

The RHEA model simulates a bilateral, heterogeneous housing market in an empirical landscape. Buyers search for properties in a seaside town where both coastal amenities and environmental risks of flooding are spatially correlated. A realtor agent serves as a mediating agent that learns about the current market trend and willingness to pay for various spatial attributes of properties [3]. At each time step, sellers of the properties advertise their ask prices while buyers select and offer bids on properties. Then, the two groups engage in dyadic (pair-wise) negotiations, which may or may not result in a successful transaction (or trade.) The ask prices are primarily market-driven: sellers (1) consult with a realtor on the appropriate price given the current trends as estimated by a hedonic analysis of the recent simulated transactions and (2) adjust the ask price if a property remains too long on a market. Buyers bid on the single property that brings them maximum utility and is affordable for their budgets. A buyer's budget constraints include their income, preference for amenities, a necessity to pay insurance premiums for flood-prone properties, and the possibility to activate an annual income growth. Spatial patterns of price dynamics and intensity of trade are the emergent outputs of the model. When studying them under various parameter settings, we noticed that the relative market power of buyers and sellers plays a role. The explanation is twofold. Firstly, a parcel that is very attractive will most likely receive several bid offers, out of which its seller chooses the highest, thus driving prices for the most desirable properties up. Secondly, sellers of the properties that are less desirable may receive only one or even no bid offers, which can result in their accepting a bid that is below their ask prices or reducing their ask prices after a number of unsuccessful trades. Thus, excess demand drives prices up while excess supply pushes them down.

For our analyses, which includes an application of the TCAG on RHEA output, we focus on exploring the dynamics of buyers and sellers count under several key primary parameter settings. The *realtor hedonic* parameter is a binary indicator of whether the realtor agents update their formula based on the evolving market prices (*adaptive*) or retain the empirically informed *static* formula [1]. The *insurance* parameter is a binary indicator for whether or not buyers consider flood insurance in their utility purchase calculation. Engaging the *growth income* indicator parameter will allow agents' incomes, their travel costs, and the insurance premium to rise over time. Therefore, insurance and income growth directly impact buyers' utilities of the properties they consider for purchase.

2 Methods and Results

2.1 Temporally Constrained Aggregated Graphs

Under TCAG, ABM states are graph vertices, and the edges indicate temporal transitions and possibly state transitions. For these analyses, we focus on the buyer and seller population sizes which rise and fall throughout the simulation. In Fig. 1, we employ TCAG on the first ten time periods (t) across 15 simulation runs.[1] These runs were executed with activation of adaptive *realtor hedonics* and no activation for *insurance* and *income growth*. In Fig. 1a, we display the TCAG of the changes in the buyer population, and in Fig. 1b, we jointly track changes in the sizes of both the buyer and seller pools. Hence, each state represents a positive, negative, or no change in the population size from the previous time step. When the state consists of changes in only one measure (buyer count), there are three distinct states. When both measures are considered, this space grows to $3 \times 3 = 9$ distinct states.

In the TCAGs, we can visually observe several features of the model's behavior. For example, the initial stages of the model appear to incur fewer distinct states especially when examining both measures jointly. This feature has implications on further development of our model. That is, if we wish to further ground the model events in reality and in real-time, then its behaviors should reflect the fact that the

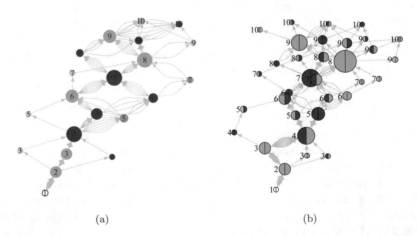

(a)   (b)

Fig. 1 Temporally constrained aggregated graphs of RHEA. *Green* indicates an increase from the previous time period; *red* indicates a decrease; and a *gray* node indicates no change. In the *right* graph, the *left* and *right* halves of each vertex, respectively, indicate changes to the buyer and seller pool counts. The *numerical labels* indicate the time step (t). The vertices have been sized by betweenness centrality (explained below). (**a**) Buyer counts only, $t_{max} = 10$. (**b**) Buyers and sellers, $t_{max} = 10$

---

[1] Our analyses here employ a subset of the output as longer and more runs render the current visualization less effective.

events occur in a pre-existing market, and at least the count of initial states should not be so distinct barring the occurrence of exogenous forces such as a major flood prior to model execution.

We also observe states that serve as *gateways* through which all or many trajectories pass. A pure, isolated gateway occurs when there exists a single vertex for a time step; all trajectories pass through such a vertex. In Fig. 1a, the sole vertices corresponding to times $t = 2$ and 4 are pure gateways. That is, all of the examined simulation runs incur an initial increase in buyer count and a decrease two periods later. These pure gateways reveal an almost deterministic aspect of the model's behavior. In this case, the buyer population will always initially grow and, more often than not, continue to grow until $t = 3$ after which point the market always reacts with a decrease in the buyer pool as indicated by the red, gateway vertex at $t = 4$. In Fig. 1b, the only pure gateway also occurs at $t = 2$ and corresponds to an initial growth in both the buyer and seller populations. We surmise that there exists a potential or gradient in these agent populations set forth by the model's initial conditions leading to the inevitable growth.

Alternatively, a non-isolated gateway is defined by the number and pattern of simulation trajectories that pass through the vertex. One might simply maintain a tally of passing trajectories, but this measure would fail to capture the bridging role a vertex plays in the overall set of state changes. A more robust measure is *betweenness centrality*, a measure widely used in social network analysis. Betweenness centrality (C_B) measures the extent to which a vertex lies in the pathways between all pairs of vertices, and thus captures the extent to which a vertex is a gateway while accounting for the global graph structure.[2] In order to account for higher betweenness for those vertices that receive or emit many edges, we substitute multiple edges with a single edge having a weight of the count of edges. These weights are inverted (i.e., 1/weight) so that the weighted edge between two vertices consecutively traversed in many simulation runs will be considered a shorter path than if they were traversed by few simulations. Given that C_B of vertices near the extremities will be biased downward, we scale the betweenness score by the temporal positions of the vertices such that those in the middle of the time span are penalized:

$$C_B^t(i) = \frac{C_B(i)}{(t_i - 1)(t_{\max} - t_i)}.$$ (2)

---

[2]More precisely, betweenness centrality is the sum of the proportions of shortest paths a vertex lies on between each pair (out of all shortest paths for each pair) [4, 5]. For betweenness centrality, we identify all the shortest paths in a graph, such that $\sigma_{s,t}$ is the number of shortest paths between vertices s and t. Betweenness centrality for vertex i then:

$$C_B(i) = \sum_{s \neq i \neq t} \frac{\sigma_{s,t}(i)}{\sigma_{s,t}}$$ (1)

Under this scaling, a simple TCAG with one vertex per time step would exhibit a C_B^t of 1 for all of the non-terminal vertices. We have sized the vertices in Fig. 1 proportional to C_B^t.

Not surprisingly, the gateway in Fig. 1a at $t = 4$ has high betweenness partly due to the expansion of states in the previous periods. Its betweenness would be lower had there been only one state at $t = 3$. Thus, the state change at that time plays a more significant role in the model's behavior. In Fig. 1b, we observe at $t = 8$ a green/green vertex (increase in both buyer and seller counts) as having the highest relative C_B, despite the existence of other states at that time. Multiple vertices at any time step potentially dilute the betweenness of the vertices in that step so a high betweenness here is particularly salient. In TCAG, a high C_B for a vertex indicates not only a relatively large number of passing trajectories but also high complexity in vertices and trajectories before and after itself. Thus, the gateway vertex at $t = 8$ exhibits high betweenness not simply because many trajectories pass through it but also because the structure of the subgraphs in the time steps prior to and after itself has significant heterogeneity. If, on the other hand, the trajectory structure (either before or after this gateway) was a single sequence, the gateway's betweenness centrality would be much lower.

2.2 Measuring Graph and ABM Complexity

Furthermore, the very count of vertices of aggregated states may describe the complexity of the model given that the states of similarly behaving runs would aggregate into a single chain of events. A perfectly random, or maximally complex, model will yield transition pathways that are less aggregable across simulation runs. The upper bound of vertices would then be $(t_{max} - 1) \times n_{states} + 1$ vertices where t_{max} is the maximum time in the TCAG and n_{states} is the number of distinct states. While the upper bound for Fig. 1a is $9 \times 3 + 1 = 28$, our TCAG contains only 21 vertices. The upper bound for Fig. 1b is $9 \times 9 + 1 = 82$ while the TCAG contains 34 vertices. The lowest complexity occurs when all model outputs exhibit identical states at each step though the state from one period to the next could vary[3]; the complexity for such graphs is $t_{max} - 1 = 9$. Our *vertex complexity* score (α) then is the vertex count as a proportion of the range defined by these upper and lower bounds:

$$|V(G_l)| = t_{max} \tag{3}$$

$$|V(G_u)| = (t_{max} - 1) \times n_{states} + 1 \tag{4}$$

$$\alpha = \frac{|V(G)| - |V(G_l)|}{|V(G_u)| - |V(G_l)|} \tag{5}$$

---

[3] An alternative formulation would differentiate between a graph comprising homogeneous states and one with heterogeneous ones.

where G_l and G_u are graphs corresponding to minimal and maximal complexity, respectively, and $|V(G)|$ is the count of vertices of some graph G. A naïve inference would place our model's complexity as being neither minimally or maximally complex: the univariate measure (buyers only) yields an $\alpha = 0.61$ while a bivariate analysis (buyers and sellers) yields $\alpha = 0.33$. Hence, depending on the complexity of the analyzed output measures, the RHEA simulation's complexity is either $\frac{1}{3}$ or $\approx \frac{2}{3}$.

We also consider the complexity of the sequence of state changes across model runs using data compression as a benchmark. As a straightforward test, we use the extent of compression of the output state change data as our secondary measure of compression: the greater the data compression, the less complex the output. We base this compression factor on the performance of the Unix utility Gzip which employs both LZ77 (Lempel–Ziv) and Huffman coding compression schemes [7, 10]. Specifically, we select the minimum of the compression ratios afforded by the matrix of states, ordered by time step (row) by simulation run (column), and its transpose. In Fig. 2, we chart both vertex and Gzip complexities for the four combinations of activation of two key parameters that impact buyer utility: *insurance* and *income growth*.

Firstly, we notice the inclusion of insurance and income growth markedly decreases the complexity of the model in its initial phase while the complexities are less variable in the later stages. This pattern appears to hold, though somewhat diminished, in the bivariate analysis. Furthermore, the data compression (Gzip) complexity roughly produces similar trends as the vertex complexity. However, further investigation into how and why they are different is warranted.

We visually confirm the effect of the insurance and income growth parameters on simplifying the graphs in Fig. 3. While the graphs in Fig. 3 appear similar to those of Fig. 1, there are differences particularly in the initial time steps, which now incur fewer vertex states. Furthermore, this decrease in complexity is accompanied by

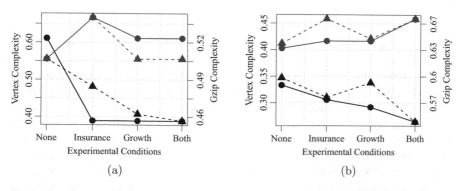

$$(a) \qquad\qquad\qquad (b)$$

Fig. 2 Complexity of TCAGs. The *black* and *red lines*, respectively, denote the complexities for the first ten periods and the last ten periods (50–60) of the simulation run. The *solid* and *dashed lines* denote the vertex and Gzip complexities, respectively. (**a**) Univariate (buyers only). (**b**) Bivariate (buyers and sellers)

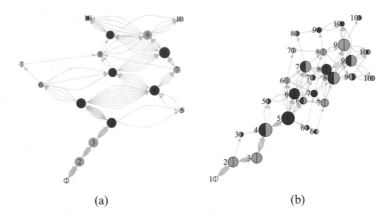

Fig. 3 Temporally constrained aggregated graphs of RHEA for insurance and income growth conditions. (**a**) Univariate (buyers only). (**b**) Bivariate (buyers and sellers)

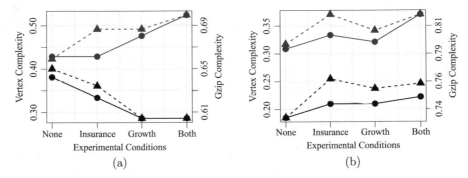

Fig. 4 Complexity of TCAGs for Static Hedonics. The *black* and *red lines*, respectively, denote the complexities for the first ten periods and the last ten periods. The *solid* and *dashed lines* denote the vertex and Gzip complexities, respectively. (**a**) Univariate (buyers only). (**b**) Bivariate (buyers and sellers)

fewer gateway vertices indicated by the lack of largely sized vertices in both Fig. 3a and b. In fact, higher betweenness centrality scores can only occur in more complex TCAGs. Given that the betweenness measure for a given vertex scales with both the trajectories that pass through it and possibly the count of vertices that occur before and after itself, the measure itself encapsulates some portion of vertex complexity.

In the outputs above, we had activated another key parameter: the use of *adaptive realtor hedonics* to determine market prices. So next, we examine the effect of a market price determination through *static* hedonics on RHEA's graph complexity. In Fig. 4, we plot the various complexities in the univariate and bivariate cases.

The impacts of the static hedonic condition are modest. While the initial TCAG complexity for the univariate graph decreases here as it does for the adaptive hedonic setting, the magnitude of the decline (due to activation of both insurance and income growth) is smaller. Secondly, the TCAG based on the final stages of the model runs exhibits an increase in complexity through the activation of the two parameters; earlier, we observed no distinct trend in the final stages for the *adaptive* realtor hedonic condition. The bivariate situation departs from the declining trend: a mild increase is visible (Fig. 4b). All this points to the subtle interaction between the market evolution (or lack thereof) and complexity in the buyer utility function.

3 Discussion

The portrayal of aggregated state transitions in the RHEA ABM offers an alternative view of its behavior from traditional methods of visualization (e.g., line graphs) and statistics (e.g., regression models); this approach may provide additional insights. The application of network analytic methods to understand the complexity of a model (here, RHEA) at various stages further develops our understanding of the model's subtle behavior particularly in reaction to varying parametric conditions. In fact, we can associate these conditions to the complexity in the model's behavior.

This approach warrants further development. Specifically, the metrics employed offer more information for subsets of the model's time span rather than in its entirety due to the limited heterogeneity in the outputs imposed by a small number of states. Naturally, we would need to next adapt the measurements to allow for a larger number of states. Furthermore, accounting for the heterogeneity of states across time (e.g., oscillation vs. homogeneous sequences) can lead alternative, possibly more robust, complexity measures. The betweenness centrality measure may be further exploited for these purposes as it accounts for some aspect of vertex complexity in exposing key, gateway states. Alternative scaling for the betweenness measure is worth investigating. For example, scaling not just by the temporal position but also by the vertex counts before and after the measured vertex would allow the measure to be strictly based on the structural features of the trajectories. Finally, the complexity induced by graphs drawn from random state sequences would be quite informative, allowing us to assess if our analyzed outputs have some equivalence to noise or not.

Acknowledgements This material is based on work supported by NWO DID MIRACLE (640-006-012) NWO VENI grant (451-11-033), and EU FP7 COMPLEX (308601).

References

1. Bin, O., Kruse, J.B., Landry, C.E.: Flood hazards, insurance rates, and amenities: evidence from the coastal housing market. J. Risk Insur. **75**(1), 63–82 (2008)
2. Crouser, R.J., Freeman, J.G., Winslow, A., Chang, R.: Exploring agent-based simulations in political science using aggregate temporal graphs. In: IEEE Pacific Visualization Symposium (PacificVis). IEEE, New York (2013)

3. Filatova, T.: Empirical agent-based land market: integrating adaptive economic behavior in urban land-use models. Comput. Environ. Urban. Syst. **54**, 397–413 (2014)
4. Freeman, L.C.: A set of measures of centrality based on betweenness. Sociometry **40**(1), 35–41 (1977)
5. Freeman, L.C.: Centrality in social networks: conceptual clarification. Soc. Netw. **1**(3), 215–239 (1979)
6. George, B., Shekhar, S.: Time-aggregated graphs for modeling spatio-temporal networks. In: Roddick, J.F., Benjamins, V.R., Si-said Cherfi, S., Chiang, R., Claramunt, C., Elmasri, R., Grandi, F., Han, H., Hepp, M., Lytras, M., Misic, V., Poels, G., Song, I.Y., Trujillo, J., Vangenot, C. (eds.) Advances in Conceptual Modeling - Theory and Practice. Lecture Notes in Computer Science, vol. 4231, pp. 85–99. Springer, Berlin (2006)
7. Huffman, D.: A method for the construction of minimum-redundancy codes. Proc. IRE **40**(9), 1098–1101 (1952)
8. Lee, J.S., Filatova, T., Ligmann-Zielinska, A., Hassani-Mahmooei, B., Stonedahl, F., Lorscheid, I., Voinov, A., Polhill, G., Sun, Z., Parker, D.: The complexities of agent-based modeling output analysis. J. Artif. Soc. Soc. Simul. **18**(4), 4 (2015)
9. Shekhar, S., Oliver, D.: Computational modeling of spatio-temporal social networks: a time-aggregated graph approach. In: 2010 Specialist Meeting – Spatio-Temporal Constraints on Social Networks (2010)
10. Ziv, J., Lempel, A.: A universal algorithm for sequential data compression. IEEE Trans. Inf. Theory **23**(3), 337–343 (1977)

Network Influence Effects in Agent-Based Modelling of Civil Violence

Carlos Lemos, Helder Coelho, and Rui J. Lopes

Abstract In this paper we describe an agent-based model of civil violence with network influence effects. We considered two different networks, 'family' and 'news', as a simplified representation of multiple-context influences, to study their individual and joint impact on the size and timing of violence bursts, the perceived legitimacy, and the system's long term behaviour. It was found that network influences do not change either the system's long term behaviour or the periodicity of the rebellion peaks, but increase the size of violence bursts, particularly for the case of strong 'news impact'. For certain combinations of network influences, initial legitimacy, and legitimacy feedback formulation, the solutions showed a very complicated behaviour with unpredictable alternations between long periods of calm and turmoil.

Keywords Agent-based model • Civil violence • Network influences

1 Introduction

The study of social conflict phenomena, and civil violence in particular, is an important topic in political science, sociology, social psychology, and social simulation studies. Social context factors can increase the potential for violence [9], whereas

---

C. Lemos (✉)
CISDI and Instituto de Estudos Superiores Militares, Lisboa, Portugal

BioISI and Faculdade de Ciências da Universidade de Lisboa, Lisboa, Portugal

Instituto Universitário de Lisboa (ISCTE-IUL), Lisboa, Portugal
e-mail: cmrso@iscte-iul.pt

H. Coelho
BioISI and Faculdade de Ciências da Universidade de Lisboa, Lisboa, Portugal

R.J. Lopes
ISCTE - Instituto Universitário de Lisboa, Lisboa, Portugal

Instituto de Telecomunicações, Lisboa, Portugal

© Springer International Publishing AG 2017
W. Jager et al. (eds.), *Advances in Social Simulation 2015*, Advances in Intelligent Systems and Computing 528, DOI 10.1007/978-3-319-47253-9_21

widespread access to information and communication technologies (ICT) and Social Networks (SN) can trigger gradual (e.g. escalation) or sudden (e.g. revolution) uprisings [11], which in turn change the social context.

Epstein et al. [3] (see also Epstein [2]) introduced a very successful agent-based model (ABM) of rebellion against a central authority (Model I) and ethnic violence between two rival groups mediated by a central authority (Model II), in an artificial society with two types of agents ('agents' and 'cops' for representing citizens and policemen, respectively). The success of Epstein's ABM derives from the simplicity and soundness of the action rule for 'agents', the relevance of the dependent variables used and the capability for representing mechanisms of collective violence. Epstein's ABM has been extended and refined by several authors for studying different conflict phenomena [4, 10] and mechanisms of violence uprisings [12].

In this paper we present an ABM of civil violence with network influence effects by considering two forms of network influence, (1) 'family' represented by a union of small undirected cliques (individual families), and (2) 'news' represented by a union of directed star networks with agents of a new type (called 'media') as hubs. These networks provide an abstract representation of two important influence modes in a society, one associated with highly cohesive small scale communities connected by strong undirected links (two-way influence) with local information and high internal homogeneity [7], and another associated with (weaker, one-way) directed links through which influential agents shape global perceptions [7, 15]. The purpose of the present work is to seek answers to the following questions:

- How can network influences due to 'family' and 'news' networks be included in an ABM of civil violence while preserving its simplicity?
- What is the impact of network influences on the nature of the solutions (equilibrium or complex)?
- What is the relative importance of network influences with respect to other mechanisms such as imprisonment delay and legitimacy feedback?

The remainder of this paper is organized as follows. In Sect. 2 we present the theoretical background, with emphasis on Epstein's ABM of civil violence and an extension of that model which includes small scale memory effects (imprisonment delay), media influence, and legitimacy feedback. Section 3 contains a description of the present ABM. In Sect. 4, we present the results of the model and their discussion for different combinations of 'family' and 'news' network influences as well as for homogeneous or heterogeneous legitimacy perception. Section 5 contains a summary of the conclusions.

2 Theoretical Background

2.1 Epstein's ABM of Civil Violence

Epstein's ABM of civil violence [2, 3] simulates rebellion against a central authority in an artificial society with two types of agents, 'agents' and 'cops' for representing citizens and policemen, respectively, moving in a homogeneous 2D torus space. Both types of agents have one movement Rule M: 'move to a random empty site within the agent's vision radius'; and one action rule, called Rule A for 'agents' and Rule C for 'cops', as described below. 'Agents' can be in one of three possible states, 'quiet', 'active' (rebellious), or 'jailed'. 'Agents' that are not 'jailed' switch between 'quiet' and 'active' according to the following action rule:

Rule A: if $G - N > T$ be 'active'; otherwise be 'quiet'

where $G = H \cdot (1 - L)$ is the grievance, $H \sim U(0, 1)$ is the perceived hardship, L is the perceived legitimacy of the central authority assumed equal for all agents, $N = R \cdot P$ is the net risk perception, where $R \sim U(0, 1)$ is the risk aversion, P is the estimated arrest probability, and T is a threshold (assumed constant for all 'agents'). The form of the arrest probability presented in Epstein's model is

$$P = 1 - \exp(-k \cdot (C/A)_v) \tag{1}$$

where C and A are the number of 'cops' and 'active' agents within the agent's vision radius v and $k = 2.3$ is the arrest constant [2, 3]. Implementations of Epstein's ABM often replace $(C/A)_v$ by $\lfloor C_v/(A_v + 1) \rfloor$ in Eq. (1) which leads to a drop of P from 0.9 to zero when $C = A$, avoids divide-by-zero errors (the 'agent' counts itself as 'active' when estimating the arrest probability) and produces complex solutions with intermittent bursts of rebellion [4, 12, 17]. 'Cop' agents have one action Rule C: Inspect all sites within v' and arrest a random 'active' citizen, where v' is the 'cop' vision radius (which may be different from v). Arrested citizens are removed from the simulation space ('jailed') for $J \sim U(0, J_{max})$ cycles (jail term), where J_{max} is set as an input variable.

The strength of Epstein's model lies in its simplicity (just two types of agents with two simple rules for each type), the relevance of the variables chosen for representing the social context (legitimacy) and individual attributes (grievance, hardship, and risk aversion), and its explanatory power (intermittent bursts of rebellion, effects of sudden or gradual variation of legitimacy or deterring capability of the central authority, etc.).

2.2 The Effects of Imprisonment Delay, 'News Impact', and Legitimacy Feedback

The model used in the present work is based on an extension of Epstein's ABM of civil violence (Model I) that includes (1) a time delay for imprisonment, (2) a third type of agent called 'media' for representing the 'news impact effect' of the system, and (3) endogenous legitimacy variation [12]. In the ABM developed herein, we combined the imprisonment delay and improved legitimacy feedback with a formulation of network influence effects, in which 'family' influence is modelled via a network of undirected and unconnected cliques (families) and 'news impact' is modelled using a third type of agents, called 'media', working as hubs of a directed star network. This allows a better representation of information propagation and collective behaviour processes related to civil violence in real societies.

3 Model Description

3.1 Synopsis

The ABM used in this work was implemented in NetLogo [16], using the 'Rebellion' NetLogo Library Model example [17]. Table 1 shows a summary of the model characteristics, using a subset of the 'Overview, Design Concepts and Details' (ODD) protocol [8]. The details of the implementation are described below.

3.2 Model Entities

The model entities are the agents, the scenario (spatial domain), and the networks. The scenario is a 2D homogeneous torus space, which is appropriate for an 'abstract' ABM [5]. Figure 1 shows the class diagram for all agents in the NetLogo implementation. The 'observer' (i.e. model user) box shows the global parameters and the model's main procedures (setup and go). The initial densities for 'citizen' and 'cop' agents, number of 'media' agents, simulation duration, vision radius, initial (reference) government legitimacy, maximum jail term, 'fight duration', 'media audience', 'family size', and the influence weights for 'family' and 'news' networks are numeric parameters. The 'legitimacy-feedback' variable F_L is a list with three strings, 'none', 'global', and 'agents', used to define the legitimacy feedback mechanism (see Fig. 1).

Table 1 Simplified ODD description of the ABM of civil violence with network influence effects

ODD item	Description
Purpose	Introduce network influence effects in an extended version of Epstein's ABM of civil violence with imprisonment delay and legitimacy feedback
Entities, state variables, and scales	Agents: 3 types of agents, 'citizen', 'cop', and 'media' with one 'move' and one 'behave' rule Networks: 2 networks, one consisting of a union of directed star networks with 'media' agents as central hubs ('news coverage') and another consisting of a union of unconnected cliques ('family')
Scenario	Homogeneous 2D torus space
Scales	Whole artificial society, undefined time step, and patch size
	Spatial scales in units of patch size: vision radius
	Time scales in units of time step size: 'fight duration', 'jail term'
Process overview and scheduling	All agents activated once per cycle in random order
Submodels	Legitimacy feedback
	Aggregation of network influences

'Citizen' Agent Specification

'Citizen' agents have one move rule and one action rule, and can be in one of the following states: 'quiet', 'active' (rebellious), 'fighting', or 'jailed'. Agents that are not 'fighting' or 'jailed' change state between 'quiet' and 'active' according to their action rule. The move rule is the same as Rule M in the original model.

To formulate the action rule, we need to specify how the agent's own perception is to be aggregated with the information conveyed by the 'family' and 'news' networks, which agent attributes are affected by the network influences, and how the agent's final decision is made. Our proposed solution is based on two conjectures. The first is that an individual decides by aggregating basic (raw) elements instead of information processed by others. This implies, for example, that the number of rebellious agents seen by 'family members' and 'media' is more relevant in forming the legitimacy perception than the legitimacy perceptions of these individuals. The second conjecture is that (1) the legitimacy percept (a 'latent concept' [6]) is affected by the own perception and network influences, (2) the state ('quiet' or 'active') is

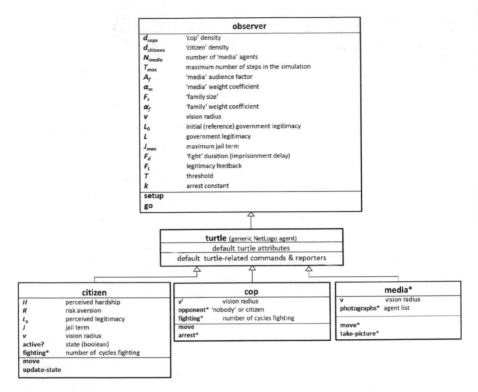

Fig. 1 Class diagram for all agent types in the NetLogo implementation. In NetLogo, agent types are implemented as subclasses of a generic 'turtle' class via the `breed` primitive. Agent types, attributes, and methods that are extensions of Epstein's model are marked by an *asterisk*

affected by the own perception and 'family' influence, and (3) the estimated arrest probability is affected only by the individual's own perception. This is consistent with the idea that in dangerous situations individuals rely on themselves and their family and when survival is at stake they act on their own. This leads to the formulation of the following two-step action rule (somewhat similar to the two-step rule of the Standing Ovation model of Miller and Page [14]):

Rule A1: if $G - N > T$ be 'active'; otherwise be 'quiet'

Rule A2: if more than 50 % of the 'family members' are 'active', be 'active'

where $G = H \cdot (1 - L_p)$ is the level of grievance, $N = R \cdot P$ is the net risk perception, T (constant exogenous variable) is a threshold, $H \sim U(0, 1)$ is the (endogenous) perceived hardship, $L_p \in [0, 1]$ is the 'perceived government legitimacy', $R \sim U(0, 1)$ is the (endogenous) risk aversion, and P is the estimated arrest probability computed using the expression

$$P = 1 - \exp(-k \cdot \lfloor C_v/(A_v + 1) \rfloor) \tag{2}$$

where $k = 2.3$ and C_v and A_v are the numbers of 'cops' and 'active' citizens within the agent's vision radius, respectively. 'Active' agents engaged by one 'cop' agent change state to 'fighting' if $F_d > 0$, or 'jailed' if $F_d = 0$. 'Fighting' agents are immobilized for F_d cycles before they are 'jailed'. 'Jailed' agents are removed from the simulation space for J cycles, after which they are reinserted in a random empty site within the simulation space with their state set to 'quiet'.

'Cop' Agent Specification

'Cop' agents can be in two states, 'non-fighting' and 'fighting', 'Non-fighting' cops have the same move and action rules as in Epstein's model. In the simulations reported herein, we used that same vision radius for 'citizens' and 'cops' ($v = v'$). If $F_d = 0$ 'cop' agents immediately arrest one 'active' citizen; if $F_d > 0$ they seek one 'suspect', mark it as 'opponent' and start 'fighting' with it for F_d cycles. During the 'fight' both enforcing 'cop' and its opponent are immobilized and at the end of the 'fight' the 'active' citizen is 'jailed' for $J \sim U(0, J_{max})$ cycles.

'Media' Agent Specification

'Media' agents have the following two rules:

Rule M': If there are any 'fighting' agents within the vision radius v, movetotheempty

site that is closest to the nearest 'fighter'; otherwise follow Rule M

   Rule P: Take one 'picture' of a 'fighter' within the vision radius

Rule M' is a departure from the use of random movement and torus geometry in 'abstract' ABM (interaction probabilities independent of position, no clustering emergent patterns), and was used to represent in a very simplistic way the 'agenda setting bias' towards showing violence. In the present version of the model Rule P does not influence the dynamics since neither the legitimacy update nor the 'news influence' depend on the number of 'pictures' recorded by 'media' agents, but this rule is still useful to get information about how efficient the 'news coverage' is.

Networks' Specification

The 'family' network is set by forming cliques of undirected links between citizens using the undirected-link-breed primitive. The clique size is defined by the family-size parameter. The 'news' network is set by connecting each 'media' agent to a proportion of 'citizens' defined by the audience-factor parameter, via directed links created using the directed-link-breed NetLogo primitive. One 'citizen' can be connected to more than one 'media' agent. Both networks remain fixed during the whole simulation.

3.3 Process Overview and Scheduling

The model is implemented in two main procedures, setup and go, which initialize the simulation and run the main cycle, respectively. The setup procedure clears all variables from the previous simulation, initializes the global variables, creates the agents list, builds the 'news' and 'family' networks (if there are any 'media' agents and the family size is greater then one, respectively), displays the simulation space, and opens the output file used for post-processing. The go procedure implements the main cycle, which consists of the following operations: (1) test for termination and closing of the output file; (2) initialization of global variables that are reset at each time cycle (number of arrests and 'pictures' taken by 'media' agents); (3) update the legitimacy; (4) run the move and action rules for all 'non-fighting' agents; (5) decrement F_d for all 'fighting' agents; and (6) print the cycle information to the output file.

3.4 Legitimacy Feedback

Legitimacy feedback is formulated by expressing the legitimacy as a function of three variables, 'legality' (L_{leg}), 'justification' (L_{just}), and 'acts of consent' (L_{cons}) [6]. The form of the legitimacy function is a key but unsolved question in political science [1, 6]. In the present ABM we considered the following expression:

$$L = L_0 \cdot \left(\frac{1}{4} \cdot (L_{\text{leg}} + L_{\text{cons}}) + \frac{1}{2} L_{\text{just}} \right) \tag{3}$$

in which

$$L_{\text{leg}} = \frac{n_{\text{quiet}}}{N} \tag{4}$$

$$L_{\text{just}} = \frac{1}{2} \cdot \left(1 - \frac{n_{\text{active}} + n_{\text{fighting}}}{N} \right) \tag{5}$$

$$+ \frac{1}{2} \cdot \left(1 - \exp\left[-\frac{\ln(2)}{2} \cdot \lfloor \frac{N}{n_{\text{active}} + n_{\text{fighting}} + n_{\text{jailed}} + 1} \rfloor \right] \right) \tag{6}$$

$$L_{\text{cons}} = L_{\text{leg}} \tag{7}$$

where N is the population size and n_{quiet}, n_{active}, n_{fighting}, and n_{jailed} are the total number of 'citizens' in each state. For the theoretical foundations and formulation of these functions, see [6] and [13], respectively.

If F_L is set to 'none' or 'global', the value of L_p is set equal to the value of the global variable L. If F_L is set to 'agents', L_p is computed for each 'citizen' agent using Eqs. (3)–(7) with n_{quiet}, n_{active}, n_{fighting}, and n_{jailed} replaced by aggregate values obtained using

$$n^*_{\text{active}} = \alpha \cdot A_v + \alpha_f \cdot \overline{A_f} + \alpha_m \cdot \overline{A_m} \tag{8}$$

and analogous expressions for n^*_{quiet}, n^*_{fighting}, and n^*_{jailed}. In Eq. (8), A_v, A_f, and A_m denote the numbers of 'active' citizens that are 'visible', 'visible by family members', and 'visible in news', respectively; α_f and α_m are the influence weights for the 'family' and 'news' networks; and $\alpha = 1 - \alpha_f - \alpha_m$.

4 Results and Discussion

We performed three sets of simulations and compared the results with a reference case run using Epstein's original ABM (Run 2 in Appendix A of [3]). In the first set we investigated the effect of varying the network influences without introducing imprisonment delay and legitimacy feedback. In the second set we combined network influence effects with imprisonment delay and legitimacy feedback, for the same value of initial legitimacy of the reference case ($L_0 = 0.82$). In the third set, we studied the effect of increasing the initial legitimacy ($L_0 = 0.89$). We considered family sizes 3, 4, and 6 and three combinations of 'media' audience and influence factor, to simulate two types of society: 'rural' with numerous families and low 'media' impact, and 'technological' with opposite characteristics. In all cases, we analysed the impact of the newly introduced effects on the system's long term behaviour, and in the cases with punctuated equilibrium or large oscillating peaks of rebellion we studied the waiting time and size of the rebellion peaks. We used a 40×40 torus space, 1120 'citizens' (70 % density), 64 'cops' (4 % density) and maximum jail term $J_{\max} = 30$. Tables 2, 3, and 4 show the parameters for the three sets of simulations. Legitimacy feedback was computed using Eqs. (3)–(7). We performed ten simulations for each case, with a duration of 2000 cycles in the first and second sets and 5000 cycles in the third set (due to the difficulty in determining the long term behaviour).

Figure 2 shows plots of the simulation space for two runs of different cases. These plots allow a suggestive visual interpretation of the spatial distribution of the 'news' network coverage (weak in the first case, strong in the second) and imprisonment delay ('cops' and 'agents' involved in temporary fights, one 'media' agent near 'fighting' agents, plotted in a larger size).

Table 5 shows the mean value and standard deviation of the waiting times and peak sizes of the rebellion peaks (maximum number of 'active' agents in large bursts of violence) for the simulations of the first set.

It can be concluded that in Epstein's ABM the periodicity of the rebellion peaks is determined by the jail term parameter. Introduction of network influences does not change either the system's long term behaviour (see Table 2) or the waiting times between rebellion peaks, but significantly increases the peak sizes. This is

Table 2 Parameters and system's long term behaviour for the first set of simulations

	E1	E1F	E1N	E1-LF-WM	E1-SF-LM
L_0	0.82	0.82	0.82	0.82	0.82
num. Media	2	2	2	1	2
m. audience	0	0	20 %	10 %	20 %
α_m	0.0	0.0	0.1	0.1	0.2
Family size	0	4	4	6	3
α_f	0.0	0.4	0	0.4	0.1
Behaviour	Punctuated equilibrium	Punctuated equilibrium	Punctuated equilibrium	Punctuated equilibrium	Punctuated equilibrium

E1 is the reference case. Case E1F includes only a 'family' network and case E1N only a 'news' network. Cases E1-LF-WM and E1-SF-LM simulate societies with large family size and influence and poor 'media' coverage and small family size and influence and large exposure to 'media', respectively

particularly notorious for the case of small 'family' and large 'news' influence, such as in modern technological societies where people tend to stick to TV and SN in detriment of family contact.

Imprisonment delay and legitimacy feedback had a larger impact and changed the system's long term behaviour in several ways (Table 3). Increasing the initial legitimacy (Table 4) lead to complex solutions. Figure 3 shows that in case F2L089, the system was near a tipping point, with indefinite long term behaviour, alternating between long periods of calm and turmoil. Such alternations occurred after hundreds of cycles in an unpredictable way. In real societies, apparently stable authoritarian regimes may suddenly face large rebellions and in democratic regimes we often observe alternating periods of calm and protests.

5 Conclusions

We presented an extension of Epstein's ABM of civil violence with network influence effects associated with two different networks, 'family' and 'news', including two other effects, imprisonment delay and endogenous legitimacy variations. We performed three sets of simulations to study the effects of (1) network influence for different network sizes and influence factor, (2) legitimacy feedback and imprisonment delay, and (3) variation of the initial legitimacy, combined with network influences and legitimacy feedback, and compared the results with a reference case run using Epstein's model.

The results from the first set showed that network influences did not change either the system's long term behaviour or the periodicity of violence bursts, but increased their size, particularly for small 'family' and large 'news' influence. The simulations of the second set showed that the introduction of legitimacy feedback changed the

Table 3 Parameters and system's long term behaviour for the second set of simulations

	F2	F3	N1	N2	NF1	NF2
L_0	0.82	0.82	0.82	0.82	0.82	0.82
F_d	0	0	0	1	0	0
feed. mech.	Global	Agents	Agents	Global	Agents	Global
num.media	0	0	2	2	2	2
m. audience(%)	0	0	20	20	20	20
α_m	0.0	0.0	0.1	0.1	0.1	0.1
family size	4	4	0	4	4	4
α_f	0.4	0.4	0.0	0.4	0.4	0.4
Behaviour	Violence peaks no calm periods	Violence peaks no calm periods	Violence peaks no calm periods	Permanent rebellion	Permanent rebellion	Permanent rebellion

Cases F2 and F3 include only 'family' networks, cases N1 and N2 only 'news' networks, and cases NF1 and NF2 include both types of influences, considering homogeneous (global) and heterogeneous (agents) legitimacy feedback

Table 4 Parameters and system's long term behaviour for the third set of simulations

	F2L089	F3L089	NF1L089	NF2L089
L_0	0.89	0.89	0.89	0.89
feed. mech.	Global	Agents	Agents	Global
num.media	2	2	2	2
m. audience(%)	0	0	20	20
α_m	0.0	0.0	0.1	0.1
family size	4	4	4	4
α_f	0.4	0.4	0.4	0.4
Behaviour	Tipping point indefinite	Violence peaks no calm periods	Violence peaks no calm periods	Tipping point indefinite

Cases F2L089 and F3L089 include family influence only, whereas cases NF1L089 and NF2L089 include both types of influence ('family' and 'news')

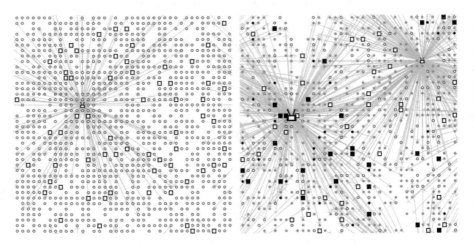

Fig. 2 Plots of the simulation space for runs of cases E1-LF-WM (*left*) and N2 (*right*). 'Quiet', 'active', and 'fighting' agents are represented by *white (hollow)*, *grey*, and *black circles*, respectively. 'Jailed' agents are hidden from view. 'Fighting' and 'non-fighting' cops are represented by *white (hollow)* and *black squares*, respectively. 'Media' agents are represented by small TV icons, which are larger when they are 'taking pictures'. 'News' links are represented in *light grey* and 'family' links are hidden from view

system's long term behaviour from punctuated equilibrium to bursts of violence with no calm periods or permanent rebellion. For the third set of simulations the solutions showed a very complicated behaviour with unpredictable alternations between long periods of calm and turmoil occurring after several hundreds of cycles. These results reinforce the conjecture that network influences by themselves do not trigger revolutions, but amplify their size (first set of simulations). Legitimacy variations and their relationship with the model's parameters and dependent variables (i.e. the social context in real situations) are more important, for they determine the system's behaviour in very complicated and sensitive ways. Thus, networks are important for

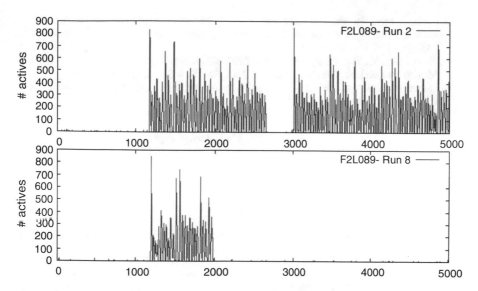

Fig. 3 Time history of number of 'active' citizens, case F2L089, run 2 and run 8

Table 5 Mean value and standard deviation of the waiting time and peak size, for the first set of simulations

	Wait. Time μ	Wait. Time σ	Peak size μ	Peak size σ
E1	29.2	8.7	252	85
E1F	30.2	7.7	348	117
E1N	31.6	8.0	360	109
E1-LF-WM	31.0	8.7	311	110
E1-SF-LM	32.1	7.48	504	153

triggering uprisings only if their existence contributes for changing the perceived legitimacy (second and third sets of simulations), which in real situations depends on their size, structure, influence, and information content.

Acknowledgements Support by the CISDI - Instituto de Estudos Superiores Militares - Lisbon, Portugal to one of the authors (Carlos Lemos) is gratefully acknowledged. Support by centre grant (to BioISI, Centre Reference: UID/MULTI/04046/2013), from FCT/MCTES/ PIDDAC, Portugal, to Carlos Lemos and Helder Coelho is also acknowledged.

References

1. Dogan, M.: Encyclopedia of Government and Politics, vol. I, chap. 7 - Conceptions of legitimacy, pp. 116–126. Routledge (1992)
2. Epstein, J.M.: Modeling civil violence: an agent-based computational approach. Proc. Natl. Acad. Sci. USA **99**, 7243–7250 (2002)

3. Epstein, J.M., Steinbruner, J.D., Parker, M.T.: Modeling civil violence: an agent-based computational approach. Center on Social and Economic Dynamics, Working Paper No. 20 (2001)
4. Fonoberova, M., Fonoberov, V.A., Mezic, I., Mezic, J., Brantingham, P.J.: Nonlinear dynamics of crime and violence in urban settings. J. Artif. Soc. Soc. Simul. **15**(1), 2 (2012)
5. Gilbert, N.: Agent-Based Models (Quantitative Applications in the Social Sciences). SAGE Publications, Thousand Oaks (2007)
6. Gilley, B.: The Right to Rule. How States Win and Lose Legitimacy. Columbia University Press, New York (2009)
7. Granovetter, M.S.: The strength of weak ties. Am. J. Sociol. **78**(6), 1360–1380 (1973)
8. Grimm, V., Bergern, U., DeAngelis, D.L., Polhill, J.G., Giskee, J., Railsback, S.F.: The odd protocol: a review and first update. Ecol. Model. **221**, 2760–2768 (2010)
9. Gurr, T.R.: Why Men Rebel. Fortieth Anniversary Paperback Edition, 446 p. Routledge, London/New York (2016)
10. Kim, J.W., Hanneman, R.A.: A computational model of worker protest. J. Artif. Soc. Soc. Simul. **14**(3), 1 (2011)
11. Kuran, T.: Sparks and prairie fires: a theory of unanticipated political revolution. Public Choice **61**, 41–74 (1989)
12. Lemos, C., Lopes, R.J., Coelho, H.: An agent-based model of civil violence with imprisonment delay and legitimacy feedback. In: 2014 Second World Conference on Complex Systems (WCCS), Agadir, Morocco, 10–12 Nov, pp. 524–529 (2014)
13. Lemos, C., Lopes, R.J., Coelho, H.: On legitimacy feedback mechanisms in agent-based models of civil violence. Int. J. Intell. Syst. **31**(2), 106–127 (2015)
14. Miller, J.H., Page, S.L.: Complex Adaptive Systems. Princeton University Press, Princeton (2007)
15. Watts, D., Dodds, P.S.: Influentials, networks and public opinion. J. Consum. Res. **34**, 441–458 (2007)
16. Wilensky, U.: NetLogo. Center for Connected Learning and Computer-Based Modelling, Northwestern University, Evanston, IL (1999). http://ccl.northwestern.edu/netlogo/
17. Wilensky, U.: NetLogo Rebellion model. Center for Connected Learning and Computer-Based Modeling, Northwestern University, Evanston, IL (2004). http://ccl.northwestern.edu/netlogo/models/Rebellion

Modeling the Evolution of Ideological Landscapes Through Opinion Dynamics

Jan Lorenz

Abstract This paper explores the possibilities to explain the stylized facts of empirically observed ideological landscapes through the bounded confidence model of opinion dynamics. Empirically left-right self-placements are often not normally distributed but have multiple peaks (e.g., extreme-left-center-right-extreme). Some stylized facts are extracted from histograms from the European Social Survey. In the bounded confidence model, agents repeatedly adjust their ideological position in their ideological neighborhood. As an extension of the classical model, agents sometimes completely reassess their opinion depending on their ideological openness and their propensity for reassessment, respectively. Simulations show that this leads to the emergence of clustered ideological landscapes similar to the ones observed empirically. However, not all stylized facts of real world ideological landscapes can be reproduced with the model.

Changes in the model parameters show that the ideological landscapes are susceptible to interesting slow and abrupt changes. A long term goal is to integrate models of opinion dynamics into the classical spatial model of electoral competition as a dynamic element taking into account that voters themselves shape the political landscape by adjusting their positions and preferences through interaction.

Keywords Continuous opinion dynamics • Bounded confidence • European social survey • Stylized facts • Homophile adaptation • Random reconsideration • Consensus • Polarization • Plurality

1 Introduction

Voters' ideological preferences are the basis of electoral behavior in spatial models of electoral competition. This holds for the classical rational choice model of Downs [3] as well as for agent-based models of multidimensional and multiparty competition [8, 11, 12]. In these models parties or candidates follow optimizing

J. Lorenz (✉)
Jacobs University Bremen, Focus Area Diversity and Bremen International Graduate School of Social Sciences (BIGSSS), Campus Ring 1, 28759 Bremen, Germany
e-mail: post@janlo.de

© Springer International Publishing AG 2017
W. Jager et al. (eds.), *Advances in Social Simulation 2015*, Advances in Intelligent Systems and Computing 528, DOI 10.1007/978-3-319-47253-9_22

or satisficing strategies and heuristics to win as many voters as possible, while
the voters have stable never changing preferences on which basis they vote non-
strategically for the party which manifesto is closest to their ideological ideal point.
Thus, only the choice of a party is modeled not the choice of the preferences of
voters. Laver [11, p. 280] acknowledges that "Almost all observers of real politics
believe voter preferences to evolve dynamically in response to the development
of political competition, yet static models of party competition find this type of
feedback difficult to handle." This paper shall serve as a first attempt to model
dynamics of voter preferences based on opinion dynamics between voters.

Opinions are subjective statements which summarize and communicate emotions, attitudes, beliefs, and values to others. Thus, on the one hand, opinions are
a manifestation of individual preferences because of their summarizing nature. On
the other hand, opinions can trigger changes of opinions of others because they
are an act of communication. While political science has a large literature on the
mechanism of opinion formation within the individual and how it can be shaped
towards public opinion through mass media and political communication [see, e.g.,
7, 23], not so much attention has been spent to study systemic effects triggered by
the mechanisms of interaction between voters.

In this context, systemic effects are effects which are triggered by behavior on
the individual level but which go beyond the scope of the individual. Thus, we look
at the evolution of the landscape of opinions based on local interaction to detect
possible hidden driving forces which shape the evolution of opinion landscapes.
We do this partly instead of studying in detail the individual process of opinion
formation for contextual information or the aggregation of public opinion from
individual opinions.

One suitable model is the bounded confidence model of continuous opinion
dynamics [1, 9, 10, 21]. The model triggered a large stream of literature within
the scientific communities of social simulation and the physics of socio-economic
systems [see 13]. This paper is to explore the usability of the model as a model of
preference formation of many agents in ideology spaces based on some stylized
facts of ideological landscapes based on left-right self-placements in different
countries in the European Social Survey (ESS).

The paper is organized as follows. First, empirical opinion landscapes will
be presented and stylized facts will be derived. Second, the bounded confidence
model of continuous opinion dynamics will be presented. Third, the main dynamics
triggered on the level of opinion landscapes will be shown. Fourth, the matching
between empirical landscapes and modeling results will be discussed.

2 Opinion Landscapes: Empirical Observations

We use data from the European Social Survey (ESS rounds 1–6) to get a glimpse of
opinion landscapes and their characteristics. Figure 1 shows the left–right ideology
landscapes of France and Sweden over all available ESS rounds from 2002 to 2012.

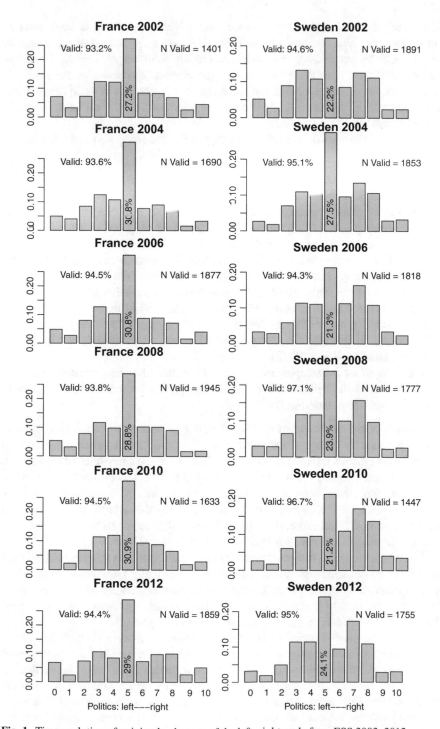

Fig. 1 Time evolution of opinion landscapes of the left–right scale from ESS 2002–2012

Answering the survey, people self-place themselves on a left–right scale with 11 grades ranging from zero (left) to ten (right). Shown are weighted histograms over the full samples per country in rounds. Individuals were weighted using the design weights delivered with the ESS dataset. This corrects for biases of the sampling design. Accompanying each opinion landscape, the percentage of valid answers and the number of valid answers is reported. Valid answers typically exclude "Refusal," "Don't know," and "No answer."

Looking at the empirical opinion landscapes the following **stylized facts** can be extracted:

1. Distributions never visually fit any "standard distribution." It is far from being uniform or normal distributed.
2. The largest peak is always at the center (bin 5). It almost always exceeds neighboring peaks by far giving a "non-continuous" impression.
3. Multiple peaks are almost ubiquitous.
4. Very often extremal peaks exist on both sides. The next-to-extremal bins 1 and 9 are almost always lower than both of their neighboring bins.
5. Peaks at 3 (moderately left) and 7 (moderately right) are frequent. The next-to-center bins at 4 and 6 are often low compared to both of their neighbors. There seems to be a tendency that moderate left and moderate right clusters might form. Especially on the right wing the peak of the cluster might also lie at 8 instead of 7 (e.g., Israel or France 2012).
6. The "shape" of landscapes away from the center and the extremes often has a "smooth" look without abrupt jumps giving the impression that a smooth continuous curve underlies the bins.

Looking at the evolution of opinion landscapes over time shows that opinion landscapes usually change slowly over time. Focusing on the off-center moderate left and right clusters around the bins 3 and 7 shows some tendencies of movement. Clusters might slightly drift—sometimes to the left, sometimes to the right. Sometimes the whole opinion landscape seems to drift in some direction (e.g., Sweden seems to drift slightly to the right). Sometimes the left and the right clusters seem to drift in different directions (e.g., France 2010 seems to have left and right clusters closer to the center than in 2004 as well as in 2012).

Two conclusions can be drawn from these stylized facts. First, empirical opinion landscapes do not follow simple distributions, in particular they are not at all close to be normally distributed. Second, empirical opinion landscapes nevertheless show some smoothness which points to some process which structures them in typical clustered shapes which are not random or externally triggered.

How do these non-simple but structured opinion landscapes evolve? Numerous factors might of course play a role. For example, there might be psychological reason why certain opinions are more attractive than others (e.g., the central or the extremal bins). In this paper we will focus on reasons which might lie in the interaction and adaptation of individuals. In the following section we will look at the bounded confidence model of opinion dynamics which produces clustered opinion landscapes.

As a side note, multimodality of ideological landscapes has also been used by the father of the spatial model [3, Fig. 8] as an explanation why two parties might not fully converge to the median voter's position and as a reason for the emergence of new parties at newly evolving modes.

3 Bounded Confidence Model of Continuous Opinion Dynamics

Model Definition Let us consider a society of N agents in which agent i is characterized by an *ideological position* x_i (also called *opinion* in the following) in the range [0,1] which is subject to change over time while the agent communicates with others. All agents are further equipped with a homogeneous *bound of confidence ε* also from the range [0,1] which is not subject to change but determines the maximal ideological distance to others which an agent is willing to take into account to revise its ideological position. Thus, the ideological positions of all agents at time t form the vector $x(t) \in [0, 1]^N$ which defines an opinion landscape. As ideological position changes over time, $x(t)$ is the dynamic variable of the model of opinion dynamics, while the bound of confidence ε is a static variable. The bound of confidence somehow models homophily, which is the tendency to interact positively only with agents which are similar to oneself [6, 18].

We consider agents to change their ideological positions because of two processes: *homophile adaptation* and *random reconsideration*.

Homophile Adaptation Consider that at each time step an agent i meets another agent j picked at random and interacts with it only when their distance in opinion is less than its bound of confidence $|x_i(t) - x_j(t)| < \varepsilon$. If agent j is close enough agent i changes its ideological position to the average of the two positions $x_i(t + 1) = (x_i(t) + x_j(t))/2$. If agent j is too far away agent i's position remains unchanged $x_i(t + 1) = x_i(t)$.

Reconsideration Sometimes agents reconsider their ideological position from scratch. Each agent reconsiders its ideological position at each time step with probability p (which is thought to be small, e.g., $p = 0.1$). When an agent reassesses its ideological position it chooses a random value from the interval [0, 1] with equal probability independently of the positions of others. Thus, the agent's new opinion is sampled from a uniform distribution. The parameter p is the second global static parameter of the model.

To start a run of the model each agent starts with a step of reconsideration to setup the initial ideological positions.

This model resembles the bounded confidence model proposed by Krause [10] and Deffuant et al. [1] independently from each other. Both models differ in their communication regime [20]. Here, we concentrate on the communication regime of Deffuant et al. [1] where agents only engage in pairwise interaction as described.

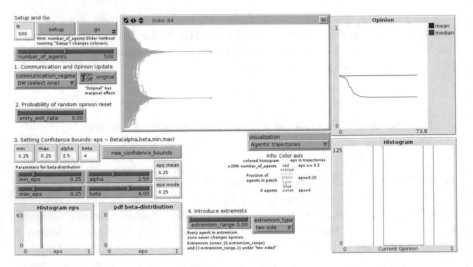

Fig. 2 Full screen of the Netlogo model of Lorenz [16]. A run is shown for 500 agents with homogeneous bound of confidence $\varepsilon = 0.25$ and without reconsideration ($p = 0$)

Originally, the model was formulated without the process of reconsideration, which was introduced as noise by Pineda et al. [19]. It is called uniformly distributed opinion noise by Mäs et al. [17]. Several other modifications of the models such as multidimensional opinion spaces [14], heterogeneous bounds of confidence [15], relative agreement, and extremism [2] were introduced and analyzed.

The version with pairwise communication and random reconsideration (and some other options) is implemented in Netlogo [22] for easy use [16]. This version is used in the following to produce the following figures.

Figure 2 shows the full screen of the Netlogo model (in a slightly changed design compared to [16]). The main panel in the upper center shows the trajectories of opinions up to the current time step (here $t = 64$) moving from left to right. For longer runs this panel turns into a "rolling" display showing only the latest time steps. The interface (left-hand side and bottom of the screen) includes besides the core controls on for the number of agents and the "setup" and "go" buttons the following functions:

1. controls for the communication regime. In this paper "DW (select one)" with original "On."
2. a slider to set the probability of reconsideration p.
3. a module to initialize random distributions of heterogeneous bounds of confidence, not used here.
4. options to introduce extremists, not used here.

On the right-hand side of the model screen there are two observers which accompany the central panel of trajectories. The top panel shows the evolution of the mean and the median opinion over the full time the simulation has been running.

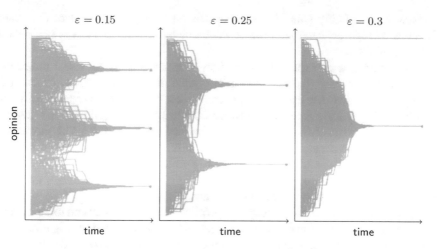

Fig. 3 Trajectories for $N = 500$ agents with homogeneous ε without reconsideration leading to plurality (*left*, $\varepsilon = 0.15$), polarization (*center*, $\varepsilon = 0.25$), and consensus (*right*, $\varepsilon = 0.3$)

The bottom panel shows the histogram of the current opinions in eleven equidistant bins, resembling the discretization of empirical opinion landscapes from the ESS.

Model Dynamics Let us first look at homogeneous bounds of confidence without random reconsideration.

Bound of Confidence ε Large bounds of confidence (approximately $\varepsilon > 0.27$) lead to *consensus*. Intermediate bounds of confidence (approximately $0.18 < \varepsilon < 0.27$) lead to *polarization* into two equally large opinion clusters, one moderately left-wing and one moderately right-wing. Small bounds of confidence ($\varepsilon < 0.18$) lead to *plurality* of three or more clusters. Figure 3 shows trajectories for different bounds of confidence to demonstrate consensus, polarization, and plurality. The dynamical process towards stable polarization over time runs as follows: (1) Starting with initially uniformly distributed opinions, central agents (with opinions more than ε away from the boundary) are equally likely to find a close enough communication partner at either side. Extremal agents close to the boundary of the opinion space instead can only find close enough communication partners slightly more central. Thus they will move closer to the center, but at most $\varepsilon/2$ as they ignore others when they are farther than ε away. (2) After a few time steps this leads to a slightly higher concentration of agents in the region $[\varepsilon/2, \varepsilon]$ and the opposing region on the other side of the opinion space. The concentration of agents remains the same in the center and declined at the extremes. Now, an agent between the center and ε has a higher chance to find a close enough communication partner at the other side of the center because the concentration of agents is higher there. Thus, it is more likely that it moves from the center towards the region of higher concentration. The same happens on the other side of the opinion space. (3) Regions of slightly higher concentration quickly gain higher concentration through this kind of attraction. Regions of high concentration finally converge to opinion clusters which have a distance of about 2ε.

Similar arguments explain the evolution of more clusters for lower bounds of confidence. Of course, the random initial conditions and the random selection of communication partners can lead to different outcomes, but even for relatively low numbers of agents the separation between consensus and polarization as attractive final opinion landscapes is quite sharp at the critical value $\varepsilon = 0.27$. This means that random influence decides which of the two attractive landscapes will be reached when the bound of confidence is at such a critical value. When the bound of confidence is far away from a critical value it is almost sure that always the same pattern of clustered opinion landscape is reached.

The typical locations bifurcating at the critical values into more and more clusters are shown in a bifurcation diagram in [13]. Note that in this study we ignore minor clusters of minimal size which structurally emerge at the extremes and between the major clusters, because they usually contain only zero to two agents in the sample sizes we use.

From a political science perspective the view on the mean and the median opinion is interesting (see upper right panel in the model's full screen in Fig. 2). In uniformly distributed opinions mean and median are of course almost the same. Interestingly, in the ε range of polarization the median quickly starts to deviate from the mean when opinion dynamics starts. The mean remains almost central but due to the depopulation of the center the median finally ends up to be in one of the two off-central clusters far away from the compromising position in the center. Which of the two clusters ends up slightly larger than the other is subject to random fluctuation.

In summary, the model with homogeneous bounds of confidence and without reconsideration is thus able to produce clustered opinion landscapes, but they are way to clustered to match the empirical opinion landscapes from the ESS.

Random Reconsideration with Probability p The introduction of random individual reconsideration of opinions with probability p blurs the peakedness of clusters such that they visually come closer to the opinion landscapes observed empirically in the ESS data. Random reconsideration also triggers slight movements of clusters.

Figure 4 shows three example trajectories in heatmap visualization and a snapshot of the histogram. As can be inferred from the trajectories, the histogram fluctuates but keeps its typical shape. The values $\varepsilon = 0.15, 0.25, 0.3$ are taken from Fig. 3 and the values $p = 0.12, 0.2, 0.09$ are chosen in an attempt to visually match some characteristics of ESS opinion landscapes. The matching will be discussed in the final section.

Finally, it shall be demonstrated how the model with random reconsideration reacts close to critical values of the bound of confidence. Without randomness the critical bound of confidence between polarization and consensus is $\varepsilon = 0.27$. Figure 5 shows snapshots from the model running with $\varepsilon = 0.28$ and a probability of random reconsideration $p = 0.09$.

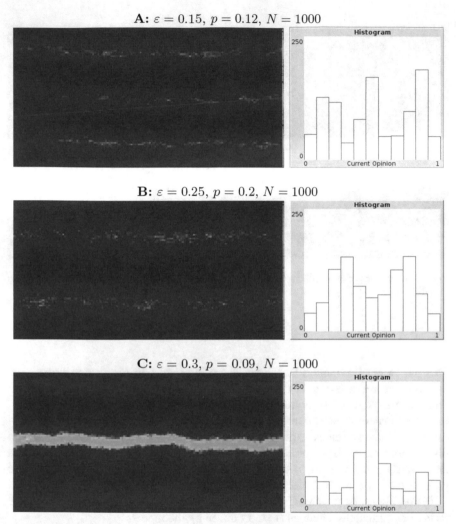

Fig. 4 Trajectories of $N = 1000$ agents with closed-minded (**a**), open-minded (**c**), and intermediate bounds of confidence (**b**) and different probabilities of random reconsideration p. The bounds of confidence are the same as in Fig. 3

The snapshots show that two meta-stable states exist—a polarized one with two clusters and a consensual one with a central clusters which position moves considerably in the central range. The system switches between these states through rare random events. The transition from polarization appears to be a slow one (panel top right) through a random drift of the central cluster and the evolution of the second cluster in the other half of the opinion space. In contrast, the transition from polarization to consensus is rather abruptly triggered by a random fluctuation which suddenly builds a critical mass to connect the two sides.

$\varepsilon = 0.28,\; p = 0.09,\; N = 1000$

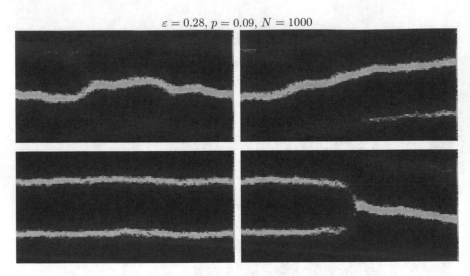

Fig. 5 Trajectories of $N = 1000$ agents with bounds of confidence at a critical value in heat map visualization

4   Discussion

To what extent do opinion landscapes from the model in Fig. 4 match empirical opinion landscapes in Fig. 1? Generally, no model specification seems to be able to reproduce all stylized facts from empirical distributions totally satisfactory. Thus, the development of a parameter fitting procedure is left for future models and the further discussion is exploratory and qualitatively. Nevertheless, the model distributions match the empirically observed distributions visually much better than any specification of a (a, b)-beta distribution. As the model has also only two parameters (ε, p) this is already a small success.

Going along the increasing bounds of confidence from Fig. 4 different stylized facts of empirical distributions are reproduced by the model.

Closed-minded agents (Fig. 4a $\varepsilon = 0.15$): Under a relatively high probability of reconsideration ($p = 0.12$) an opinion landscape with three peaks is produced—a central, a moderate left, and a moderate right cluster as we see it in many empirical landscapes, e.g., in the left–right landscape of Sweden 2002 or Germany's European integration landscape 2012; but empirical landscapes of this type usually also have extremal peaks, the moderate peaks lie a bit closer to the center and never on the bins 1 and 9 and the central peak is much more pronounced.

Intermediate agents (Fig. 4b $\varepsilon = 0.25$): Under a high probability of reconsideration ($p = 0.2$) the opinion landscape is still polarized as for the non-noisy case, but clusters are blurred such that their shape and location at bins 2,3, respectively, 7,8 matches moderate off-center empirical clusters. In this specification a dominant central cluster as observed empirically misses completely. If it would exist it would absorb the moderate clusters quickly. Also extremal peaks as empirically found are missing.

Open-minded agents (Fig. 4c $\varepsilon = 0.3$): Under an intermediate probability of reconsideration ($p = 0.09$) the opinion landscape shows a large central peak as observed empirically, also small peaks at the extremes match the empirical landscapes, although empirical extremal peaks lie more pronounced at 0 and 10 and not on 0,1, respectively, 9,10 as in the model. The weak point here is that empirically we observe much more mass on the bins close to the center.

In conclusion, several stylized facts can be reproduced by this two parameter model much better than classical distributions can, but the model cannot produce a large central peak, and tendencies for small and blurred moderate off-center peaks and small or tiny extremal peaks at the same time in its current form.

Agent-based models for the evolution of ideological landscapes are still in its infancy and it remains to show if they can add interesting insight to political dynamics. At least the possibility for counterfactual analysis at critical values seems promising for studying and understanding self-driven abrupt changes in political landscapes.

Acknowledgements Presented at ESSA@Work Groningen 2015. The author thanks Klaus Troitzsch, Nigel Gilbert, and Geeske Scholz for comments and advice. A former version of the paper was presented at ECPR General Conference Sept 3–6, 2014, Glasgow in the Panel P289 "Preference Formation and Formal Models of Politics." Part of this research was funded by the German Research Council (DFG) grant no. LO2024/2-1 "Opinion Dynamics and Collective Decision: Procedures, Behavior and Systems Dynamics."

References

1. Deffuant, G., Neau, D., Amblard, F., Weisbuch, G.: Mixing beliefs among interacting agents. Adv. Complex Syst. **3**, 87–98 (2000)
2. Deffuant, G., Neau, D., Amblard, F., Weisbuch, G.: How can extremism prevail? a study based on the relative agreement interaction model. J. Artif. Soc. Soc. Simul. **5**(4), 1 (2002)
3. Downs, A.: An Economic Theory of Democracy. Harper & Row Publishers, New York (1957)
4. ESS Round 1–5: European Social Survey Cumulative Rounds 1–5 Data. Data file edition 1.1. Norwegian Social Science Data Services, Norway – Data Archive and distributor of ESS data (2010)
5. ESS Round 6: European Social Survey Round 6 Data. Data file edition 2.0. Norwegian Social Science Data Services, Norway – Data Archive and distributor of ESS data (2012)
6. Feld, S., Bernard, G.: Homophily and the focused organization of ties. In: Hedström, P., Bearman , P. (eds.) The Oxford Handbook of Analytical Sociology. Chap. 22, pp. 521–543. Oxford University Press, Oxford (2009)
7. Feldman, S.: Structure and consistency in public opinion: the role of core beliefs and values. Am. J. Polit. Sci. **32**(2), 416–440 (1988)
8. Fowler, J.H., Laver, M.: A tournament of party decision rules. J. Confl. Resolut. **52**(1), 68–92 (2008)
9. Hegselmann, R., Krause, U.: Opinion dynamics and bounded confidence, models, analysis and simulation. J. Artif. Soc. Soc. Simul. **5**(3), 2 (2002)
10. Krause, U.: A discrete nonlinear and non-autonomous model of consensus formation. In: Elyadi, S., Ladas, G., Popenda, J., Rakowski, J. (eds.) Communications in Difference Equations, pp. 227–236 Gordon and Breach Pub, Amsterdam (2000)

11. Laver, M.: Policy and the dynamics of political competition. Am. Polit. Sci. Rev. **99**(02), 263–281 (2005)
12. Laver, M., Schilperoord, M.: Spatial models of political competition with endogenous political parties. Philos. Trans. R. Soc. B Biol. Sci. **362**(1485), 1711–1721 (2007)
13. Lorenz, J.: Continuous opinion dynamics under bounded confidence: a survey. Int. J. Mod. Phys. C **18**(12), 1819–1838 (2007)
14. Lorenz, J.: Fostering consensus in multidimensional continuous opinion dynamics under bounded confidence. In: Helbing, D. (ed.) Managing Complexity: Insights, Concepts, Applications. Understanding Complex Systems, pp. 321–334. Springer, Berlin (2008)
15. Lorenz, J.: Heterogeneous bounds of confidence: meet, discuss and find consensus! Complexity **15**(4), 43–52 (2010)
16. Lorenz, J.: Netlogo Model: Continuous Opinion Dynamics under Bounded Confidence. NetLogo User Community Models (2012). See also http://janlo.de/applets/bc.html
17. Mäs, M., Flache, A., Helbing, D.: Individualization as driving force of clustering phenomena in humans. PLoS Comput. Biol. **6**(10), e1000959 (2010)
18. McPherson, M., Smith-Lovin, L., Cook, J.M.: Birds of a feather: homophily in social networks. Annu. Rev. Sociol. **27**, 415–444 (2001)
19. Pineda, M., Toral, R., Hernandez-Garcia, E.: Noisy continuous-opinion dynamics. J. Stat. Mech: Theory Exp. **2009**(08), P08001 (18pp.) (2009)
20. Urbig, D., Lorenz, J., Herzberg, H.: Opinion dynamics: The effect of the number of peers met at once. J. Artif. Soc. Soc. Simul. **11**(2), 4 (2008)
21. Weisbuch, G., Deffuant, G., Amblard, F., Nadal, J.-P.: Meet, discuss, and segregate! Complexity **7**(3), 55–63 (2002)
22. Wilensky, U.: NetLogo.Center for Connected Learning and Computer-Based Modeling. Northwestern University, Evanston, IL (1999)
23. Zaller, J.R.: The Nature and Origin of Mass Opinion. How Citizens Acquire Information and Convert It into Public Opinion, pp. 40–52. Cambridge University Press, Cambridge (1992)

Changing Habits Using Contextualized Decision Making

Rijk Mercuur, Frank Dignum, and Yoshihisa Kashima

Abstract In this paper we aim to present a new model in which context influences a combination of both habitual and intentional environmental behavior using the concept of social practices. We will illustrate the model using meat-eating behavior while dining out.

Keywords Agent-based simulations • Context • Habits • Meat eating

1 Introduction

Despite both governmental and scientific interventions, greenhouse gas (GHG)-emissions continue to rise. Climate change is primarily driven by green house (GHG)-emitting human behavior and could therefore be largely mitigated by interventions in human behavior. For example, a global transition to a healthy low meat-diet would have a substantial impact on lowering GHG-emissions [9, 15]. However, human behavior, including meat eating, is the least understood aspect of the climate change system [8].

One possible major defect is the focus of interventions (and theory) on *intentional* (i.e., voluntary) behavioral change [1, 6]. The underlying idea is that to change behavior one must change ones intention. However, this does not adequately account for the extend to which behavior is influenced by *context* and, consequently, *routines*. Context is defined as the setting in which something happens, including location, people, and past behavior [2, 18]. Repeated behavior in the same context can lead to habits (i.e., routines) [18]. Habits are often contrasted with intentions, capturing the fast-thinking, and automatic side of behavior. Interventions predominantly focus on changing intentions, for example, by providing more information, not acknowledging that one behaves, at least partly, because one behaved similarly

R. Mercuur (✉) • F. Dignum
Utrecht University, Utrecht, The Netherlands
e-mail: rijkmercuur@gmail.com; f.p.m.dignum@uu.nl

Y. Kashima
University of Melbourne, Parkville, VIC, Australia
e-mail: ykashima@unimelb.edu.au

© Springer International Publishing AG 2017
W. Jager et al. (eds.), *Advances in Social Simulation 2015*, Advances in Intelligent Systems and Computing 528, DOI 10.1007/978-3-319-47253-9_23

before in the same context [1]. A possible solution might lie in using a notion of sociology called *social practice*. Sociologist from the social practice tradition focuses more on the practice than the individual highlighting its relation with context [14].

In this paper we aim to present a new model in which context influences a combination of both habitual and intentional environmental behavior using the concept of social practices. We will illustrate the model using meat-eating behavior while dining out.

2 Background

One widely studied primary predictor of environmental behavior is the notion of values. Values represent what a person finds important in life. We see them both as reasons that trigger an action as well as post-action criteria for evaluation [16]. In line with Schwartz we model a person's characteristics in terms of four different value types: openness, conservation, self-transcendence, and self-enhancement [12].

One of the most influential theories that tries to capture the complex interaction of influences (e.g., values) on intentions is Azjen and Fishbein's reasoned action approach [5]. However, we believe their negligence of habits accounts for the reason that interventions based on this approach (e.g., [13]) have had little or no effect on largely habitual behavior such as meat eating.

Habits are learned dispositions to repeat past responses. Their automaticity emerges from patterns of covariance between context and behavior. They are often the residue of goal pursuit (i.e., intentions), but once formed, perception of contexts triggers the associated response without a mediating goal [18]. Note that everyday vernacular implies a dichotomous view of habits and intentions, but behavior is more realistically conceived as on a continuum between habitual and intentional each requiring different interventions [10].

Sociologist from the social practice tradition focuses less on the individual and more on the practice itself. They study how practices emerge, persist, or disappear in society by securing or losing "carriers." We adopt this idea of a practice as a distinct (epistemological) entity that is "a temporally and spatially dispersed nexus of doing and saying things" [11, p. 89]. In our model, each individual and each enactment thus relates to a shared notion of the practice of "dining out."

We will base our model on a notion of Shove et al. [14] called *social practice*, originally consisting of three *elements*: materials (covering physical aspects), competences (skills and knowledge needed for the practice), and meanings (referring to the mental and social associations of practices).

The social practice of eating, for example, can be divided into its materials—tableware, cutlery, and food—its required competences—etiquette and usage of cutlery—and the meaning of pleasure, health or the achievement of a necessary chore [7]. On enactment one combines these distinct elements into the single practice of eating. The practice of eating, however, includes more than simply the intake of food. It is a social activity consisting of selecting food, preparing dishes, making social arrangement for meals, and judgments about taste [17]. In our model we compress this complex practice in the tangible form of *dining out*.

3   Model

3.1   Social Practices and Agents

Based on the work of Dignum et al. [2–4] we present a model that uses the macroscopic notion of social practices in microscopic deliberation. A social practice is constituted by the following aspects:

Physical context   describes the physical environment that can be sensed (e.g., meat venue, vegetarian venue, or mixed venue) including affordances that describe the natural physical conditions to enact the practice (e.g., meat venue affords eating meat/vegetarian venue affords eating vegetarian) and Triggers that describe physical context that (often) co-varied with the practice (e.g., 5× eating meat in venue 1, 10× eating vegetarian in venue 2, etc.)

Social context   describes the other agents that can be sensed (fellow customers in venue, e.g., agent 1, agent 2, etc.) including social triggers that describe the social context that (often) co-varied with the practice (e.g., 5× eating vegetarian with agent 1, 10× eating vegetarian with agent 2, etc.)

Values   refers to the meaning of the practice, i.e., the values it furthers

Embodiment   refers to the set of possible actions, plans, roles, and norms that the agent will use to guide its behavior within this practice (e.g., eating meat, eating vegetarian)

Evaluation   captures agents own judgment about past enactments of practice.

The agents all have their own image of the social practice of dining out. Some of the aspects will be equal for all agents (e.g., affordances, values), but each agent also records its own experience with that practice (e.g., by updating triggers and evaluation). In addition the agents heterogeneity is ensured in their difference in values. Each agent attributes a different *level of importance* to openness, conservation, self-transcendence, and self-enhancement.

The social practice as described above is used in the deliberation of the agents as depicted in Fig. 1.

In the first stage of the deliberation an agent filters the impossible behavior out by comparing its current environment (i.e., context) with the social practice. The context can afford the behavior of meat eating, vegetarian eating, or both.

In the second stage of the deliberation an agent can decide on an embodiment based on habitual triggers and past evaluation. The agent filters the least salient embodiment out by comparing its current environment (i.e., context) with the social practice. The trigger of a practice represents how frequent some embodiment co-varied with a context. The evaluation variable represents how well past enactments

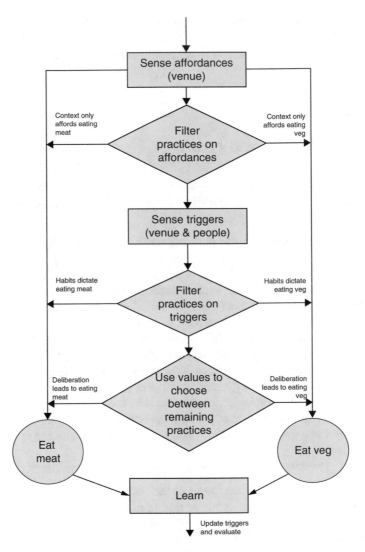

Fig. 1 Schematic overview of the deliberation cycle of an agent

went in this context. The salience of an embodiment depends thus on past success and frequency in this context. How salient certain behavior has to be to pass this filter depends on how high an agent values openness and conservation.

In the third stage of the deliberation an agent will use his values to make an intentional choice out of the remaining possible embodiments. The values of self-transcendence and self-enhancement have a threshold (corresponding to the importance they attribute to the value) and a current satisfaction level. The satisfaction level slowly decreases over time, but an action can increase the satisfaction level. The agent will choose the embodiment that furthers the value with the lowest satisfaction compared to its threshold.

After the enactment an agent will learn. Firstly, by updating its history of enactment by keeping track of the past performance context in the trigger variable of the social practice. Secondly, it will evaluate if it wants to do the behavior again. The agent evaluates by (1) comparing the level of importance of its values to the value of the enacted behavior and (2) by comparing its own behavior to the behavior of the other agents. The higher the agent values self-enhancement the more important it finds the first, the higher it values conservation and self-transcendence the more important it finds the latter.

4   Hypotheses, Experiments, and Further Work

The presented model will allow us to test hypotheses revolving around the continuum between habitual and intentional behavior. For example, one might want to motivate an agent to change its physical context (e.g., to a new restaurant) in order to move an agent from a "bad" habit to a "good" intentional action. In this model this intervention will not necessarily work. The agent's social context might activate the old habit or reinforce the norm. Furthermore, the target agent forms the social context of other agents. The individual change of an agent might have a domino effect on the other agents, possibly resulting in a global change. This model gives more insight in what combination of intervention might bring about such a global change. One idea is to motivate the agents not only to change their physical context, but also to invite like-minded agents such that subcultures that divert from the norm can grow and develop. This paper represents research in progress, but we claim that these kinds of combined agent deliberation models can shed new insights in when and how the behavior of people can change in light of habits and social contexts.

References

1. Abrahamse, W., Steg, L., Vlek, C., Rothengatter, T.: A review of intervention studies aimed at household energy conservation. J. Environ. Psychol. **25**, 273–291 (2005)
2. Dignum, V., Dignum, F.: Contextualized planning using social practices. In: International Workshop on Coordination, Organizations, Institutions, and Norms in Agent Systems, Springer International Publishing, 2014
3. Dignum, F., Prada, R., Hofstede, G.J.: From autistic to social agents. In: Autonomous Agents and Multi-Agent Systems (AAMAS), pp. 1161–1164 (2014)
4. Dignum, V., Jonker, C., Prada, R., Dignum, F.: Situational deliberation getting to social intelligence. In: Workshop-Computational Social Science and Social Computer Science: Two Sides of the Same Coin, Guildford, 23–24 June 2014 (2014)
5. Fishbein, M., Azjen, I.: Predicting and Changing Behavior The Reasoned Action Approach. Taylor & Francis, New York (2011)
6. Gifford, R.: Environmental psychology matters. Ann. Rev. Psychol. **65**, 541–79 (2014)
7. Halkier, B.: A practice theoretical perspective on everyday dealings with environmental challenges of food consumption. Anthropol. Food (S5) (2009)

8. Intergov. Panel Climate Change. Climate Change 2014 Synthesis Report Summary Chapter for Policymakers (2014)
9. Jones, C.M., Kammen, D.M.: Quantifying carbon footprint reduction opportunities for U.S. households and communities. Environ. Sci. Technol. **45**(9), 4088–4095 (2011)
10. Moors, A., De Houwer, J.: Automaticity: a theoretical and conceptual analysis. Psychol. Bull. **132**(2), 297–326 (2006)
11. Schatzki, T.R.: Social Practices. A Wittgensteinian Approach to Human Activity and the Social. Cambridge University Press, Cambridge (1996)
12. Schwartz, S.H.: An overview of the Schwartz theory of basic values an overview of the Schwartz theory of basic values. In: Online Readings in Psychology and Culture, vol. 2, pp. 1–20 (2012)
13. Seigerman, M.: Manipulating Meat-Eating Justifications: A Novel Approach to Influencing Meat Consumption (2014)
14. Shove, E., Pantzar, M., Watson, M.: The Dynamics of Social Practice: Everyday Life and How it Changes. SAGE Publications, London (2012)
15. Stehfest, E., Bouwman, L., Van Vuuren, D.P., Den Elzen, M.G.J., Eickhout, B., Kabat, P.: Climate benefits of changing diet. Clim. Change **95**(1–2), 83–102 (2009)
16. van der Weide, T.L.: Arguing to motivate decisions. SIKS Dissertation Series (2011).
17. Warde, A.: What sort of practice is eating? In: Shove, E., Spurling, N. (eds.) Sustainable Practices: Social Theory and Climate Change, p. 221. Routledge, London (2013)
18. Wood, W., Neal, D.T.: A new look at habits and the habit-goal interface. Psychol. Rev. **114**(4), 843–863 (2007)

SocialSIM: Real-Life Social Simulation as a Field for Students' Research Projects and Context of Learning

Larissa Mogk

Abstract SocialSIM is an educational project that won the Instructional Development Award of Freiburg University in 2014. The project is experimenting with real-life simulation as didactic tool in social sciences. A team of sociologists, cultural anthropologists, economists and political scientists designed a one-day social simulation on a macroscale. For the participants of the one-day real-life simulation, SocialSIM was base to experience and shape the dynamic interplay of complex social, political and economic processes. For students it was furthermore a platform to develop skills in qualitative and quantitative research methods and learn about the different perspectives of the disciplines taking part in the project.

Keywords Social simulation • Research • Education

SocialSIM is an educational project that won the Instructional Development Award of Freiburg University 2014. The project is experimenting with real-life simulation as didactic tool in social sciences. A team of Sociologists, Cultural Anthropologists, Economists and Political Scientists designed a 1-day social simulation on a macro scale. The simulation treats societal processes covering the management of common pool resources and environmental challenges as focal points of interest. The project used the 1-day simulation game with over 100 participants as platform for experiencing social, political and economic dynamics and as research field for students. Creating an atmosphere where real-life processes could be openly reproduced and redefined in a simplified setting without the necessity for players to follow up a certain strategy to win the game. Instead it was the aim of the simulation design to give a frame in which players feel free to practise their ideal of social interaction, as well as political and economic organisation. This simulated environment served as field of research for students researching on social practise.

L. Mogk (✉)
Department of Social Anthropology, Freiburg University, Freiburg im Breisgau, Germany
e-mail: Lariss.Mogk@gmail.com

© Springer International Publishing AG 2017
W. Jager et al. (eds.), *Advances in Social Simulation 2015*, Advances in Intelligent Systems and Computing 528, DOI 10.1007/978-3-319-47253-9_24

In the following sections, the simulation game, being the centre of action, is described, before I explain the structure of the project and introduce the participants and their learning experiences.

In the first phase of the simulation game "Build your own society!" players (recruited mostly in the academic surroundings) were separated in three communities and let to the playground where they had to agree on a political system, on the property rights of their resources and a set of values for their own community. These decisions were obligatory for each community to continue the game but could be revised at any time. Besides players were animated to symbolically collect and produce resources that could be inserted to develop each community subsequently. The aim of the first phase (being introduced by a short introductory film in the beginning of the social simulation) was to build a bridge to start the interaction with the other communities.

All actions in the simulation game were design to have a certain structural function in the game mechanics and furthermore represent the emotional involvement linked with that action. For instance mining at SocialSIM required crawling through dark and narrow tunnels extracting pieces of black coal. Coal was necessary to produce goods at the factory but (as players found out during the game) caused serious emissions that affected the environment and players "everyday life". To overcome these obstacles became the challenge in the second phase of the simulation game.

Besides the political organisation and the productive tasks players were able to solve quests to built up infrastructure for their community and investigate on technological solutions for the upcoming environmental changes in the second phase of the simulation game. Furthermore they had the option to instal universities and leisure areas, a market place and internet. As the participants were free to move and act between the communities, decisions addressing the environment were also posed to that level. On a common conference the political representatives decided about future actions.

In a great showdown at phase three players were asked to decide if they would like to stay in their communities to search for a better life or leave with a spaceship to the next planet. The latter was only offered to a limited number of people and thus caused serve inconsistencies. The final task for the remains was than to work on the balance of the CO_2-emissions. After the game the SocialSIM experience ended with a two hours official reflection phase and an informal part.

The participants benefitted from the 1-day social simulation experiencing the dynamics of social, political and economic processes, training certain skills and reflecting their own role taken in the simulation. Due to simplification they were able to capture the essential characteristics of societal systems and managed to compare them with real-life processes. For instance, players experienced the benefits of representative democracy, but also learned about the constraints and disadvantages that this form of political empowerment involved. Concerning the environmental challenges players observed, how developments between the communities emerged and how social bonds, politics and economics were intertwined. To give an example, two of the communities decided not to invest in nuclear energy or fossil fuels, which were far more productive than renewable energy but caused CO_2-emissions.

Being scarce in energy resources, they additionally bought energy resources from the third community, which had a great income with their oil well. According to the complexity of the simulation, these effects where reflected afterwards in the reflection phase, where different experiences were shared to capture the whole picture of the occurrences. Finally players discussed analogies with real-life processes and the different levels of abstraction.

In terms of skills, participants were practising leadership, solving conflicts and developed debating competencies in between their own community and in correspondence with the others. They arranged and negotiated political rules with people they did not know before and they were organising the complex economic processes that were linked to political questions on the management of resources in general and the fish pond as one example for a common-pool resource. Furthermore players selected their specific role in the simulation and were animated to reflect the certain positions taken and the self-identifying attributes linked. In comparison with their everyday routines they were able to see personal potentials.

Besides the players of the simulation game and their learning experiences, there was a group of students from sociology, cultural anthropology and political sciences accompanying the social event. They used the simulated reality as research field. As assistants of the organisational team they were provided with detailed insights and had access to the simulation game. Students taking part were at the end of their Bachelor degree or in their Master studies having some or no experiences and training with methodologies.

During the whole simulation day students collected data for their research using interviews, participant observation and certain types of questionnaires for this approach. The simulation was an ideal surrounding to experience research methods being accompanied by teaching staff and fellow students. All students had worked on their own research design beforehand, guided by 2 weekly seminars, preparing the social simulation from an academic viewpoint. One was organised within each discipline, the other was held in interdisciplinary sessions. Both seminars were taking part for one semester with the simulation day at the end and a final feedback session in the week after the event. The seminar aimed at preparing students' research projects and providing a preliminary with regards to content and sensitivity to simulations. Students were trained to analyse human behaviour, social phenomena and political structures within the simulation and link their results to theoretical approaches. Furthermore the aim was to recognise the own disciplinary perspective and differentiate it from the others.

Some Master students even took part in the simulation design team combining their own research question with the simulation design. Two students elected SocialSIM for their one-year research project in sociology, which allowed them to get into the designing processes more deeply. One of the research projects contained simulating the management of common-pool resources simulated fishing with a fish pond. The need to be cautious about the regeneration rate and the amount of fish remaining in the pool and the rules set and controlled were observed by the student. During tests of the simulation design that took place beforehand he had the chance to observe different models and work out principles for his own research.

To guarantee that the prepared research topics could be observed during the social simulation, a team of sociologists, cultural anthropologists, political scientists, economists and a cultural studies academic created the simulation design. The designing process took 9 months and included the game mechanics, the productive tasks and quests, the economic core and the equipment. The designing team closely worked together with the lecturers to agree on certain topics for the academic approach.

All the outcomes are presented in a bilingual handbook titled "Lernen und Forschen mit sozialen Simulationen. Das real-life Simulationsspiel SocialSIM" which was published by Waxmann in 2015. For further information visit our SocialSIM webpage (www.socialsim2014.wordpress.com).

Simulating Thomas Kuhn's Scientific Revolutions: The Example of the Paradigm Change from Systems Dynamics to Agent Based Modelling

Georg P. Mueller

Abstract Based on evolutionary game theory, this paper presents a model that allows to reproduce different patterns of change of the main paradigm of a scientific community. One of these patterns is the classical scientific revolution of Thomas Kuhn (The Structure of Scientific Revolutions. University of Chicago Press, Chicago 1962), which completely replaces an old paradigm by a new one. Depending on factors like the acceptance rate of extra-paradigmatic works by the reviewers of scientific journals, there are however also other forms of change, which may e.g. lead to the coexistence of an old and a new paradigm. After analysing the different types of paradigm-changes and the conditions of their occurrence by means of EXCEL based simulation runs, the article explores the applicability of the model to a particular case: the spread of agent based modelling at the expense of the older systems dynamics approach. For the years between 1993 and 2012 the model presented in this article reproduces the observed bibliometric data remarkably well: it thus seems to be empirically confirmed.

Keywords Kuhn's scientific revolutions • Multi-paradigmatic science • Evolutionary game theory • Agent based modelling • Systems dynamics

1 Background and Overview

This chapter refers to the famous book "The Structure of Scientific Revolutions", which Thomas Kuhn published the first time in 1962 (see [1]). The book describes the life cycle of so-called paradigms, which starts with the introduction of new

This is a substantially enlarged and improved version of an article in German language, published by the same author: G. Mueller, Die Krise der wissenschaftlichen Routine. In: Verhandlungen des 37. Kongresses der Deutschen Gesellschaft für Soziologie. http://www.publikationen.soziologie.de. Bochum, 2015.

G.P. Mueller (✉)
University of Fribourg, Pérolles 90, Fribourg CH-1700, Switzerland
e-mail: Georg.Mueller_Unifr@bluewin.ch

scientific theories, an agenda of problems to be solved, and methods that are rapidly accepted by the scientific community at the expense of a previously used paradigm. This revolutionary stage is followed by a period of "normal science", where the problems — i.e. the "puzzles" in Kuhn's terminology — of the new paradigm are treated and successfully solved by means of its specific methods. At the end, the limitations of the paradigm become more and more visible since many of its remaining "puzzles" turn out to be unsolvable. This is the time, when the scientific community is ready for another scientific revolution by abandoning the existing paradigm in favour of a new one. Thus, science according to Kuhn is a sequence of paradigms, which in the stage of normal science monopolistically dominate the activities of a scientific community.

In spite of its excellent reputation, Kuhn's book has two major shortcomings: First, it is mainly based on historical examples and thus neglects the institutional framework of contemporary science like peer-reviewing or the publish-or-perish rule for academic careers. Second, Kuhn's book mainly deals with sequences of mutually exclusive paradigms and thus does not really come to grips with multi-paradigmatic situations, which are so typical for the humanities and social sciences (see [2]). In order to tackle these difficulties, we present in the following sections a simulation model, which is based on game theoretical premises. As proposed by [3], it takes the institutional settings of modern science better into account and offers the possibility to reproduce the coexistence of paradigms.

Obviously there are other simulation models of scientific revolutions (see [4]), the most prominent ones being developed by Sterman [5] and Sterman/Wittenberg [6], who directly refer to Kuhn's work. Whereas these two authors used a systems dynamics approach (see [7]: Chap. 3), the present chapter is based on evolutionary game theory (see e.g. [8]), which we consider as much more appropriate to the study of competition between paradigms. Moreover, the cited works [5] and [6] of Sterman and Wittenberg are purely theoretical, whereas this paper attempts to corroborate the theoretical simulations with empirical data. The respective analyses in Sect. 4 demonstrate that our model is able to grasp not only the complete replacement of successive paradigms, as described by Kuhn [1], but also the more complex reality of multi-paradigmatic scientific communities. This again is a major advantage over the older model of Sterman/Wittenberg, which seems to explain only the total replacement of an exhausted paradigm by a new one (see [6]: 329, Fig. 7a).

2 A Game Theoretical Model of the Competition Between Paradigms

In its simplest form, evolutionary game theory (see [8], [9]: Chap. 8, [10]) departs from the idea of two randomly interacting species and an associated matrix of $2 \times 2 = 4$ pairs of possible payoffs, which determine the so-called fitness of the two species as well as their reproduction and death rates: the higher the mentioned fitness of the first species as compared to the second, the higher its population growth at the expense of the other.

These basic ideas from theoretical biology have successfully been used for the analysis of dynamic social processes (see e.g. [11–14]). Thus we are going to tackle the modelling of Kuhn's scientific revolutions on the basis of evolutionary game theory. Obviously, the two interacting species are in this case the supporters of the old and the new paradigm, which we describe by:

$$S_n = \text{Share of the supporters of the \underline{n}ew paradigm,} \tag{1a}$$

$$S_o = 1 - S_n = \text{Share of the supporters of the \underline{o}ld paradigm.} \tag{1b}$$

The arenas where these two "species" encounter are editorial boards of scientific journals, search committees for filling academic posts, or institutions for funding research projects. In each of these arenas academics appear in the role as suppliers and requesters of publication space, posts at universities, or research money. Due to the exclusiveness of their paradigms, interactions in the mentioned arenas are rather hostile for encounters of different paradigms and relatively friendly between representatives of the same paradigm. This has consequences for the academic careers of the requesters, which we are going to analyse in the following paragraph for the case of the submission of articles to scientific journals.

If we assume that the composition of reviewers of journals by paradigm corresponds to the paradigm-orientation of the general population of scientists, it is possible to calculate the total *acceptance rates* A_o *of the old paradigm* and a respective value A_n for the *new paradigm*. Both are the sums of the *acceptance rates* A_i *of intra-paradigmatic* and A_e *of extra-paradigmatic works*, weighted by the population shares S_n and S_o. For reviewers supporting the old paradigm, intra-paradigmatic works are authored by members of the old paradigm and extra-paradigmatic works by supporters of the new one. For reviewers representing the new paradigm, the definitions of intra- and extra-paradigmatic works are just the reverse. Thus, according to Table 1:

$$A_n = S_o * A_e + S_n * A_i \tag{2a}$$

$$A_o = S_o * A_i + S_n * A_e \tag{2b}$$

Table 1 The acceptance rates of the old and the new paradigm

Reviewer's paradigm:	Share	Author's paradigm:	
		Old paradigm	New paradigm
Old paradigm	S_o	A_i	A_e
New paradigm	S_n	A_e	A_i
Total acceptance rates of old/new paradigm		$A_o =$ $S_o * A_i + S_n * A_e$	$A_n =$ $S_o * A_e + S_n * A_i$

A_i = acceptance rate of intra-paradigmatic articles; A_e = acceptance rate of extra-paradigmatic articles; A_o = acceptance rate of articles based on old paradigm; A_n = acceptance rate of articles based on new paradigm. Source: [22]: Table 1.

Implicitly we are postulating in the formulas (2a) and (2b) that the result of the reviewing process is influenced by the *randomness* of the assignment of reviewers to manuscripts, as demonstrated by [15]. If we assume in addition that there is a publication bias against new ideas (see [16]: p. 71 and [17]: Chap. 3) such that

$$A_e < A_i \tag{3}$$

the Eqs. (2a) and (2b) imply that new paradigms have at the *beginning* of their existence a lower acceptance rate A_n than the dominating old paradigm, since in this situation $S_o \approx 1$ and $S_n \approx 0$.

At the beginning, however, new paradigms have the advantage of offering to ambitious scientists a lot of easy-to-solve new puzzles such that the *ease of discovery E_n* is at this stage for the *new paradigm* much higher than the *ease of discovery E_o* of the old paradigm. As mentioned by Kuhn (see [1]: Chap. 7), the latter is in its final stage often confronted with insurmountable difficulties in solving its own scientific puzzles. Thus the ease of discovery obviously has consequences for the scientific productivity F_o of the supporters of the old and F_n of the new paradigm, which modify the effects of the initial non-acceptance of the new paradigm in the following way:

$$F_n = E_n * A_n = E_n * (S_o * A_e + S_n * A_i) \tag{4a}$$

$$F_o = E_o * A_o = E_o * (S_o * A_i + S_n * A_e) \tag{4b}$$

Since the above-mentioned productivity in terms of accepted and published papers determines the careers of the respective scientists, we are using in (4a) and (4b) as left-hand-terms the letters F_n and F_o, which stand for the *fitness* of the two groups of scientists. Hence we hypothesise in accordance with the general assumptions of evolutionary game theory (see e.g. [10]: Chap. 3) that the *growth of the supporters of the new paradigm* is

$$\Delta S_n = \delta * (F_n - F_o), \text{ if } 0 < S_n < 1, \text{ else } \Delta S_n = 0, \tag{5a}$$

where δ is a constant laps of time. Similarly we assume that the *growth of the supporters of the old paradigm* equals

$$\Delta S_o = \delta * (F_o - F_n), \text{ if } 0 < S_o < 1, \text{ else } \Delta S_o = 0. \tag{5b}$$

Both equations are conceptualised in such a way that the shares S_o and S_n do not leave their definition interval [0,1] and always sum up to 1.[1] The changes which they describe are partly due to the transitions of established scholars between paradigms

---

[1]From (5a) and (5b) follows $\Delta S_n = -\Delta S_o$ such that the sum $S_n + S_o$ is *time-invariant* and always yields 1 (see formula (1b)).

and partly to the rational choice of young scientists, who start their careers with the paradigm that promises the more successful professional future.

The advantage of the new paradigm in terms of a higher ease of discovery E_n tends to decrease by the number of newly solved scientific puzzles, which is proportionate to $F_n * S_n$, i.e. the product of the relative size S_n of the population of scientists and its productivity F_n. Similar things hold true for the dynamics of the ease of discovery with the *old* paradigm. Consequently we postulate:

$$\Delta E_n = -\delta * F_n * S_n, \text{ with initial value } E_n = 1 \text{ and } \delta = \text{constant laps of time.}$$
$$(6a)$$

$$\Delta E_o = -\delta * F_o * S_o, \text{ with initial value } E_o \leq 1 \text{ and } \delta = \text{constant laps of time.}$$
$$(6b)$$

Hence, after some time, both paradigms are depleted and may be replaced by a third paradigm, which is however not considered in the simulations that follow.

3 Model Simulation

3.1 Introductory Remarks

This section pursues two related goals:

(i) We want to look for an inventory of the different *types of population dynamics* that can be reproduced by the model. Of special interest are on the one hand the empirically observed coexistence of two paradigms and on the other the complete replacement of the old paradigm by a new one, as described by Kuhn [1].

(ii) We attempt to analyse the *determinants* of the mentioned patterns of population-dynamics. Given the limited number of exogenous model parameters, we focus on the acceptance rates of extra-paradigmatic works A_e and the initial ease of discovery E_o by the old paradigm. For reasons of standardisation we set for the start of the simulations the ease of discovery of the new paradigm $E_n = 1$ and the acceptance rate of intra-paradigmatic works $A_i = 1$. This way E_o and A_e become *relative* values, i.e. fractions of the former ones.

In view of the complexity of our model we tackled the goals (i) and (ii) by *simulation* experiments: they allowed to study the effects of parameter changes on the population dynamics of the supporters of the old and the new paradigm in a rather easy way. This method obviously required the translation of the model into a computer program. We used for this purpose an EXCEL spread-sheet with columns being defined as time-dependent variables, like e.g. S_o and S_n and rows representing subsequent time-points with a laps of time $\delta = 0.1$. The rows are linked in such a way that changes of variable-values on one line are propagated to the next, as described by the difference equations (5a, 5b) and (6a, 6b). This process always started under the assumption that between t = 0 and t = 1 a new paradigm showed up and lowered the share of the supporters of the old paradigm from an initial de facto monopole $S_o = 1.0$ to $S_o = 0.95$. By simulation of the subsequent population dynamics it was e.g. possible to analyse, under which conditions the new paradigm is crowded out by the old or alternatively further spreads and finally becomes the mainstream of the scientific community.

3.2 Simulated Types of Population Dynamics of Scientists

As a matter of fact, a relatively small number simulation experiments with randomly selected parameter values E_o and A_e show the population dynamics that correspond to the *classical revolutions* described by Kuhn [1]. Figure 1a is an example for these rather rare situations, where the new paradigm immediately attracts a growing number of scientists until it completely replaces the old one.

Much more frequent than the classical "perfect" revolutions are in our simulation experiments the *incomplete* ones, as exemplified by Fig. 1b: the new paradigm immediately starts to grow at the expense of the old. The latter however recovers after some time and leads to a multi-paradigmatic situation, which is often observed in the social sciences. A closer look at Fig. 1b explains this fluctuation in the support for the two paradigms: the rapid start of the new paradigm leads to its early exhaustion and soon lowers its ease of discovery E_n. Between time t=30 and t=100, the E_n of the new paradigm is already smaller than the E_o of the old, which this way gets a chance for a revival (see Fig. 1b). This dynamic of E_o and E_n is in sharp contrast to the classical scientific revolution, depicted in Fig. 1a: here the ease of discovery of the new paradigm is for a much longer time, i.e. until t = 50, above the old one and thus leads to its complete victory.

In about half of all simulation experiments with randomly selected parameter values E_o and A_e, the change of paradigm is *delayed:* the new paradigm is available, but for some years the old is still vigorous enough to exert monopolistic control of the scientific community. Only after a latent period of further depletion, the old paradigm breaks down and triggers either a *complete* (Fig. 2a) or an *incomplete revolution* (Fig. 2b).

Last but not least there is the rather rare possibility that the outbreak of a scientific revolution is not only temporarily but even infinitely delayed and consequently ends in a *failed revolution.* Hence, from the perspective of evolutionary game theory there are particular conditions (see Sect. 3.3), under which paradigms can be *evolutionarily stable.*

In sum, this model is able to reproduce not only the scientific revolutions of Kuhn [1] but many other phenomena of scientific change like delayed revolutions, where new ideas come too early to be accepted by the scientific community, or incomplete revolutions that lead to multi-paradigmatic science. Thus in view of the last-mentioned category of changes, the model fulfils one of the major goals of this chapter (see Sect. 1).

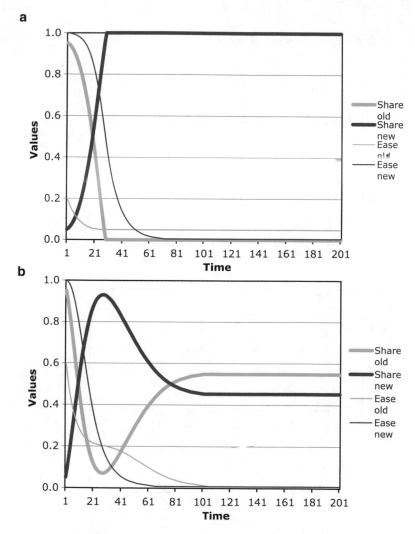

Fig. 1 (**a**) Kuhn's classical revolution: immediate and complete change from the old to a new paradigm (initial parameter values: $E_o = 0.2$, $A_e = 0.2$). (**b**) An incomplete revolution: coexistence of the old and a new paradigm (initial parameter values: $E_o = 0.6$, $A_e = 1.0$)

3.3 The Determinants of the Stability and Long-Term Dominance of Paradigms

The previously encountered types of paradigm-changes differentiate mainly with regard to the following two dimensions:

(i) The *stability of the old paradigm* in the case of the arrival of a new one: it may be *immediately unstable, temporarily stable,* or *permanently stable.* In the first

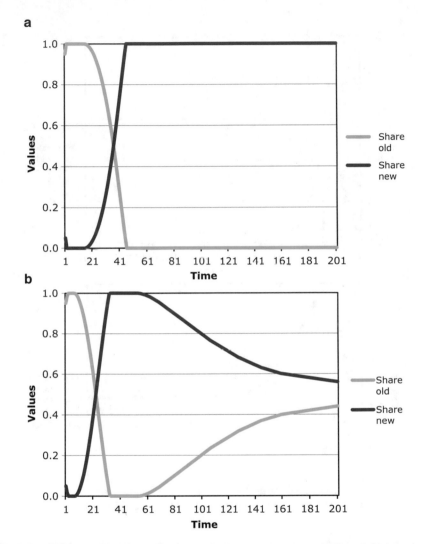

Fig. 2 (**a**) A delayed classical revolution: complete change to a new paradigm (initial parameter values: $E_o = 0.8$, $A_e = 0.2$). (**b**) A delayed incomplete revolution: transition to the coexistence of the old and a new paradigm (initial parameter values: $E_o = 0.9$, $A_e = 0.5$; source: [22]: Fig. 1)

case we expect a classical or an incomplete revolution, in the second delayed changes, and in the third a failed revolution.

(ii) The *paradigm*, which finally *dominates* after the changes induced by the arrival of a new paradigm have fully developed. In the long run the dominating model of science may be the *new paradigm*, the *old paradigm*, or *both paradigms*. The first case corresponds to the effect of a classical revolution, the second of a failed revolution, and the third of an incomplete revolution.

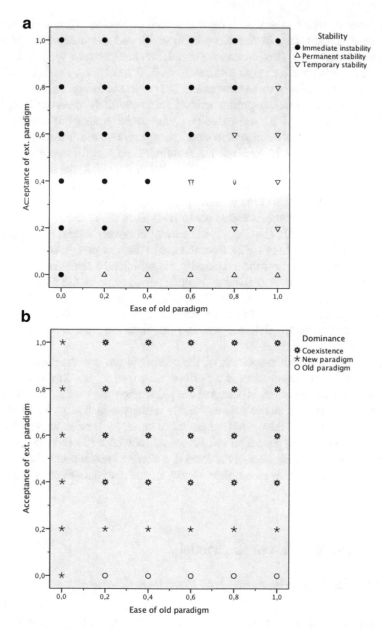

Fig. 3 (**a**) The stability of the old paradigm, by values of A_e and E_o (source: [22]: Fig. 2a). (**b**) Long-term dominance of different paradigms, by values of A_e und E_o (time horizon: 400 units of time; source: [22]: Fig. 2b)

The outcome of the model with regard to both dimensions (i) and (ii) depends on the ease of discovery of the old paradigm E_o and the acceptance of extra-paradigmatic works A_e. Thus we ran the simulation model over a span of 400 units of time with $\delta = 0.1$ and for varying values $E_o = 0., 0.2, 0.4, \ldots, 1$ and $A_e = 0., 0.2, 0.4, \ldots, 1.$[2] The results of these simulations are presented in Fig. 3a, b.

As Fig. 3a demonstrates, scientific revolutions *immediately* break out if $A_e \geq E_o$. Thus, if the reviewers of the old paradigm, who initially have full control of the editorial boards, are too indulgent to new extra-paradigmatic ideas, a change of paradigms is very likely. If $A_e > 0.2$ the revolution remains *incomplete* and leads to a multi-paradigmatic compromise (see Fig. 3b). If $A_e \leq 0.2$, the revolution ends with the dominance of the new paradigm, as described by Kuhn [1] (see Fig. 3b). It is important to note that this kind of a *complete* classical revolution only occurs for a small minority of randomly selected parameter values of A_e and E_o.

Alternatively, if $A_e < E_o$, but $A_e > 0$, the ease of discovery with the old paradigm is too high for an immediate swing from the old to the new paradigm (see Fig. 3a). Nevertheless, after some time the old paradigm is sufficiently depleted and a *delayed revolution* breaks out. For $A_e > 0.2$ it ends again with a compromise between the old and the new paradigm (see Fig. 3b). If $A_e \leq 0.2$, the growth of the supporters of the new paradigm is slower but finally leads to a classical revolution, where the dominance of the scientific field completely shifts from the old to the new paradigm (see Fig. 3b).

Finally, if $A_e = 0$, the supporters of the old paradigm use their initial control of the scientific production to exert a perfect "censorship": no extra-paradigmatic work from the new paradigm is accepted for publication. As Fig. 3a demonstrates, this kind of censorship is an evolutionary stable strategy, which turns the invasion of the field by supporters of the new paradigm into a failure, at least as long as there are any puzzles from the old paradigm left that can be solved by its representatives. In the very moment when E_o reaches the level 0, a classical revolution is immediately triggered, which ends with a complete victory of the new paradigm, as Fig. 3a, b show for $E_o = 0$.

4   An Empirical Test of the Model

4.1   The Explanandum and Its Operationalization

This paper aims at an explanation of the rise and fall of two paradigms of social simulation: systems dynamics simulation (see [7]: Chap. 3) and agent based modelling (ABM) (see [18], [7]: Chap. 8). The former was introduced by Forrester [19] and dominated the simulation literature of the 1970s and 1980s. The latter has

---

[2]By definition $A_e = 0$ and $E_o = 0$ are the *lowest* possible values of these two parameters. Similarly, since $A_e \leq A_i = 1$ and $E_o \leq E_n = 1$, A_e and E_o *cannot exceed* the value 1.

its roots in the work of Schelling [20] and spread after 1990 at the expense of the older systems dynamics approach. However, agent based modelling was never able to crowd the other competitive approaches out. According to Table A1 (see Data Appendix), this rise of ABM was rather an incomplete than a complete revolution, and thus resembles Fig. 1b more than Fig. 1a. Hence the case under consideration is an interesting test on whether our model is able to reproduce also the quantitative aspects of an incomplete scientific revolution.

Unfortunately, the shares of scientists S_o and S_n adhering to the *old* and the *new* paradigm are much more difficult to measure than the *shares of their respective publications* P_o and P_n, which can easily be extracted from bibliographies. Therefore we tested our model by explaining the publication shares P_o and P_n, which we hypothesise to be the standardised products of the shares of scientists and their fitness related productivity:

$$P_n = (F_n * S_n) \,/\, (F_n * S_n + F_o * S_o) \tag{7a}$$

$$P_o = (F_o * S_o) \,/\, (F_n * S_n + F_o * S_o) \tag{7b}$$

Obviously the shares of the old- and the new-paradigm publications P_o and P_n sum up to 1.[3]

In order to measure P_n, we used the electronic bibliography of Scholar Google [21] as a basic resource that allowed us to count for each year between 1993 and 2012 the *absolute* number of articles with the keyword "agent based" in the title. For measuring P_o we utilised the same bibliography and determined the number of articles with the title-words "system dynamics" or alternatively "systems dynamics". Subsequently we calculated the relative shares P_o and P_n by dividing the number of articles in the old, respectively in the new paradigm through the number of both types of articles. The intermediate and final results of this procedure are presented in the annex in Table A1 (see Data Appendix). The figures are obviously only a rough approximation to reality, with many erroneous omissions and inclusions of articles. Its also important to keep in mind that the data refer not only the social sciences but to any scientific activity covered by Scholar Google, thus e.g. including engineering.

4.2 An Empirical Tests with Preliminary Results

As shown in the previous Figs. 1a and 2b, the dynamics of the model depend very much on the values of its "free" parameters like e.g. the acceptance of extra-paradigmatic works or the ease of discovery with different paradigms. Thus, these parameters have the advantage that they can be used in order to fit the model to

---

[3] $P_n + P_o = (F_n * S_n) / (F_n * S_n + F_o * S_o) + (F_o * S_o) / (F_n * S_n + F_o * S_o)$
$\qquad = (F_n * S_n + F_o * S_o) / (F_n * S_n + F_o * S_o) = 1$

the data. Ideally they should be determined with regression-like statistical methods. However, this is for the present model rather difficult, among others because of the missing time series data for the ease of discovery. Hence the author changed the values of the parameters δ, E_o, and A_e by trial and error, until there was for the whole analysed period between 1993 and 2012 a good correspondence between the outcome of the respective simulation run and the observed shares P_n of publications, referring to the new agent based modelling paradigm. This ad hoc method has yielded the following results:

$$\delta = 0.0191 \tag{8a}$$

$$A_e = 1 \tag{8b}$$

$$S_o = 0.872 \text{ at time } t = 1 \tag{8c}$$

$$E_o = 0.892 \text{ at time } t = 1, \tag{8d}$$

where for reasons of standardisation all simulation-experiments started at time t=1 with the parameter values $A_i = 1$ and $E_n = 1$. The resulting model-fit,[4] defined as the mean difference between the observed and the simulated share of publications equals 0.0041 and thus appears to be quite ok: the simulated trajectory of the publication share P_n is on the average less than half of a per cent away from true share of these publications. This positive evaluation of the model is further corroborated by Fig. 4, which shows a good correspondence between the real and the simulated temporary evolution of P_n, especially with regard to the geometrical properties of the two curves, like e.g. the peaks or the phases of acceleration. However, it has to be kept in mind that a more profound assessment of the model is only possible on the basis of additional examples of paradigm changes, preferably with other types of revolutionary dynamics.

The parameter-estimates (8a) to (8d) are not only useful for a good model fit but also help to understand the modelled processes of science: especially striking in this respect is the estimate $A_e = 1$ (see (8b)),[5] which means that the extra-paradigmatic papers are treated by the journal reviewers in a very similar way as the intra-paradigmatic papers with the same value $A_i = 1$. This is probably due to the fact that the representatives of the new ABM-paradigm had even at the beginning of the simulated period enough opportunities to publish in journals, which were not under control of the older systems dynamics paradigm. Of similar interest as $A_e = 1$ is the strikingly high ease of discovery $E_o = 0.892$ of the old paradigm at the initial time-point t=1. This probably reflects the fact that the systems dynamics paradigm was not really in crisis, when agent based modelling entered the scientific scene. Obviously this is a different situation from the one described by Kuhn

---

[4]Model-fit = Square root of (Sum of squares between observed and simulated P_n/20) = 0.0041

[5]As explained earlier in Sect. 3.1, $A_e = 1$ is a relative and not an absolute acceptance rate.

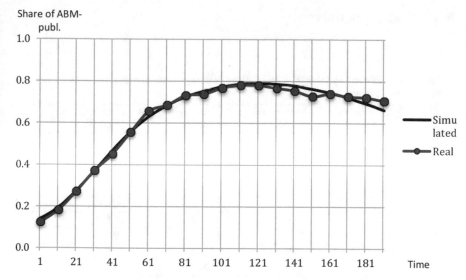

Fig. 4 The temporary evolution of the observed and the simulated share P_n of publications based on agent based modelling (ABM) (Time $= 1 \approx 1993$, Time $= 11 \approx 1994$, ..., Time $= 181 \approx 2011$, Time $= 191 \approx 2012$)

[1] at the outbreak of a scientific revolution. Consequently we cannot expect the disappearance of the old paradigm, as suggested by Kuhn, and the coexistence of two paradigms seems to be an intuitively plausible result of the model.

5 Summary and Conclusions

In this chapter we present a model that allows to simulate the classical scientific revolution of Th. Kuhn [1] as well as many other forms of paradigmatic changes like the stable coexistence of an old and a new paradigm. According to Fig. 3b, Kuhn's revolution seems to be a possible but rather *special* event that can only occur if the acceptance A_e of external paradigms is rather *low*. Given the large number of scientific journals, the difference between the intra- and extra-paradigmatic acceptance $A_i = 1$ and A_e is probably often only small. Thus, A_e too is for many cases close to 1 such that Kuhn's revolution becomes according to Fig. 3b a *rare* event. Moreover, due to the mentioned high values of A_e, the old paradigm need not really be depleted in order to enable the *immediate* start a new one: as shown in Fig. 3a, the triggering of this kind of paradigmatic change simply requires that the old paradigm has an ease of discovery $E_o < A_e$. The high acceptance rate A_e of new external paradigms makes this a *likely* event, which leads according to Fig. 3b to *multi-paradigmatic* science — in reality not only with two, but often *several* paradigms coexisting in parallel. Obviously, the model presented in this paper is not made for situations with *more than two* simultaneous paradigms and thus requires in the future an additional modification.

A.1   Data Appendix

Table A.1 Numbers and shares of publications in the agent based modelling paradigm and the systems dynamics paradigm

Year	Sys. dynamics: number	Agent based: number	Both paradigms: number	Agent based: share P_n
1993	183	26	209	0.124
1994	202	45	247	0.182
1995	198	74	272	0.272
1996	247	146	393	0.372
1997	297	244	541	0.451
1998	282	351	633	0.555
1999	285	543	828	0.656
2000	357	775	1132	0.685
2001	366	994	1360	0.731
2002	411	1160	1571	0.738
2003	450	1480	1930	0.767
2004	472	1680	2152	0.781
2005	541	1920	2461	0.780
2006	601	1980	2581	0.767
2007	644	1970	2614	0.754
2008	736	1970	2706	0.728
2009	787	2240	3027	0.740
2010	845	2250	3095	0.727
2011	856	2220	3076	0.722
2012	905	2180	3085	0.707

Source: own calculations, based on [21]

References

1. Kuhn, T.: The Structure of Scientific Revolutions. University of Chicago Press, Chicago (1962)
2. Tracy, J., et al.: Tracking trends in psychological science. In: Dalton, T., Evans, R. (eds.) The Life Cycle of Psychological Ideas. Kluwer, New York (2004). Chapter 5
3. Edmonds, B., et al.: Simulating the Social Processes of Science. J. Artif. Soc. Soc. Simulat. http://jasss.soc.surrey.ac.uk/14/4/14.html (2011)
4. Sobkowicz, P.: Simulations of opinion changes in scientific communities. Scientometrics **87**, 221–232 (2011)
5. Sterman, J.: The growth of knowledge: testing a theory of scientific revolutions with a formal model. Technol. Forecast. Soc. Change **28**, 93–122 (1985)
6. Sterman, J., Wittenberg, J.: Path dependence, competition, and succession in the dynamics of scientific revolution. Organ. Sci. **10**, 322–341 (1999)

7. Gilbert, N., Troitzsch, K.: Simulation for the Social Scientist. Open University Press, Maidenhead (2011)
8. Maynard Smith, J.: Evolution and the Theory of Games. Cambridge University Press, Cambridge (1993)
9. Webb, J.: Game Theory: Decisions, Interaction and Evolution. Springer, London (2007)
10. Weibull, J.: Evolutionary Game Theory. MIT Press, Cambridge, MA (1996)
11. Axelrod, R.: The Evolution of Cooperation. Penguin, London (1990)
12. Hanauske, M.: Evolutionary game theory and complex networks of scientific information. In: Scharnhorst, A., et al. (eds.) Models of Science Dynamics. Springer, Berlin (2012). Chapter 5
13. Mueller, G.: Universities as producers of evolutionarily stable signs of excellence for academic labor markets? Semiotica **175**, 429–450 (2009)
14. Mueller, G.: The dynamics and evolutionary stability of cultures of corruption: theoretical and empirical analyses. Adv. Complex Syst. **15**(6) (2012)
15. Bornmann, L., Daniel, H.-D.: The luck of the referee draw: the effect of exchanging reviews. Learned Publ. **22**, 117–125 (2009)
16. Daniel, H.-D.: Guardians of Science: Fairness and Reliability of Peer Review. VCH Verlagsgesellschaft, Weinheim (1993)
17. Shatz, D.: Peer Review: A Critical Inquiry. Rowman & Littlefield, Lanham (2004)
18. Gilbert, N.: Agent-Based Models. Sage, Los Angeles (2007)
19. Forrester, J.: Industrial Dynamics. MIT Press, Cambridge, MA (1961)
20. Schelling, T.: Dynamic models of segregation. J. Math. Sociol. **1**, 143–186 (1971)
21. Scholar Google: Title-word entries "Agent based", "System dynamics", "Systems dynamics". http://scholar.google.de/. Accessed 30 Oct 2014
22. Mueller, G.: Die Krise der wissenschaftlichen Routine: Computer-Simulationen zu Kuhns "Structure of Scientific Revolutions." In: Lessenich, S. (ed.) Routinen der Krise — Krise der Routinen. Verhandlungen des 37. Kongresses der Deutschen Gesellschaft für Soziologie in Trier 2014. http://www.publikationen.soziologie.de/. Bochum (2015)

Urban Dynamics Simulation Considering the Allocation of a Facility for Stopped Off

Hideyuki Nagai and Setsuya Kurahashi

Abstract In this paper, we propose an agent-based urban model in which the relationship between a central urban area and a suburban area is expressed simply. Allocation and hustle of a public facility where residents stop off in daily life are implemented in the model. We clarify that transportation selection and residence selection of residents make an effect to change the urban structure and environment. We also discuss how a compact urban structure and a reduction in carbon dioxide emissions are achieved with urban development policies and improvements on attractiveness of the facility for pedestrians and cyclists. In addition, we conduct an experiment of the exclusion of cars from the center of the city. The experimental results confirmed that the automobile control measure would be effective in decreasing the use of automobiles along with a compact urban structure.

Keywords Compact city • Urban sprawl • Household relocation • Facility location problem • Traffic policy

1 Research Background and Purpose

Urban structure sprawl coming along with urban development has been one of the large themes related to urban issues for decades [5, 10], and recently decline of urban central areas has also been considered to be an issue [15, 16]. In Japan, from the beginning of the twentieth century to the high economic growth period post-World War II, many "Newtowns" modeled after "Garden City" that Ebenezer Howard was proposed had been developed in the suburbs of major cities. Public institutions, private railways company, and real estate company had led to the developments. Many of these "Newtowns" were commuter towns rather than the

H. Nagai (✉)
Department of Risk Engineering, Graduate School of Systems and Information Engineering, University of Tsukuba, Tokyo, Japan
e-mail: s1530156@u.tsukuba.ac.jp

S. Kurahashi
Graduate School of System Management, University of Tsukuba, Tokyo, Japan
e-mail: kurahashi.setsuya.gf@u.tsukuba.ac.jp

© Springer International Publishing AG 2017
W. Jager et al. (eds.), *Advances in Social Simulation 2015*, Advances in Intelligent Systems and Computing 528, DOI 10.1007/978-3-319-47253-9_26

autonomous city with the place of residence, employment, and school attendance. Residents had been assumed to commute to major cities by rail. But thereafter, by popularization of people's lifestyle based on private cars as main transportation means, expansion of low-density urban area and hollowing of city centers have been under progress up to the present [9, 12]. There is concern about this situation because of the possible consequences such as the decline in living convenience due to weakened public services, loss of neighboring communities along with deterioration of public security. Additionally, another concern is a problem caused by the excessive dependence on automobiles. This results in problems such as increase of energy consumption and air pollution. It is also obvious as the population declines and becomes older in the future, while such problems just mentioned will become more serious amid the trend for the population to concentrate in large cities. Therefore, a countermeasure for this situation based on transformation into a system of a compact city has been explored similarly to many European cities. But as we take a position from this point of view that considers the city where we live to be a system structured by autonomous actions made by a wide variety of agents as individuals and organizations, such as families and companies [1], the difficulty in the direct control of the dynamism of the city is highlighted. Furthermore, experiments on major urban policies in the real cities are almost impossible to perform since there are major constraints about cost and time.

In this study we conducted a multi-agent simulation based on simplified urban models. This was done in order to verify whether various measures related to public space can indirectly bring about a compact city.

2 Related Studies

As for the theme of location of residential areas, there exist previous experiments based on statistical models focusing on the relationships between the characteristics of transportation means and the location of residential areas [11, 17]. In this study, we adopted multi-agent simulation. In multi-agent simulation, interactions between individual agents and the environment are reflected. For this reason, this simulation is expected to give contributions to measure the effectiveness of politics in complicated environments [8, 14].

A number of multi-agent simulation experiments that focus on urban structure sprawl exist [2, 20]. One of them was done by Taniguchi et al. They constitutively verified the fact that a soft measure that controls the use of automobiles changes in city formation through transportation selection and residence selection of individuals [18]. As a result, they suggested the possibility that cities can be guided indirectly to a desirable formation by adding changes in transportation behavior, namely commuting, which is one of the basic activities of humans. On the other hand, recently, there have been indications that town streets can serve as the most important public space that produces urban diversity and vitality [7], and "the third place" other than homes ("the first place") and workplaces ("the second place") becomes essential to revitalize local democracy and communities [13]. These points have been reevaluated. The importance of informal public spaces has also been recognized, while such attempts are applied to the actual urban and regional planning [4, 21]. With this in mind, in this study, we examined the qualitative benefits that can be obtained by the existence of informal public spaces other than residences and working regions and by being in such places.

Isono et al. presented a method to calculate the most optimum location and scale of facilities where people can easily stop off or visit momentarily. By doing case studies, they demonstrated that locating a library near a train station would be effective in order to achieve a high level of convenience [6]. A general optimum location problem is basically handled by modeling the direct access to a facility from residential areas as the starting point. Yet in this model they focused on access to the facility influenced by the main transportation behavior of people on a daily basis. In this study, we considered feedback that a highly attractive public facility could change the distribution of residential areas.

With those in mind, in this study we used an urban model based on what was used in the study done by Taniguchi et al. [18] based on an assumption that transportation selection of residents. This assumption was to do with their movement as they leave home, commute to work or school, as they might stop off in a regional facility, such as the library, as mentioned by Isono et al. [6], and they return home. With that, we performed experiments by using multi-agent simulation in order to verify a possibility as to whether changes in the allocation of such facility where residents could visit along with bustle around such places could really indirectly change the urban structure into a compact city through the intermediary of transportation selection and residence selection of residents. Additionally, we similarly verified changes in urban structures in cases where transport measures to control use of automobiles with the purpose of urban environmental improvement as one of the mobility managements. We also discussed the structure of the changes we verified.

3   Simulation Model

3.1   Urban Model

Figure 1 shows the schematic of the urban model in this study. This schematic is a simplified model of the real urban areas initial state of the central urban areas and the suburban commuter towns that are connected to central areas by railway and highway.

Considering one simulation space, two domains are set, the residential zone and the destination zone. In the residential zone, homes for each individual resident agent are allocated as the base point of each of agent. In the residential zone, destinations that correspond to each resident agent are allocated as the halfway point. In addition, one facility is allocated as a place where all residential agents

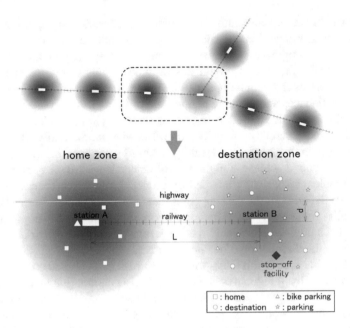

Fig. 1 Simple urban model

stop off in or around the destination zone. Recently, complex facilities, that provide public functions such as libraries and a museums featuring other supplementary public functions and commercial functions, are increasing. Many of them were built and maintained by the cooperation of both public and private sectors, and various residents can casually stop off there. The stop-off facility in this urban model is assumed such complex facilities. A train station is allocated at the center of each of the residential zone and the destination zone. These stations are connected by a railway. Station A is located in the residential zone, while station B is located in the destination zone. Station A is allocated to the west and station B is allocated to the east at the same latitude with distance L. As the default value, residents of the same number as the number of resident agents, n, is randomly allocated based on the normal distribution centering of station A. Similarly, destinations of the number of n are also allocated centering on station B. With the assumption that uniform and high-density sidewalks and main roads are allocated. Resident agents move on this continuous planar space by walking, by bicycle, or by car. Additionally, a highway is allocated to the north of the railway tracks from the west edge to the east edge of the model plane. Resident agents moving on this road by car can move faster than those in other spaces. Station A is also equipped with a bicycle parking space that can hold a sufficient number of bicycles. In the destination zone, bicycle and car parking spaces of the same number as the number of destinations is allocated in a similar distribution. The facility where resident agents stop off is also equipped with bicycle and car parking spaces that can hold a sufficient number of bicycles and cars.

3.2   Transport Behavior

Each resident agent leaves the home for the destination every day. Depending on the experiment, the resident agent goes home directly or stops off in a stop-off facility. The transportation behavior of each agent from the departure place to the final destination by using a single or multiple transport means is referred to as the linked trip. The linked trip is achieved by either of the following main transport means: by walking, by bicycle, by train, or by car.

Cost

The total movement cost of the i-th linked trip for each resident agent is calculated as follows:

$$C_i = \omega_T C_T + \omega_M C_M + \omega_F C_F - \omega_B B$$

$$\omega_T, \omega_M, \omega_F, \omega_B \geq 0$$

$C_T, C_M, C_F,$ and B indicate time cost, fee cost, fatigue cost, and bustle bonus, respectively. Similarly, $\omega_T, \omega_M, \omega_F,$ and ω_B indicate each preference bias. Total movement cost C_i is calculated every time each resident agent comes back home. The resident agent changes the value V_i of the i-th linked trip as shown below according to total movement cost C_i.

$$V_i \longleftarrow \alpha(-C_i) + (1 - \alpha)V_i$$

The transportation selection of each resident agent is decided by the ϵ-greedy method based on this value, V_i, every day. By the ϵ-greedy method agents select behavior randomly from all options at probability ϵ, while selecting the most valuable behavior at probability $1 - \epsilon$. Probability ϵ gradually declines according to the equation below ($\gamma < 1$), and this converges to 0 gradually through trials.

$$\epsilon \longleftarrow \gamma \epsilon$$

Influence of Bustle Around the Stop-Off Facility

In this study, the bustle bonus is considered when each resident agent moving by walking or by bicycle moves within a radius of R_F centered on the stop-off facility which was set as an influential area related to the stop-off facility. Here, it is assumed that this bustle bonus can be obtained when one or more resident agents that move by walking or by bicycle within the radius of r_{bust} where the relevant resident agent exists. Bustle bonus B is determined as shown below.

$$B = \min(\eta_{\text{bust}} D_{\text{bust}}, B^{\text{max}})$$

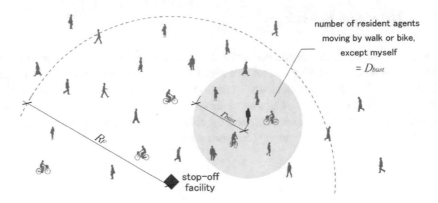

Fig. 2 Mechanism of gainning bustle bonus

Here, D_{bust} indicates the number of the resident agents moving by walking or by bicycle within the radius of r_{bust} around the relevant resident agent, while η_{bust} indicates the exchange coefficient from density to the bustle bonus (Fig. 2).

The exchange coefficient of the bustle bonus, η_{bust}, indicates the implementation level of urban development to produce further bustle by enhancing the attractiveness of the place by means of public–private measures such as developing sidewalks and cycle roads and enriching stores, according to the accumulation of pedestrians within the influential area of the stop-off facility. Hereinafter, this measure is referred to as the bustle promotion measure.

Movement

In experiments performed in this study, all resident agents leave their homes simultaneously. When all resident agents reach their destinations, all of them then leave their destinations for the stop-off facility. Afterwards, those resident agents that reached the stop-off facility leave for their homes.

3.3 Residence Selection

When a specified number of days have passed, some resident agents change their residences. As for those resident agents that are subject to changing their residents, a specified number of resident candidates are presented within a range where homes are not significantly off the resident distribution. Resident agents select the new residences where the total living cost C_i^l of each of them is at a minimum. The total living cost C_i^l is the sum of the total movement cost C_i and the land rent R_i.

$$C_i^l = C_i + R_i$$

Movement Cost

The total movement cost C_i is calculated by virtually transferring to the presented resident candidate and commuting from that location. At that time, the total movement cost while experiencing traffic jams or the bustle bonus is calculated when other resident agents move simultaneously. Based on the study made by Fujii et al. [3] their transportation means should be the same as the one taken before the transfer. In other words, the residence selection of each resident agent depends on the regular transportation selection of the relevant resident agent.

Land Rent

The land rent R_i of each resident candidate should follow the equation below by referring to the equation used for Togawa's study [19] that expresses the relationship between the consumption rate of land use and the land rent.

$$R_i = \eta_R^h I_i^h \left(\frac{A^h}{A} \right) + \eta_R^d I_i^d \left(\frac{A^d}{A} \right)$$

Here, η_R^h and η_R^d indicate the exchange coefficient, I_i^h and I_i^d indicate the number of residents and destinations within the relevant area, A indicates the area of the relevant range, and A^h and A^d indicate the consumption area by the unit of residents or destinations. In other words, the land rent increases occur because of the accumulation of the population of residents, labor, and employment.

4 Experiment 1—Effect of the Allocation of the Stop-Off Facility

4.1 Experimental Outline

First, we conducted an experiment in order to examine the effectiveness of the allocation of the stop-off facility and the bustle promotion measure within the influential area of the stop-off facility. Values were based on fragmentary information from a wide variety of reference documents including experimental models used by Taniguchi et al. [18] and considered to be valid. These values were configured as model parameters. In each experiment, resident agents learned of the transportation selection for 30 days while their transportation selection was fixed so that the movement cost could be minimized. Afterwards, a loop process to select their resident areas from 10 resident candidates for 20 times, after which they determined

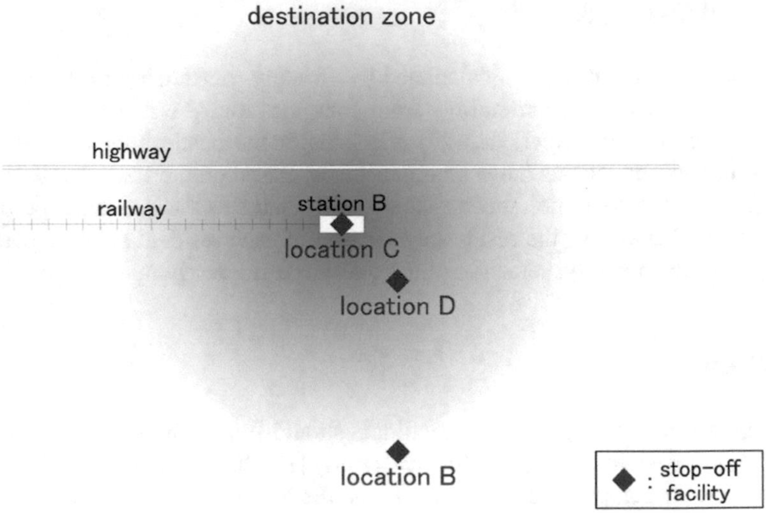

destination zone

highway

railway   station B

location C

location D

location B   ◆ : stop-off
                                              facility

Fig. 3 Stop-off facility locations

the final resident distribution and the transportation means. Additionally, the number
of resident agents that transfer to new residents simultaneously was determined to
be 1/10 of the whole number of the resident agents.

Experiments were performed under the condition where there were four ways
of allocating the stop-off facility and where four different bustle bonus exchange
coefficients were combined (Fig. 3).

- A : not allocating
- B : suburbs, 2 km south and 0.5 km east from station B
- C : central area, same place as station B
- D : central area, 0.5 km south and 0.5 km east from station B

- $\eta_{bust} = 0, 10, 20, 30$

Through these experiments, we observed changes in the ratio of transportation
means, resident distributions, total CO_2 emission, and the average moving time.
The CO_2 emission was to be expressed in percentage based on the CO_2 emission
without stopping off at the stop-off facility.

4.2 Experimental Results

In the case that no measures were implemented, residences of resident agents that
used cars were broadly distributed in both the center of the destination zone with
high land cost due to the low total movement cost and the surrounding areas of the
destination zone with low land cost (A in Fig. 4 and Table 1). In this situation, an

■ : walk   ■ : bike   ■ : train   ■ : car

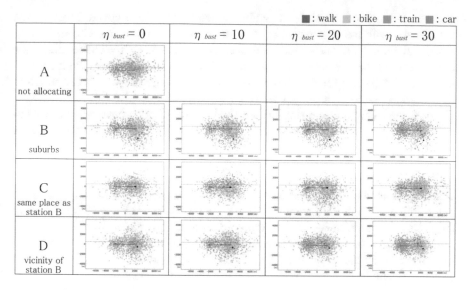

Fig. 4 Residences final distribution of experiment 1

Table 1 Result of experiment 1

| | Ratio of main transportation mean | | | | | |
	Walk (%)	Bike (%)	Train (%)	Car (%)	CO_2 emission (%)	Moving time (min)
A	1.3	3.9	4.8	90.0	100	14.2
B0	1.0	2.8	3.5	92.7	139.1	38.5
B10	0.9	2.8	3.1	93.2	141.2	38.4
B20	0.9	2.7	3.6	92.9	145.1	39.2
B30	5.1	2.2	48.8	43.9	76.7	63.5
C0	1.0	3.1	4.7	91.1	98.6	19.3
C10	1.1	2.6	5.4	90.9	100.0	19.8
C20	1.1	2.9	5.6	90.5	99.4	19.6
C30	1.1	2.7	6.1	90.1	99.7	19.6
D0	1.0	3.0	4.2	91.8	98.6	24.3
D10	1.0	3.1	4.1	91.8	100.3	24.2
D20	3.2	2.8	36.0	58.0	68.8	33.2
D30	4.5	2.9	66.8	25.8	38.3	44.5

urban structure sprawl that can be seen in the real urban areas was observed where the urban structure significantly changed from the beginning structure where zoning was conducted in order to separate residential areas from working areas to the sprawl structure based on the use of private cars. Therefore, the experiments we performed confirmed the validity of the urban model for this study.

The experimental results also confirmed that the simulation process changed very little just by adding the action of stopping off in the stop-off facility, wherever in the destination zone the stop-off facility is allocated. We discovered this by the comparison with the case where no measures were taken (A, B0, C0, and D0 in Fig. 4 and Table 1). In many cases, however, concurrent use of the bustle promotion measure would be effective in making a compact urban structure and reduction of CO_2 emission accompanied with the process of achieving a compact urban structure (B0–30, and D0–30 in Fig. 4 and Table 1).

Here, the location of the stop-off facility became advantageous when a reasonable distance was set from the destination station (D0–30 in Fig. 4 and Table 1). When the stop-off facility was allocated at the same location as the train station, the bustle bonus could not be obtained through the process of moving toward station B (C0–30 in Fig. 4 and Table 1). Therefore, this could prevent those using cars from changing to another transportation means. In other words, this result suggested the possibility that slight differences in locations of the stop-off facility could bring about significant differences in the future urban structures and environment.

Additionally, our experiments clarified that residents would be compelled to experience some inconvenience of additional travel time where a compact urban structure would be achieved and CO_2 emission would be reduced.

5   Experiment 2—Effect of the Automobile Use Control Measures

5.1   Experimental Outline

This section describes an experiment performed regarding control of driving cars in the urban central area in addition to the allocation of the stop-off facility. This experiment differed from experiment 1, in that 10 parking lots were located on the circle whose center was station B and that included major destinations, at equal intervals. Within this circle, those agents using cars moved from the nearest parking lot from their destinations on foot. In other words, experiment 2 was performed under the hypothesis where the "park and walk" measure would be implemented. Additionally, from the point of view of removing through traffic, the highway was placed on the outside of the circle. Additionally, residence candidates out of this circle were presented to those agents using cars with respect to the choice of residence.

5.2   Experimental Results

When compared to the case where only the stop-off facility was allocated, generally, the experimental results confirmed that this automobile control measure would be effective in decreasing the use of cars along with a compact urban structure and the

■: walk ■: bike ■: train ■: car

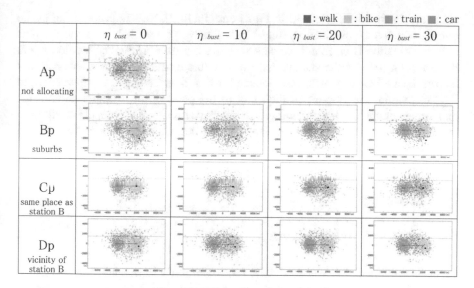

Fig. 5 Residences final distribution of experiment 2

Table 2 Result of experiment 2

	Ratio of main transportation mean				CO₂ emission (%)	Moving time (min)
	Walk (%)	Bike (%)	Train (%)	Car (%)	CO$_2$ emission (%)	Moving time (min)
Ap	1.6	13.8	22.5	62.0	85.2	28.9
Bp0	1.2	11.8	5.0	82.1	138.5	60.1
Bp10	1.1	11.1	10.9	76.8	131.3	60.5
Bp20	5.4	5.0	47.1	42.5	79.8	70.2
Bp30	5.4	3.8	64.5	26.3	55.5	77.0
Cp0	1.2	20.7	45.3	32.8	55.5	40.7
Cp10	1.7	17.6	40.7	40.0	63.6	42.7
Cp20	4.0	11.4	36.5	48.2	71.7	45.9
Cp30	6.3	7.9	31.7	54.1	80.0	47.9
Dp0	1.0	23.4	34.8	40.9	66.5	47.7
Dp10	3.1	11.5	54.4	31.1	53.3	51.0
Dp20	4.5	7.5	71.5	16.5	34.6	53.0
Dp30	4.1	5.8	80.9	9.2	26.7	54.6

reduction of CO_2 emission (Fig. 5 and Table 2). In particular, where the stop-off facility was located within the range where automobile traffic was removed, a significant effect appeared at the stage where no bustle bonus was considered (Cp0 and Dp0 in Fig. 5 and Table 2). But, as the bustle bonus exchange coefficient was increased, although the location of the stop-off facility differed by approximately only 700m, the effect of decrease in those using cars and CO_2 emission reduction were worsened in the former (Cp0–30 in Fig. 5 and Table 2) and were while enhanced in the latter (Dp0–30 in Fig. 5 and Table 2). This is because the bustle bonus, which was obtained by those using railways traveled from the stop-off facility to station B on foot, became zero since the stop-off facility was located at the same location as the station. As a result, this bustle bonus was subordinated by the bustle bonus which was obtained when those using cars traveled from the destination, the stop-off facility, and to the parking lot. In other words, this experiment suggested the possibility that the bustle promotion measure could have harmful effects on urban structures and the environment depending on the location of the stop-off facility.

6 Conclusion

6.1 Research Achievements

In this urban model, the residents' daily transportation selection is influenced by the other residents around them through the congestion and the enjoyment of bustle, and similarly residence selection of residents is influenced by the other residents through the change of rand rent. By using this model we conducted simulation experiments for validating of indirect measures as development of the stop-off facility with respect to a compact city that is an issue to be addressed in urgent. First of all by the simulation where no facility was allocated, we reproduced sprawl of the urban structure that is commonly observed in the real cities, while confirming the validity of this urban model. And the results of experiments confirmed that the results did not change only by allocating the stop-off facility when compared to the case where no facility was allocated, regardless of the location of the stop-off facility, however, concurrent use of the bustle promotion measure around the facility would be effective in achieving compact urban structures along with the reduction of CO_2 emission. In addition, these results confirmed that slight differences in the location of the stop-off facility could bring about significant differences to the effect.

We then performed other experiments in order to examine the automobile use control measure in the urban central area with the purpose of urban environmental improvement. The experimental results clarified that the synergistic effect of the bustle promotion measure and the automobile use control measure would have outstandingly effects in achieving compact urban structures and reduction of CO_2 emission in some cases. At the same time, depending on the location of the stop-off facility, the bustle promotion measure might worsen urban structures and the environment.

Any of these results suggest that when we develop the stop-off facility for the purpose of induction to a compact city we should also pay close attention to that location and development around that.

6.2 Future Perspectives

As for remaining issues, first, we need to review the consistency between the simulation model and the real cities. For example, we came to a conclusion that allocating the stop-off facility in the same location as the destination station could probably cause negative effects. When considering some cases of bustling in-station stores with commercial success, however, we need to review our model so that the sojourn time at the stop-off facility including qualitative value for staying at the facility is considered. Additionally, we assumed that all resident agents stop off in such stop-off facility every day, however, it might be better to incorporate variation of frequency of stopping off according to attraction of and distance to the facility and by actions of other residents into the simulation model.

Second, we need to examine the evaluation standard for the simulation results when the guidelines for measures that should be implemented to the real cities are provided. In addition to the standards referred to by this study including compact urban structures, CO_2 emission, and movement hours, we need to clarify the standard that are or should be focused on when the actual policies are planned.

References

1. Batty, M.: Cities and Complexity: Understanding Cities with Cellular Automata, Agent-Based Models, and Fractals. The MIT Press, Cambridge (2007)
2. Brown, D.G., Robinson, D.T.: Effects of heterogeneity in residential preferences on an agent-based model of urban sprawl. Ecol. Soc. 11(1), 46 (2006)
3. Fujii, S., Someya, Y.: A behavioral analysis on relationship between individuals' travel behavior and residential choice behavior. Infrastruct. Plan. Rev. Jpn. Soc. Civil Eng. 24, 481–487 (2007)
4. Glaeser, E.: Triumph of the City: How Our Greatest Invention Makes US Richer, Smarter, Greener, Healthier and Happier. Pan Macmillan, London (2011)
5. Haase, D., Lautenbach, S., Seppelt, R.: Modeling and simulating residential mobility in a shrinking city using an agent-based approach. Environ. Modell. Softw. 25(10), 1225–1240 (2010)
6. Isono, Y., Kishimoto, T.: Properties of utility model and optimal location of public library considering halfway stop. Pap. City Plan. 46(3), 415–420 (2011)
7. Jacobs, J.: The Death and Life of Great American Cities. Vintage, New York (1961)
8. Jager, W., Mosler, H.J.: Simulating human behavior for understanding and managing environmental resource use. J. Soc. Issues 63(1), 97–116 (2007)
9. Kaido, K.: Urban densities, quality of life and local facility accessibility in principal Japanese cities. In: Future Forms and Design for Sustainable Cities. Architectural Press, Oxford (2005)
10. Kazepov, Y.: Cities of Europe: Changing Contexts, Local Arrangement and the Challenge to Urban Cohesion, vol. 46. Wiley, Oxford (2011)
11. Kim, J.H., Pagliara, F., Preston, J.: The intention to move and residential location choice behaviour. Urban Stud. 42(9), 1621–1636 (2005)
12. Millward, H.: Urban containment strategies: a case-study appraisal of plans and policies in Japanese, British, and Canadian cities. Land Use Policy 23(4), 473–485 (2006)
13. Oldenburg, R.: The Great Good Place: Café, Coffee Shops, Community Centers, Beauty Parlors, General Stores, Bars, Hangouts, and How They Get You through the Day. Paragon House Publishers, New York (1989)

14. Railsback, S.F., Grimm, V.: Agent-based and individual-based modeling: a practical introduction. Princeton University Press, Princeton (2011)
15. Rieniets, T.: Shrinking cities—growing domain for urban planning? (2005). Retrieved 11 Dec 2007
16. Rieniets, T.: Shrinking cities: causes and effects of urban population losses in the twentieth century. Nat. Cult. **4**(3), 231–254 (2009)
17. Rouwendal, J., Meijer, E.: Preferences for housing, jobs, and commuting: a mixed logit analysis. J. Reg. Sci. **41**(3), 475–505 (2001)
18. Taniguchi, T., Takahashi, Y.: Multi-agent simulation about urban dynamics based on a hypothetical relationship between individuals' travel behavior and residential choice behavior. Trans. Soc. Instr. Control Eng. **47**, 571–580 (2012)
19. Togawa, T., Hayashi, Y., Kato, H.: A expansion of equilibrium type land use model by multi agent approach. 37th the Committee of Infrastructure Planning and Management (IP), Japan Society of Civil Engineers (2008)
20. Vega, A., Reynolds-Feighan, A.: A methodological framework for the study of residential location and travel-to-work mode choice under central and suburban employment destination patterns. Transp. Res. Part A: Policy Pract. 43(4), 401–419 (2009)
21. Zukin, S.: Naked City: The Death and Life of Authentic Urban Places. Oxford University Press, Oxford (2009)

Using ABM to Clarify and Refine Social Practice Theory

Kavin Narasimhan, Thomas Roberts, Maria Xenitidou, and Nigel Gilbert

Abstract We use an agent-based model to help to refine and clarify social practice theory, wherein the focus is neither on individuals nor on any form of societal totality, but on the repeated performances of practices ordered across space and time. The recursive relationship between social practices and practitioners (individuals performing practices) is strongly emphasised in social practice theory. We intend to have this recursive relationship unfold dynamically in a model where practitioners and social practices are both considered as agents. Model conceptualisation is based on the principle of structuration theory—the focus is neither on micro causing macro nor on macro influencing micro, but on the duality between structure (macro) and agency (micro). In our case, we conceptualise the duality between practitioners and practices based on theoretical insights from social practices literature; where information is unclear or insufficient, we make systematic assumptions and account for these.

Keywords Social practice theory • Modelling social practices • Agent-based model • Structuration theory

1 Introduction

In this paper we seek to clarify the core principles of social practice theory using an agent-based model. Existing information regarding the dynamics and growth of social practices[1] are very rich but quite dense in that they are open for multiple interpretations. For instance, Kuijer provides an interesting tabular summary of the evolution of bathing as a social practice between 500 BC (the Roman empire) and

---

[1]The terms 'practices' and 'social practices' are used interchangeably.

K. Narasimhan (✉) • T. Roberts • M. Xenitidou • N. Gilbert
Centre for Research in Social Simulation (CRESS), Department of Sociology, University of Surrey, Guildford GU2 7XH, UK
e-mail: k.narasimhan@surrey.ac.uk; t.m.roberts@surrey.ac.uk; m.xenitidou@surrey.ac.uk; n.gilbert@surrey.ac.uk

© Springer International Publishing AG 2017   307
W. Jager et al. (eds.), *Advances in Social Simulation 2015*, Advances in Intelligent Systems and Computing 528, DOI 10.1007/978-3-319-47253-9_27

the 1970s [1, p. 110]. But an exact pattern for the growth trajectory of the bathing practice is not available in current literature. It is a similar case for many other social practices—there is rich evidence suggesting that practices evolve over time but there is no analytical understanding of how this happens. We propose an agent-based model conceptualisation, a recognised method to formalise theoretical insights [2], to understand the mechanisms underlying the evolution of social practices. It is our aim that a model implemented based on our conceptualisation would be able to account for the evolution of practices in a systematic fashion. The rest of this paper is structured as follows. We provide an overview of social practice theory in Sect. 2, which leads to identifying three specific processes we have chosen to conceptualise in Sect. 3. The actual model conceptualisation is then introduced in Sect. 4, followed by conclusions in Sect. 5.

2 Overview of Social Practice Theory

Social practice theory draws on many of the core principles of Gidden's theory of structuration [3], which considers that human activity and the social structures which shape it are recursively related. Shove notes that human activities are shaped and enabled by social structures of rules, and in turn, the structures are reproduced through human action [4]. In social practice theory, the focus is neither on individuals nor on any form of societal totality but on individuals performing practices that are ordered across space and time. Individuals are still a part of the social system, but the starting point for understanding social systems is the performance of practices by individuals (also referred to as *practitioners*).

Performing a social practice refers to the routine accomplishment of what people consider to be the normal ways of life [5, p. 117]. Individuals are seen as carriers of practice; they carry out various tasks and activities that practices require [6]. This does not mean that individuals are regarded as passive beings [3], at the same time they are also not active in the sense of being involved in conscious decision-making [7]. Instead, individuals are considered to be skilled agents who actively negotiate and perform practices in the course of their daily lives. The 'social' status of a practice is then more a consequence of its stable reproduction beyond the limits of space, time and single individuals.

Practices are performed when all the relevant component elements are linked together [8, in review]. There is some debate about the nature of the different elements, but there is a growing consensus around Shove's understanding of practices being made up of three core element groups: materials, meanings and skills [5]. As a minimum, one element from each of the three categories is required for a performance of the practice to occur. Through repeated performances, practices are reproduced across space and time. Practices can also evolve and/or eventually die out as the component elements change.

Practices rarely occur in isolation; they come together as *bundles* to make up lifestyles or habitus. For example, doing the laundry combines a number of individual practices such as loading and unloading the washing machine, drying

Fig. 1 Practices contributing to the laundry bundle

the clothes and ironing and storing laundered clothes (see Fig. 1). While these all remain separate practices, they are often performed together, leading to them sharing a common label.

Bundles (or practices) are also linked to each other. As the practices and elements in one bundle change there is a knock on effect for other related bundles. This process is known as *co-evolution* [8, in review]. For example, as the number of entertainment and wireless communication devices, such as televisions, laptop computers, tablets and smart phones, increases the practices associated with the use of these devices become less spatially constrained. This has led to the dispersal of such activities around the home with implications for other bundles of practices such as heating. Whereas in the past the whole family might spend an evening watching television together in the living room only needing to heat one room, today family members could be dispersed around the house independently watching, playing or communicating on multiple devices necessitating the need to heat the whole house. Similarly, as daily lives become more dependent on ICT,[2] an increasing number of bundles of practices become connected through the skills required to operate ICT equipment—e.g., using a computer for food preparation (using the internet to find a recipe) and communication (sending an e-mail). Shared meanings also link practices together, for example, the desire for privacy while undertaking activities such as showering and using the lavatory.

3 Identifying Key Model Processes

Gilbert defines *target* as the social phenomenon or process that an agent-based model seeks to represent [9]. In our case, demonstrating the evolution of social practices is the target phenomenon. From the social practices literature, we have

---

[2]Information and Communications Technology.

identified three specific processes that contribute to this target phenomenon. First of these is the coming together of elements to enable the performances of practices. Shove suggests that meaning, material and skill elements come together to signify the performance of practices [10]. Elements initially exist in isolation waiting for links to be made, when this happens, practices come to exist. On the other hand, when elements are no longer linked, practices become obsolete and eventually cease to exist. Just as elements come together to form practices, loosely associated practices come together to form bundles, e.g., the aforementioned laundry example. In our model conceptualisation, we consider both the coming together of elements to form practices and the coming together of practices to form bundles.

The second aspect we consider as contributing to the growth of social practices is the recursive relationship between practitioners and practices. The relationship signifies how practitioners perform practices, and how with repeated performances practices become established and widespread, i.e., become norms that feedback into future performances. Shove suggests an example to illustrate this concept [10]. Showering as a social practice has rapidly evolved during the last 50 years, and in this time, it has become a societal norm for people to shower everyday. This norm has influenced the practice of showering in different ways—new materials (e.g. soaps, bath gels), new skills (e.g. ability to use shower units) and new meanings (e.g. personal hygiene standards) have emerged. Figure 2 demonstrates the coming together of elements constituting the modern day practice of showering. The recursive relationship between practitioners and practices signifies how, in one direction micro-level phenomena (coming together of elements) leads to macro-level phenomena (the performance of practices). In the other direction, it signifies how practices afford the norms for their performances.

Lastly, we consider the linked performances of practices, *aka* the co-evolution of practices. Co-evolution is the result of two or more practices evolving at the same time as a consequence of elements being shared between them. For instance, as

Fig. 2 Elements underlying the social practice of showering

noted in Sect. 2, the growth of visual entertainment and ICT devices have influenced the evolution of heating practices in households.

Existing literature does not provide adequate details for modelling the processes we have identified here. For instance, there is theoretical and illustrative clarity of the recursive relationship between practitioners and practices, but there is no analytical understanding of the actual mechanisms involved. There is no explanation of rules and conditions under which elements come together (i.e. become linked) or disaggregate to influence the performances of practices. Likewise, there is no evidence suggestive of practices influencing norms for their performance. There is also a lack of understanding of mechanisms causing the co-evolution of practices (and bundles[3]). By systematically accounting for the missing details, we intend for our model conceptualisation to:

1. Demonstrate the coming together of elements to enable the performance of practices; demonstrate the coming together of practices to signify the performance of bundles.
2. Delineate the recursive relationship between practices and practitioners in an analytical fashion.
3. Demonstrate the co-evolution of practices.

4 Conceptualisation of the Model

4.1 Basic Principles

We consider four main entities in our model: elements, practitioners, practices and bundles. The last three entities are agents in the model, while elements (i.e. meaning, material and skill) are objects used by practitioners (h_meaning, h_material and h_skill) and practices (p_meaning, p_material and p_skill), respectively. Practitioners are agents that perform practices, e.g., in the context of the laundry and showering examples considered above, households can be regarded as practitioner agents.[4] We consider practices as agents in the model to demonstrate the co-evolution relationship, which entails interaction between practices causing them to share elements. Consequently, we treat practices as agents that are capable of interacting with one another, to share and adapt their elements as an outcome of those interactions. Lastly, we consider bundles as agents that emerge in the model when two or more relevant practices come together.

---

[3]Co-evolution applies to both practices and bundles.

[4]Technically, it is individuals living within households who perform the practices. Since we like to consider several practices performed within the bounds of the household, even if different people perform them, we conceptualise households as the practitioners instead of individuals.

Through the following example we demonstrate the relationship between the agents (households as practitioners, drying clothes as the social practice and laundry as the bundle). Household agents draw at least one element each (i.e. one *h_meaning*, one *h_material* and one *h_skill*) to perform the social practice of drying clothes. The exact elements chosen to perform the practice can differ across households. For example, one can use tumble dryer as the material, while another uses cloth airer and some other household uses a radiator drying rack. This is an example of three different *h_material* elements used by three different household agents. Despite differences in the actual elements used, all households intend to perform the same practice, i.e., dry clothes. Hence it is possible to say that the social practice of drying clothes has three different *p_material* elements contributing to its performance. On any particular occasion, if a household agent performs two or more associated practices, e.g., wash clothes, dry clothes and iron clothes, it is then considered to perform the laundry bundle. The specific roles of practitioners, practices and bundle agents are further detailed in Sect. 4.3 after reviewing the general characteristics of agents and their interactions in Sect. 4.2.

4.2   General Characteristics of Agents and Their Interactions

Before we present the distinct properties of the practitioner, practice and bundle agents in Sect. 4.3, we here present their common characteristics. Per Macy and Willer's recommendations, the agents are simple, autonomous, interdependent and adaptive [11]:

– **Agents are simple** in that they follow simple rules. Practitioners follow simple rules by way of performing practices out of habit. The assumption is based on the understanding of social practice theory that actors perform practices not as a consequence of conscious decision-making processes but mostly out of habit. Likewise, practice agents also follow simple rules to influence the norms for performances. As there are no theoretical insights of how bundles influence norms for future performances, we hypothesise that bundles influence norms by aggregating (a simple additive aggregation) the individual norms imposed by contributing social practices. Consider the laundry example in Sect. 2. To determine norms imposed by laundry as a bundle, we propose computing the aggregate of norms imposed by each individual social practice contributing to the laundry bundle (loading the washing machine, unloading the washing machine, iron clothes, etc.).
– **Agents are interdependent** in the sense that their actions affect one another. Interdependency of agents in the model can be explained at two levels. At one level, the coming together of material, meaning and skill elements facilitates the performance of practices, which in turn influences the norms

affecting future performances of practices. At another level, the coming together of practices facilitates the production and reproduction of bundles; these in turn lead to norms (at the bundle level) affecting future performances of practices contributing to bundles.

- **Agents are autonomous** in the sense there is no global authority (entity) directing practitioners to perform practices; instead, they emerge purely from interactions at the micro-level. Elements referenced by practitioners come together to enable them to perform practices. Two or more practices come together to enable performances of bundles. In the other direction, practices and bundles influence norms for future performances of practices. These norms act as, in what [11] refers to as additional environmental constraints influencing the actions pursued by practitioners with regard to performing practices. However, constraints do not directly cause a practitioner to perform any practice.
- **Agents are adaptive:** Practitioners modify performances based on the influences of existing practices and bundles. Each individual practitioner utilises locally available information to perceive about and react to the norms imposed by the practice and bundle agents in its environment. Likewise, each individual practice (and bundle) agent uses locally available information to update its elements based on current performances by practitioners and the co-evolution links between practices.

There are three levels of interaction between model entities: micro-, meso- and macro-level interactions. Interactions between material, meaning and skill elements referenced by practitioners signify the **micro-level interactions**. These interactions cause elements to be updated, new links to be formed between elements or existing links to be broken. Recollect that performances of practices signify the coming together of at least one meaning, one material and one skill elements. So, when micro-level interactions cause the meaning, material and skill elements essential to perform a practice to be linked, it enables a practitioner to perform that practice. In addition to that, when micro-level interactions between elements enable a practitioner to perform two or more social practices contributing to the same bundle, it leads to the performance of that bundle.

Next, interactions that occur among practitioner agents are classified as **meso-level interactions**. For instance, a practitioner agent assesses the performance of a particular practice by other practitioners in its social circle, and uses that as a *social influence* to modify or adapt its own performance of that practice. Meso-level interactions influence practitioners to modify the elements they reference and to form new links or disintegrate existing links between elements. Lastly, **macro-level interactions** signify the interactions between practices as a result of co-evolution. These interactions cause practices to modify the elements they reference or establish new links or disintegrate existing links between elements referenced across practices.

4.3   Specific Characteristics of Model Entities

We now provide a detailed description of the distinct characteristics we have
conceptualised for each model entity by considering as an example the *performance
of social practices within households*:

- **Elements** represent three distinct categories of model entities: meaning, material
  and skill. Elements are objects used by practitioners and practices. Practitioners
  have access to any number of material, meaning and skill elements. Similarly,
  practices can be made up of any number of material, meaning and skill elements.
  But for the sake of model simplicity, we have assumed that in order to perform
  one practice, practitioners draw one element each, i.e., one material, one meaning
  and one skill elements.
- **Practitioners** are agents that are able to carry out practices. To do this, they
  have access to elements required to perform practices, but there is no evidence
  for what causes them to draw elements together. So we hypothesise that a
  variety of influences act upon practitioners and enable them to combine elements
  and carry out practices. The influences to be considered will vary depending
  upon the characteristics of the practitioners and practices being modelled. For
  instance, when households are the practitioners, we consider social influences
  and historical influences to impact the performance of practices. Social influences
  enable a household to carry out practices that are commonly performed in its
  social neighbourhood. Based on Reckwitz' (2002) observation, we consider
  historical influences to enable households to repeat their previous performances
  of practices [6]. While the influences act upon households causing them to update
  their elements (i.e. *h_material*, *h_meaning* and *h_skill*), they do not directly force
  households to perform specific practices.
- **Practices** are agents that exist from the start of the simulations, but may not
  be instantiated as performances. However, when practitioners start drawing
  elements together, practices are instantiated as performances and continue to
  exist as such, so long as practitioners keep repeating performances. If practices
  are not being performed, they become obsolete and may eventually cease to exist.
  Each practice agent maintains a list of meaning, material and skill elements
  contributing to its performance. During the simulation, these elements are
  updated based on two types of influences; the first one causes practices to update
  their list of elements based on current trends in how practices are performed
  and the elements used. For example, if households adopt a new material or skill
  element for washing clothes, then the wash clothes social practice and laundry
  bundle update their respective lists of elements to include the newly adopted
  element. The second influence referred to as update based on co-evolution causes
  practices to update their elements as a consequence of their relationship with one
  another. For instance, browsing the Internet is an ICT social practice, but at the
  same time, it can also serve as a skill element associated with the cooking practice
  for finding recipes. Given this relationship, if one practice (browsing the Internet)

adopts a new material (using tablet PCs), then the other practice (cooking) also adopts that new element (tablet PCs to find recipes).

- **Bundles:** Shove suggests that practices come together to constitute bundles [10], but there is no evidence for what causes the coming together of practices. So we hypothesise establishing pre-defined soft links,[5] and then, to check if practitioners are in a position to perform these linked practices. If they are, then a new bundle agent emerges in the model. For example, let us consider *thermal control* as a bundle that includes soft links between two social practices—use central heating and do thermal retrofits. Despite being soft linked, a household will be able to perform both practices only if influences acting upon the household should allow it. For instance, if privately rented, then households cannot perform thermal retrofits,[6] whereas households whose tenure is owned can. So in the former case, households can only perform a practice (i.e. use central heating), whereas in the latter case households can perform a bundle (i.e. thermal control).

- **Environment:** We conceptualise the model environment as a virtual society where practitioners, practices and bundles exist. The model environment has variables such as temperature, day of the week and season. A key principle of social practice theory is that the performance of practices are ordered across space and time [4]. So we hypothesise defining space-time boundaries as environmental influences acting upon practices and causing them to adapt their elements. For instance, when the scope of modelling extends between households and work places, then certain practices can be limited to performance within households (cooking, showering, etc.), while certain practices are restricted to be performed only outside the physical boundary of households (e.g. driving). Similarly, the performances of practices can also be restricted based on time— time of the day (morning, afternoon, evening and night) and season. For example, it is unlikely home heating will be required during summer months, so the seasonal influence affects the meaning, material and skill elements needed to perform the central heating practice. This in turn imposes different norms during different seasons for households to perform the central heating practice.

---

[5]We define soft links as associations between practices (i.e. each practice linked to one or more other practices) defined at the start of simulations.

[6]In reality, housing tenure impacts an household's ability to undertake a thermal retrofit—a process to improve the thermal properties of a building through the use of high levels of thermal insulation and airtightness. Such a process will be much easier for people who own their homes than for those who live in privately rented properties. But for the sake of model simplicity, we have assumed that households residing in privately rented properties will not be able to do thermal retrofits.

4.4   Conceptualising Key Processes

Conceptualising the Coming Together of Elements to Enable the Performances of Practices We hypothesise a *similarity between elements rule* that serves as a checkpoint to enable practitioners to perform practices. Since there is no evidence suggesting how elements come together to enable performing practices, we propose the following strategy as a sample approach. Compute an index of similarity between elements referenced by a practitioner and a practice, which if exceeds a desired threshold enables performing the desired practice. The strategy could be formulated in the following manner. Consider each element being represented as a bit string, e.g., [0110110111]. Compute pair-wise similarity between corresponding elements referenced by a practitioner and a practice. This will be the difference between 10 (i.e. bit string length) and the Hamming distance between elements. If the average pair-wise similarity (i.e. average of similarity measures between meaning, material and skill elements) exceeds a desired threshold, then a practitioner performs the desired practice.

Conceptualising the Coming Together of Practices to Enable the Performance of Bundles Given the similarity between elements rule, we hypothesise that practitioners are able to perform a bundle if they are able to perform all (or more than a desired threshold) number of soft linked practices contributing to that bundle. Considering the previously mentioned example of the thermal comfort bundle, this rule will allow certain practitioners to perform the bundle itself, while other practitioners will only be able to perform individual social practices such as using central heating and not the bundle itself.

Conceptualising Practices Influencing Norms for Performances We had established before that practices evolve as a consequence of environmental influences, current trends and co-evolution of practices. Evolution of practices implies the evolution of elements they reference. For example, between one time step and the next, the *p_material* of a particular practice agent might transform from 1001011001 to 1001000101 as a consequence of the three influences that acted upon the agent during that time step. This in turn impacts the calculation of similarity between elements in a future time step when a decision has to be made whether or not a practitioner is able to perform the concerned practice. This is our conceptualisation for practices influencing the norms for performances.

Conceptualising the Co-evolution of Practices (and Bundles) We hypothesise that co-evolving practices will try to mirror all or parts of their elements—equivalent to sharing elements across practices. For instance, let us consider that at a particular time step the *p_meaning* element of the watching television practice is 1000110010. If using central heating is a co-evolving practice, it will then try to mirror the *p_meaning* element of the watching television practice as a consequence of co-evolution. This example is based on an empirical finding that people preferred to turn up the heating when watching TV with an intention of creating a comfortable and cosy ambience (i.e. meaning) [8, in review]. Co-evolution rules like these

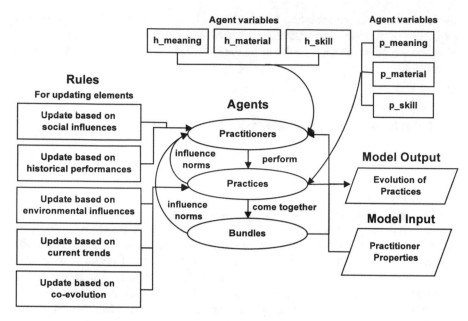

Fig. 3 Conceptualisation of households performing social practices

implemented in the model will allow practices to mirror the elements referenced by another, which in turn, will affect future performances of practices due to the similarity between elements rule.

Drawing together all the individual aspects described so far, we present our social practice theory-based model conceptualisation in Fig. 3. The flow of action between practitioner, practice and bundle agents are such that practitioners perform practices; practices come together to form bundles; both practices and bundles influence norms for the performance of practices. This is demonstrated in the middle portion of Fig. 3. We have conceptualised this part with the intention of capturing the duality between practitioners and practices, i.e., practitioners perform practices, which in turn influence the norms for future performances.

The left portion of Fig. 3 shows the rules influencing practitioner agents and practice agents, respectively. Our idea is that social influences and historical influences act upon practitioners (i.e. households) to enable them to draw elements together to perform practices. We recognise there may be other influences causing practitioners to perform practices, but here we have only considered these two. We consider three rules influencing practice agents: (1) update based on environmental influences, (2) update based on current trends and (3) update based on co-evolution.

Lastly, the right portion of Fig. 3 depicts the input to the model and the output obtained from it. We conceptualise the input to the model as being any practitioner specific parameters. For example, if households are considered as the practitioner agents, then household demographics (housing tenure and type, number of rooms, number of residents, etc.) could be an input to the model. Output obtained from the model summarises the evolution of practices. This may be visualised as a growth trajectory of practices demonstrating the evolution of material, meaning and skill elements over time, as well as measuring the spread of practices among practitioners over time.

5   Conclusion and Future Work

We have tried to achieve a three-fold outcome in this paper. Firstly, we have attempted to explicate in a systematic fashion the rich but dense concepts constituting the core principles of social practice theory. Second, to demonstrate the appropriateness of adopting a modelling position based on the theory of structuration to systematically conceptualise the key model processes. Lastly, to delineate an agent-based model conceptualisation to demonstrate the dynamics of processes associated with the production and reproduction of social practices. It is our intention that the proposed model conceptualisation will allow formulations using suitable methods for pursuing specific case studies. Social practice theory is gaining popularity in many areas of research, so it is possible to adapt the model conceptualisation we have proposed here to investigate specific real-world phenomena. For instance, the authors are implementing the model to investigate the performances of energy intensive social practices in households.

Only few attempts have been made to conceptualise social practices using agent-based models. Holtz aimed at modelling the emergence of an abstract social practice based on the level of coherence achieved when material, meaning and competence elements come together [12]. He followed up with evidence to show that the proposed model reaches *lock-in* [13] quite early on in the simulation after which further changes to the social practice were blocked [14]. Balke et al. conceptualised social practices not as outcomes intended from the model but as agents themselves within the model. Their goal was to model households, social practices and industries as entities capable of shaping one another recursively, while also affecting the evolution of material artefacts within the model [15]. None of these, however, focused on all the three model processes that we have conceptualised here.

Besides, models aimed at simulating human societies often assume either an individualistic approach or a holistic approach. The former assumes that interactions between agents lead to the emergence of societal structure, while the latter considers that societal structure governs interaction between agents. But in the present case, the inherently recursive relationship between practitioners and practices motivated pursuing an alternative modelling strategy. We have followed Gilbert's recommendation of a modelling perspective that allows for duality to exist between structure and agency [16]. In one direction, practitioners act and their actions lead to the production and reproduction of practices and bundles. In the other direction, practices both enable and constrain the actions of practitioner agents. Gilbert notes that the ability to perceive and be responsive to macro-level structures is a notable capacity of humans and that it needs to be addressed in models synthesising human societies [16]. Our conceptualisation of practitioners performing practices, which in turn, influences the norms for future performances captures the duality between structure and agency in a realistic fashion.

References

1. Kuijer, S.C.: Implications of social practice theory for sustainable design. Doctoral dissertation, TU Delft, Delft University of Technology (2014)
2. Gilbert, N., Troitzsch, K.: Simulation for the Social Scientist. McGraw-Hill International, New York (2005).
3. Giddens, A.: The Constitution of Society: Outline of the Theory of Structuration. University of California Press, Berkeley (1984)
4. Shove, E., Pantzar, M., Watson, M.: The Dynamics of Social Practice: Everyday Life and How it Changes. SAGE Publications Ltd., London (2012)
5. Shove, E. Changing human behaviour and lifestyle: a challenge for sustainable consumption? In: Consumption Perspectives Ecological Economics, pp. 111–132 . Elgar, Cheltenham (2005)
6. Reckwitz, A. Toward a theory of social practices a development in culturalist theorizing. Eur. J. Soc. Theory 5(2), 243–263 (2002)
7. Wilhite, H., Towards a better accounting of the roles of body, things and habits in consumption. In: Warde, A., Southerton, D. (eds.) The Habits of Consumption: Studies Across Disciplines in the Humanities and Social Sciences, pp. 87–99. Helsinki Collegium for Advanced Studies, Helsinki (2012)
8. Roberts, T., Balke, T., Gilbert, N.: Co-evolution of social practices: A case study of domestic energy use. [submitted] (n.d.)
9. Gilbert, G.N.: Agent-Based Models (No. 153). Sage, London (2008)
10. Shove, E.: Comfort, Cleanliness and Convenience: The Social Organization of Normality. Berg, Oxford (2003)
11. Macy, M.W., Willer, R.: From factors to actors: computational sociology and agent-based modeling. Ann. Rev. Sociol. 28, 143–166 (2002)
12. Holtz, G.: An agent-based model of social practices. In: Proceedings of the 8th Conference of the European Social Simulation Association, Salzburg (2012)
13. Arthur, W.B.: Competing technologies, increasing returns, and lock-in by historical events. Econ. J. 99, 116–131 (1989)
14. Holtz, G.: Generating social practices. J. Artif. Soc. Soc. Simul. 17(1), 17 (2014)
15. Balke, T., Gilbert, N., Roberts, T., Xenitidou, M.: Modelling energy-consuming social practices as agents. In: Social Simulation Conference (2014)
16. Gilbert, N., Conte, R.: Artificial societies: the computer simulation of social life. Taylor & Francis, Inc., London (1995)

Transition to Low-Carbon Economy: Simulating Nonlinearities in the Electricity Market, Navarre Region, Spain

Leila Niamir and Tatiana Filatova

1 Introduction

Coupled climate-economy systems are complex adaptive systems. While changes and out-of-equilibrium dynamics are in the essence of such systems, this dynamics can be of a very different nature. Specifically, it can take a form of either gradual marginal developments along a particular trend or exhibit abrupt nonmarginal shifts [1]. Nonlinearities, thresholds, and irreversibility are of particular importance when studying coupled climate-economy systems. Strong feedbacks between climate and economy are realized through energy: economy requires energy for literary every sector, while emissions need to stabilize and be even reduced to avoid catastrophic climate change [2]. Possibilities of passing some thresholds that may drive these climate-energy-economy (CEE) systems in a completely different regime need to be explored. However, currently available models are not always suitable to study nonlinearities, paths involving critical thresholds and irreversibility [3]. To be able to formulate an appropriate energy policy for this complex adaptive CEE system, policymakers should ideally have decision support tools that are able to foresee changes in energy market over the coming decades to plan ahead accordingly. Many macro models, that assume rational representative agent with static behavior, are designed to study marginal changes only. So there is a need for models that are able to capture nonlinear changes and their emergence.

ABMs are simulating human social behavior more realistically and can capture human variability and other nonlinear processes [4–9]. Since ABMs are not directly used to model climatic systems, there are no climate system thresholds considered directly. Irreversibility, however, is addressed in ABMs. The ABM of the carbon

L. Niamir (✉) • T. Filatova (✉)
Department of Governance and Technology for Sustainability, University of Twente,
P.O. Box 217, Enschede, 7500 AE, The Netherlands
e-mail: l.niamir@utwente.nl; t.filatova@utwente.nl

© Springer International Publishing AG 2017
W. Jager et al. (eds.), *Advances in Social Simulation 2015*, Advances in Intelligent
Systems and Computing 528, DOI 10.1007/978-3-319-47253-9_28

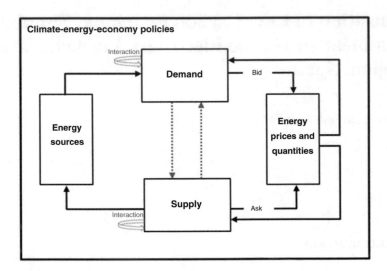

Fig. 1 Agent-based energy market—conceptual model

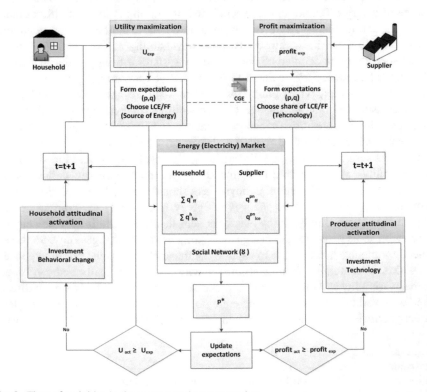

Fig. 2 Flow of activities in the agent-based energy market

emission trading impact on shifting from carbon-intensive electricity production [7] suggested that as soon as investments in new technology are made, the switch from the old technology is irreversible. Various scenarios produced by the ENGAGE ABM by Gerst et al. [10] all produce irreversible transitions to low-carbon economy. While depending on a policy, the transition can be swift or more gradual, the return back to carbon-intensive economy is unforeseeable.

2   Agent-Based Energy Market

We designed and programmed an ABM with an aim to investigate nonlinearities in energy markets. It aims to trace potential discontinuities in energy markets driven endogenously from within the economic ABM or triggered by changes in the environment. The quantities and prices of different energy sources namely low-carbon energy and fossil fuel and corresponding greenhouse gas emissions resulting from the microeconomic choices are indicators of an aggregated ABM energy market dynamics. Here we focus on the retail electricity market.

2.1   Demand

Demand side of our ABM consists of heterogeneous households with different preferences, awareness of climate change, and socioeconomic characteristics, which lead to various energy-consumption choices. Households choose a producer and energy type by optimizing utility they expect to receive (u_{exp}) given price expectations (q_{hlce}/q_{hff}) under budget constraints. Households receive utility from consuming energy (E) and a composite good (z) between which its budget is shared (Eq. 1). Moreover, households have awareness about the state of climate and environmental preferences (γ), which could potentially be heterogeneous and change over time.

$$U = z^{\alpha} \times E^{(1-\alpha)} \times C^{\gamma} \tag{1}$$

Later on we plan to implement various energy saving actions selecting from the following pool: switching to energy-efficient equipment, installing solar panels, energy saving bulbs, or change in electricity usage habits (e.g. switching off the lights).

2.2 Supply

The supply side is presented by heterogeneous energy providers, which may deliver either electricity based on low-carbon energy sources (LCE) or on fossil fuels (FF). The ABM model is being integrated with a macro-economic CGE model [11]. Thus, at this stage we do not go into the details of modeling the various energy producers where ABM can be instrumental in simulating the potential diffusion of alternative energy technologies. Instead, we simulated suppliers with different share of LCE and FF electricity production. In retail electricity market, form expectations are calculated regarding to prices (q_{plce}/q_{pff}), and share of LCE vs. FF, to deliver next time step in order to optimize their profits.

New energy prices (p^*_{lce}/p^*_{ff}) and market shares of green and grey energy are an emergent outcome of this agent-based energy market. After the market clearing, households update their price expectations and utility when comparing them to the actual market outcomes. If the total energy spending for a household are more than was expected, it stimulates a household to reconsider either an energy provider and a type of energy source, or an investment leading to energy savings, or a change in energy-consumption pattern.

2.3 Market Clearing

Due to the reasons widely discussed in the literature [12–15] agent-based markets try to distance from the traditional Walrasian auctioneer. Thus, the equilibrium price determination is replaced with alternative market structures. Different methods of market clearing evolved in the agent-based computational economics practice, which can be categorized in four main groups [14, 16].

The first category, which can be labeled "gradual price adjustment," assumes a simple price which the market-maker announces, and the demands are submitted at this price. Then if we have an excess demand, the price is increased, and if there is an excess supply the price is decreased. The price is often changed as a fixed proportion of the excess demand as in Eq. (2) [14].

$$pt + 1 = pt(1) + \alpha \left(D\left(pt \right) - S\left(pt \right) \right) \qquad (2)$$

This price adjustment method is used in Alvarez-Ramirez et al. [17]; Dieci and Westerhoff [18]; Farmer [19]; Farmer and Joshi [20]; Martinez-Echevarria [21]; Zhu et al. [22] models.

In second approach is temporary market clearing which the price is determined so that the total demand equals the total number of shares in market [12, 14, 16, 23, 24]. The advantage of this approach when compared to the "gradual price adjustment," is that there is no need to deal with market-maker. However, two critical problems are mentioned for this approach. It may impose too much market clearing, and it may not well represent the continuous trading situation of a financial market. Also, it is often more difficult to implement. It either involves a computationally costly procedure of numerically clearing the market, or a simplification of the demands of agents to yield an analytically tractable price [14].

The third category, which is the most realistic approach and is labeled "order book" market structure, simulated where demand and supply are crossed with using a certain well-defined procedure. One of the most common examples within this category of price formation mechanism is a double-auction market [14, 16, 25–27].

The fourth approach is bilateral trade and it assumes that agents bump into each other randomly and trade if it benefits them. It would appear realistic. However it may not be very natural in places where trading institutions are well defined, and function to help buyers meet sellers in a less-than random fashion [14].

We choose the first approach "gradual price adjustment" as the price determination of agent-based electivity market model, as it seems to represent the retail electricity market more accurately [28].

3 Results and Future Work

We present a work in progress with an application of the retail electricity market ABM to the Navarre region of Spain. Currently the demand and supply sides of energy (electricity) market are simulated using NetLogo with GIS and R extensions. We explore the dynamics of market shares of low-carbon electricity in the scenario where a household's choice on the type of electricity (grey or green) is driven exclusively by preferences vs. when market-clearing mechanisms is explicitly modeled. We also contrast the results for a population of household with homogeneous vs. heterogeneous preferences and awareness of climate change as well as incomes.

The future work will go on in constrain two directions. First, we aim at integrating the ABM with the CGE model to assure direct feedbacks between behavioral change with consequent changes in market shares of LCE vs. FF and impacts of these on other sectors of economy (ABM=>CGE), as well as accounting for nonresidential electricity demand and changes in households incomes as economy evolves (CGE=>ABM). Secondly, we plan to study behavioral changes and socioeconomic characteristics of households via a survey. The main goal of the survey is to elucidate the information on behavioral changes, which includes change not only in choices but also in preferences and opinions, potentially affected by social influence on the demand side (households) to feed it into the ABM.

Acknowledgment Funding from the EU FP7 COMPLEX project 308601 and the Netherlands Organization for Scientific Research (NWO) VENI grant 451-11-033 is gratefully acknowledged.

References

1. Filatova, T., Polhill, G., van Ewijk, I.: Regime shifts in coupled socio-environmental systems: review of modelling challenges and approaches. Environ. Model. Software (2015)
2. IPCC.: Fifth assessment report: climate change (2014)
3. Stern, N.: The structure of economic modeling of the potential impacts of climate change: grafting gross underestimation of risk onto already narrow science models. J. Econ. Lit. **51**(3), 838–859 (2013)
4. Arto, I., et al. Review of existing literature on methodologies to model non-linearity, thresholds and irreversibility in high-impact climate change events in the presence of environmental tipping points (2013)
5. Bonabeau, E.: Graph multidimensional scaling with self-organizing maps. Inform Sci **143**(1-4), 159–180 (2002)
6. Castel, C., Crooks, A.: Principles and Concepts of Agent-Based Modelling for Developing Geospatial Simulations. Centre for Advanced Spatial Analysis, University Collage of London, London, UK (2006)
7. Chappin, E.J.L., Dijkema, G.P.J.: An agent based model of the system of electricity production systems: exploring the impact of CO2 emission-trading. 2007 IEEE Int. Conf. Syst. Syst. Eng. **1 and 2**, 277–281 (2007)
8. Nigel, G.: Agent-Based Models. Sage Publications, Los Angeles (2008)
9. Tran, M.: Agent-behaviour and network influence on energy innovation diffusion. Comm. Nonlinear Sci. Numer. Simulat. **17**(9), 3682–3695 (2012)
10. Gerst, M.D., et al.: Agent-based modeling of climate policy: an introduction to the ENGAGE multi-level model framework. Environ Model Software **44**, 62–75 (2013)
11. Filatova, T., et al.: Dynamics of Climate-Energy-Economy Systems: Development of a Methodological Framework for an Integrated System of Models. (2014)
12. Arthur, W.B.: Complexity and the economy. Science **284**(5411), 107–109 (1999)
13. Kirman, A.P.: Complex Economics: Individual and Collective Rationality. Routledge, London (2011)
14. LeBaron, B.: Agent-based computational finance. In: Tesfatsion, L., Judd, K.L. (eds.) Handbook of Computational Economics. North-Holland, Amsterdam, The Netherlands (2006)
15. Tesfatsion, L.: Agent-based computational economics: a constructive approache to economic theory. In: Tesfatsion, L., Judd, K.L. (eds.) Handbook of Computational Economics. North-Holland, Amsterdam, The Netherlands (2006)
16. Rekike, Y.M., Hachicha, W., Boujelbene, Y. Agent-based modeling and investors's behavior explanation of asset price dynamics on artificial finanacial market. In: TSFS Finance Conference, vol 13. Sousse, Tunisia: Procedia Economics and Fianance (2014)
17. Alvarez-Ramirez, J., Suarez, R., Ibarra-Valdez, C.: Trading strategies, feedback control and market dynamics. Phys. A-Stat. Mech. Appl. **324**(1–2), 220–226 (2003)
18. Dieci, R., Westerhoff, F.: Interacting cobweb markets. J. Econ. Behav. Organ. **75**(3), 461–481 (2010)
19. Farmer, J.D.: Market force, ecology and evolution. Ind. Corp. Change **11**(5), 895–953 (2002)
20. Farmer, J.D., Joshi, S.: The price dynamics of common trading strategies. J. Econ. Behav. Organ. **49**(2), 149–171 (2002)
21. Martinez-Echevarria, M.A.: The market between sociability and conflict. Anuario Filosofico **40**(1), 175–186 (2007)
22. Zhu, T., Singh, V., Manuszak, M.D.: Market structure and competition in the retail discount industry. J. Market. Res. **46**(4), 453–466 (2009)
23. Brock, W.A., Hommes, C.H.: Heterogeneous beliefs and routes to chaos in a simple asset pricing model. J. Econ. Dynam. Contr. **22**(8–9), 1235–1274 (1998)
24. Levy, M., Levy, H., Solomon, S.: Microscopic simulation of the stock-market—the effect of microscopic diversity. J. Phys. I **5**(8), 1087–1107 (1995)

25. Chiarella, C., Iori, G., Perello, J.: The impact of heterogeneous trading rules on the limit order book and order flows. J. Econ. Dynam. Contr. **33**(3), 525–537 (2009)
26. Farmer, J.D., Patelli, P., Ilija, Z.O.: The predictive power of zero intelligence in financial markets. Proc Natl Acad Sci U S A **102**(6), 2254–2259 (2005)
27. Lux, T., Marchesi, M.: Volatility clustering in financial markets: a micro-simulation of interacting agents. Comput. Econ. Finance Eng. Econ. Syst. **7–10** (2000)
28. Federico, G., Vives, X.: Competition and Regulation in the Spanish Gas and Electiricity Markets. Spain Public-Ptivate Sector Research Center, IESE Bussiness School, Madrid (2008)

Statistical Verification of the Multiagent Model of Volatility Clustering on Financial Markets

Tomasz Olczak, Bogumił Kamiński, and Przemysław Szufel

Abstract Volatility clustering and leptocurtic, heavy tailed distribution of financial asset returns have been puzzling economists for decades. Ghoulmie, Cont, and Nadal (2005) proposed an agent-based model attempting to reproduce these stylized facts by means of the threshold switching behavior of investors. We investigate properties of the model following principles of the design of simulation experiments. We find the results to be only partially consistent with properties of empirical time series. This suggests the model to be an insightful but incomplete description of the phenomena under study.

Keywords Simulation model analysis • Volatility clustering • Distribution of asset returns • Heavy tails

1 Introduction

Volatility clustering and leptocurtic heavy tailed distribution of returns are the well-known facts manifesting in time series of financial instruments [11]. Despite much attention and research the long standing debate on origins of the phenomena is still open. Among the plausible causative mechanisms the following are suggested in the literature:

1. heterogeneity in time horizons of economic agents and rates of information arrival [1, 7];
2. instability of trading strategies due to agent learning or behavioral switching [3, 6, 9];
3. investor inertia resulting from a threshold behavior of agents trading only when magnitude of a market signal reaches a certain level [4].

In an attempt to quantify influence of the last mechanism on properties of a time series Ghoulmie et al. [4] proposed an agent-based model allowing to study its effect in isolation. The simulation results reported there confirmed capability of the model

T. Olczak (✉) • B. Kamiński • P. Szufel
Warsaw School of Economics, Al. Niepodległości 162, 02-554 Warsaw, Poland
e-mail: tolczak@gmail.com; bkamins@sgh.waw.pl; pszufe@sgh.waw.pl

© Springer International Publishing AG 2017
W. Jager et al. (eds.), *Advances in Social Simulation 2015*, Advances in Intelligent Systems and Computing 528, DOI 10.1007/978-3-319-47253-9_29

to generate time series with the desired properties, namely leptocurtic distribution of returns with semi-heavy tails and positive autocorrelation of absolute returns over many time lags (symptomatic for volatility clustering). This general conclusion, however, has been based on only two arbitrarily selected examples.

In this article we present results of the more systematic investigation of properties of the model proposed by Ghoulmie et al. [4]. We verify the following two hypotheses:

H1 the excess kurtosis of distribution of asset returns generated from the model is positive ($\kappa > 0$);
H2 the one period autocorrelation of absolute returns generated from the model is positive ($\rho_{\tau=1} > 0$).

Following the approach of [4] we use hypothesis **H1** as a proxy for detecting heavy tails. This is justified by the fact that in case of symmetric, unimodal distributions positive kurtosis indicates heavy tails [2], while at the same time empirical distributions of returns are known to be symmetric with high kurtosis, fat tails, and a peaked center as compared with the normal distribution [11, p. 93]. In a similar way we use hypothesis **H2** as a proxy for detecting volatility clustering. For empirical time series autocorrelations of both absolute and squared daily returns are known to be positive for many time lags and to be always positive at a lag of 1 day [11, pp. 90–91] —a direct consequence of volatility clustering [11, p. 313].

The rest of the paper is divided into three sections. In Sect. 2 we describe the model structure. In Sect. 3 we present the experiment design and results of the analysis. Section 4 concludes.

2 Model Description

The description of the model presented in this section is a direct adaptation of the original model by [4]. We consider a market with a single asset traded by $n \in \mathbf{N}$ agents. Trading takes place in discrete time intervals $t \in \{1, 2, \ldots\}$. At each time period agents decide to buy or sell a unit of the asset or remain idle. By denoting demand of agent i at time t by $\phi_i(t)$ this will be represented as $\phi_i(t) = 1$, $\phi_i(t) = -1$ and $\phi_i(t) = 0$, respectively. The buy or sell decisions are triggered by the market signal representing public expectations about a future return rate and modeled as a sequence of IID random variables $\varepsilon_t \sim N(0, \sigma^2)$. Agents place orders when the signal strength crosses their individual "sensitivity" threshold $\theta_i(t)$, a sell order when $\varepsilon_t < -\theta_i(t)$ or a buy order when $\varepsilon_t > \theta_i(t)$. They remain dormant when $\varepsilon_t \leq |\theta_i(t)|$. The resulting excess demand is given by $Z_t = \sum_{i=1}^{n} \phi_i(t)$ and produces a change in the asset price given by $r_t = Z_t/(\lambda n)$, where λ quantifies "market depth" and r_t can be interpreted as log rate of returns. After orders are placed and market price determined every agent independently with probability q changes its sensitivity threshold θ_i to $|r_t|$.

In addition to the analysis of the model we use the following observation showing the link between σ and λ in the model (this fact was not observed in [4]). Assume that we have some state of the model $\theta_i(t)$ for parameters σ and λ. Now consider that we scale all these parameters by $\alpha > 0$. Thus the state of our model is $\alpha\theta_i(t)$ for parameters $\sigma\alpha$ and λ/α. If we use the same random number to generate ε_t, then $\phi_i(t)$ will be identical in both cases. Therefore r_t will be α times larger in the new parameterization case. But this means that $\theta_i(t+1)$ will be also α times larger so in period $t+1$ the parameters of the model maintain the same transformation relation as in period t. This shows that the dynamics of the model is exactly the same for all parameterizations where the product $\sigma\lambda$ is constant

3 Simulation Experiment Results

We simulate our model assuming that we have $n = 100,000$ agents. The simulation is run for $110,000$ periods of which the first $10,000$ are discarded as burn-in phase. These values are much larger than assumed in [4] to ensure that we obtain information about the behavior of the model in a steady state. We keep the original ranges of values for the design parameters q, σ, and λ as chosen by [4] so that r_t can be interpreted as a daily return rate: $q \in [0.01, 0.1]$, $\sigma \in [0.001, 0.01]$, and $\lambda \in [5, 20]$. We probe the design space in 4710 random points with uniform probability of choosing a combination $(q, \sigma \cdot \lambda)$ from the set $[0.01, 0.1] \times [0.005, 0.2]$ and run the simulation 32 times in each point.

The left plot in Fig. 1 depicts dependence of kurtosis of the distribution of returns on the design parameters. It is easily seen that the model generates leptocurtic distribution of returns only in the narrow area of the design space, while in the

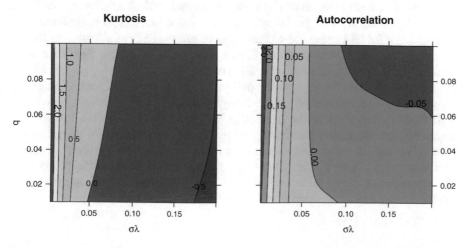

Fig. 1 Estimated response surfaces of kurtosis of distribution of returns and one period autocorrelation of absolute returns

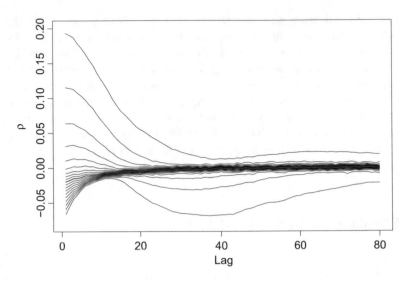

Fig. 2 Vigintiles for distribution of autocorrelation of absolute returns ρ for lags beyond one period; median value marked in *red*

remaining area the distribution of returns is mezocurtic or even platocurtic. The right plot in Fig. 1 depicts dependence of one period autocorrelation of absolute returns on the design parameters. In case of empirical time series this measure is observed to be always positive [11, pp. 90–91]. One can easily see that for the simulated time series there are significant regions of the design space where autocorrelation is either not present or negative.

Autocorrelation of absolute returns for lags beyond one period is depicted in Fig. 2. It shows the median value (in red) and vigintiles for autocorrelation across the entire parameter range. It can be seen that around 70 % of simulations produced $\rho_{\tau=1} < 0$, moreover for 95 % of simulations $\rho_{\tau=1} < 0.193$, and 90 % of simulations have $\rho_{\tau=1} < 0.115$. For the lag 40 those figures drop to 95 % of simulations having $\rho_{\tau=40} < 0.013$ and 90 % of simulations having $\rho_{\tau=40} < 0.006$. Hence for most simulations the autocorrelation is negative or very close to zero.

We confirm these findings by formally testing hypothesis **H1** and **H2** with the Bayesian interval metamodel methodology proposed by [5]. Table 1 shows the calculated probability that the respective hypotheses are true in the investigated range of parameters. The probability values should be interpreted in the following way: given the simulation data we have collected if we would uniformly choose a new random parameter combination $(q, \sigma \cdot \lambda)$ from the set $[0.01, 0.1] \times [0.005, 0.2]$, then the values given in Table 1 are chances that in this new design point we would observe the given combination of autocorrelation and kurtosis.

Table 1 Probability that the hypotheses **H1** and **H2** are true in the investigated range of parameters

Hypotheses probabilities	$\kappa > 0\,(\%)$	$\kappa \leq 0\,(\%)$
$\rho_{\tau=1} > 0$	26.58	2.76
$\rho_{\tau=1} \leq 0$	4.93	65.73

4 Conclusions

By applying the systematic approach to the simulation experiment we have shown the model of Ghoulmie et al. [4] to generate entire spectrum of return trajectories with discrepant properties conditional on the initial parametrization. The properties of leptocurtic distribution of returns and positive one period autocorrelation of absolute returns are not guaranteed to emerge from the model, in fact the probability of both properties to manifest jointly is slightly more than one quarter. We therefore conclude the model to be an insightful but incomplete description of the phenomena under study.

Our findings exemplify importance of applying the principles of experiment design and analysis for credible inference on properties of agent-based models, as postulated by many authors [8, 10].

References

1. Andersen, T.G., Bollerslev, T.: Heterogeneous information arrivals and return volatility dynamics: uncovering the long-run in high frequency returns. J. Finance **52**(3), 975–1005 (1997)
2. DeCarlo, L.T.: On the meaning and use of kurtosis. Psychol. Methods **2**(3), 292–307 (1997)
3. Gaunersdorfer, A., Hommes, C.H., Wagener, F.O.: Bifurcation routes to volatility clustering under evolutionary learning. J. Econ. Behav. Organ. **67**(1), 27–47 (2008)
4. Ghoulmie, F., Cont, R., Nadal, J.P.: Heterogeneity and feedback in an agent-based market model. J. Phys.: Condens. Matter **17**, S1259–S1268 (2005)
5. Kamiński, B.: Interval metamodels for the analysis of simulation input–output relations. Simul. Modell. Pract. Theory **54**, 86–100 (2015)
6. Kirman, A., Teyssiere, G.: Microeconomic models for long memory in the volatility of financial time series. Stud. Nonlinear Dyn. Econom. **5**(4), 281–302 (2002)
7. LeBaron, B.: Evolution and time horizons in an agent-based stock market. Macroecon. Dyn. **5**(02), 225–254 (2001)
8. Lorscheid, I., Heine, B.O., Meyer, M.: Opening the black box of simulations: increased transparency and effective communication through the systematic design of experiments. Comput. Math. Organ. Theory **18**(1), 22–62 (2012)
9. Lux, T., Marchesi, M.: Volatility clustering in financial markets: a microsimulation of interacting agents. Int. J. Theor. Appl. Finance **3**(04), 675–702 (2000)
10. Marks, R.E.: Validating simulation models: a general framework and four applied examples. Computat. Econ. **30**(3), 265–290 (2007)
11. Taylor, S.J.: Asset Price Dynamics, Volatility, and Prediction. Princeton University Press, Princeton (2005)

Social Amplification of Risk Framework: An Agent-Based Approach

Bhakti Stephan Onggo

Abstract There is a paucity in the use of simulation for theory development in the content of the social amplification of risk. Simulation modeling has the advantage of making a theory more precise and including relevant factors within broader boundaries (e.g. the use of multiple actors). This paper demonstrates how an agent-based simulation model can be developed to generate or test a theory in the context of the social amplification of risk. The challenges on model's calibration and validation are highlighted.

Keywords Agent-based simulation • Social amplification of risk framework • Model calibration • Model validation

1 Introduction

Most individuals, consciously or subconsciously, seek to manage risk. The way they assess risk depends on personal characteristics and experience as well as social processes such as norms and collective beliefs. Hence, even though the risk of a real event such as a nuclear accident, food poisoning, or pandemic is real, the perception of individuals of the same risk varies. The perception of risk often affects real risk. For example, during a SARS outbreak, people reduced their travel, which decreased the probability of the spread of SARS. On the other hand, in a flash crowd, risk perception can make real risk more imminent. Because risk perception can affect real risk, managing public risk perception is important for many organizations.

The *Social Amplification of Risk Framework* (SARF) is the framework most commonly used to describe how a society responds to risk, and in particular how a society amplifies or attenuates risk through social processes [1]. The idea behind SARF is that when a risk event happens, it produces signals. These signals are processed and sometimes distorted (amplified or attenuated) by the interactions of social actors in which each actor acts as a communication station. They do not

B.S. Onggo (✉)
Department of Management Science, Lancaster University Management School,
Lancaster, LA1 4YX, UK
e-mail: s.onggo@lancaster.ac.uk

© Springer International Publishing AG 2017
W. Jager et al. (eds.), *Advances in Social Simulation 2015*, Advances in Intelligent
Systems and Computing 528, DOI 10.1007/978-3-319-47253-9_30

simply pass the signals to others but often observe and judge each other's responses. These interactions may lead to considerable distortion to the original signal. Such distortion will become a real management issue when it produces secondary effects, such as a product boycott or a loss of institutional trust.

A substantial amount of empirical work has been conducted on or around the idea of social amplification in various contexts, such as nuclear-weapons-facility accidents [2], genetically-modified foods [3], the Sudan 1 and Hatfield scandals in the UK [4], a zoonosis disease outbreak [5] and wildfire risk [6]. The literature shows that SARF has been the subject of little modeling using simulation. Hence, there is a lack of modeling that looks at the issue from a broader perspective. Examples of simulation-modeling work on SARF include understanding the spread of fear and how a community responds to a terrorist attack [7], demonstrating how a society where individuals correct for other individuals' apparent risk amplifications, instead of simply correcting their own risk beliefs, can have a polarizing effect on risk beliefs in the society, and leading to residual worry and loss of demand for the associated products and services [8] and investigating the effect of broadcast and narrowcast communication on social risk perception [9].

2 Agent-Based Model of Social Risk Amplification

Agent-based simulation (ABS) has been used to theorize a phenomenon in a society. Classic examples include theorizing how segregation might occur in a reasonably tolerant society [10], describing the emergence of wealth inequality in a simple society [11] and theorizing how the rebellion of a subjugated population emerges against a central authority [12]. Given that the heterogeneous nature of social actors, the interactions between social actors, the social network formed by social actors and the ability of social actors to learn are relevant to SARF, the use of ABS for theory development in SARF is justifiable.

Key common actors in the SARF literature have been identified in [9], namely: individuals (lay persons or public), media, risk experts, and risk managers (including authorities or government). Hence, we have developed a model that can represent the interactions between these four types of actors. Figure 1 (left) shows a summary of the communication between actors in the model. A risk manager represents an individual or an institution (including government) that has a legal responsibility or interest in managing risk perception in a society. A risk manager regularly consults experts' assessments of real risk, public-risk perception, and public consumption. Based on this information, the risk manager will make a decision about the level of risk to be communicated to the public. A risk expert makes an assessment of a real risk based on available information (e.g. the number of individuals experiencing a specific risk event and the level of public consumption of some good or service that has become hazardous). A risk expert can communicate his/her assessment to a risk manager and to the media. The media regularly retrieve information from the public, experts, and managers, process information and communicate the processed

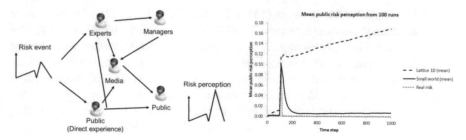

Fig. 1 Communication between actors in the simulation model (*left*); an example of the output of the simulation model (*right*)

information to the public. How the media process information depends on the role that the media assume. The individual represents a member of the public who is concerned with the risk that is being studied. An individual receives information from the media (broadcast) and his/her social network neighbors (narrow-cast). The individual will decide his/her risk perception based on these sources of information and his/her personal characteristics (including past experiences). If the risk perception of an individual is above a certain threshold, the individual will stop consuming.

The main outputs of the model are the mean and standard deviation of public-risk perception and consumption at population and individual levels. The structure of the model allows us to generate or test a theory in the context of SARF such as the impact of social network structure and role of media on the diffusion of risk perception [9]. Figure 1 (right) shows a sample output from the simulation when there is an outbreak between time 100 and 110. It shows that a society with a small-world network is more resilience than a society with a lattice network.

An important lesson learned from the development of the model is to appreciate that SARF is a framework (not a theory). Hence, the first challenge is that SARF does not provide a precise specification that can be unambiguously modeled. The second challenge is on the calibration and validation of the model. First, we need to represent the behavior of actors and the interactions between actors using a set of logical rules. It is challenging to extract this information from real-world actors. Furthermore, real-world actors are often heterogeneous. Hence, it is challenging to validate whether the rules used in the model represent the rule used by most real-world actors and whether we have represented the heterogeneity of real-world actors correctly. The parameters and rules are also contextual (for example the impact of risk perception on real risk in SARS is different from flash crowd). Hence, the data from one context may not be applicable to another context. Secondly, our model requires data about the communication between actors via their social network. Although the collection of high-fidelity quantitative data has become very common, this is not the case for qualitative data. The lack of data makes it diffult to validate a model againt empirical data. There is also a need to validate a model at the individual level and at the population level. The difficulty in validating an ABS model is reflected in the survey in [13] which found that only 35 % of models were validated conceptually and operationally. One of the potential validation approaches is discussed in [14]. In this approach, a modeler writes a validation code before a simulation model is developed. The validation code uses the result from an analytical model to determine if the simulation result is valid. The modeler then writes the simulation model until it passes the validation code. Afterwards, the modeler will write another validation code which evaluate another part of the model. This process is repeated until the simulation model is complete.

3   Conclusion and Future Work

An ABS model that enables theory generation and testing in the context of SARF has been presented. Simulation modeling has the advantage to make the definition and formulation in SARF more formal and specific in such a way that a theory can be tested. We plan to use the model in the context of a product recall after food adulteration events have been exposed. One of the main challenges is in the data collection for model's calibration and validation. We plan to use TDSM.

References

1. Kasperson, R.E., Renn, O., Slovic, P., Brown, H.S., Emel, J., Goble, R., Kasperson, J.X., Ratick, S.: The social amplification of risk: a conceptual framework. Risk Anal. **8**(2), 177–187 (1988)
2. Metz, W.C.: Historical application of a social amplification of risk model: economic impacts of risk events at nuclear weapons facilities. Risk Anal. **16**(2), 185–193 (1996)
3. Frewer, L.J., Miles, S., Marsh, R.: The media and genetically modified foods: evidence in support of social amplification of risk. Risk Anal. **22**(4), 701–711 (2002)
4. Busby, J.S., Alcock, R.E.: Risk and organizational networks: making sense of failure in the division of labour. Risk Manage. **10**(4), 235–256 (2008)
5. Busby, J., Duckett, D.: Social risk amplification as an attribution: the case of zoonotic disease outbreaks. J. Risk Res. **15**(9), 1049–1074 (2012)
6. Brenkert-Smith, H., Dickinson, K.L., Champ, P.A., Flores, N.: Social amplification of wildfire risk: the role of social interactions and information sources. Risk Anal. **33**(5), 800–817 (2013)
7. Burns, W.J., Slovic, P.: The diffusion of fear: modeling community response to a terrorist strike. J. Defense Model. Simulat. **4**(4), 298–317 (2007)
8. Busby, J.S., Onggo, B.S.: Managing the social amplification of risk: a simulation of interacting actors. J. Oper. Res. Soc. **64**, 638–653 (2012)
9. Onggo, B.S., Busby, J.S., Liu, Y.: Using agent-based simulation to analyse the effect of broadcast and narrowcast on public perception: a case in social risk amplification. In: Proceedings of the 2014 Winter Simulation Conference, pp. 322–333 (2014)
10. Schelling, T.C.: Dynamic models of segregation. J. Math. Soc. **1**, 143–186 (1971)
11. Epstein, J., Axtell, R.: Growing artificial societies: social science from the bottom up. Brookings Institution Press, Washington, DC (1996)
12. Epstein, J.M.: Modeling civil violence: an agent-based computational approach. Proc. Nat. Acad. Sci., **99**(3) (2002)

13. Heath, B., Hill, R., Ciarallo, F.: A survey of agent-based modeling practices (January 1998 to July 2008). J. Artif. Soc. Soc. Simulat. **12**(9) (2009)
14. Onggo, B.S., Karataz, M.: Agent-based model of maritime search operations: a validation using test-driven simulation modelling. Proceedings of the 2015 Winter Simulation Conference (2015)

How Precise Are the Specifications of a Psychological Theory? Comparing Implementations of Lindenberg and Steg's Goal-Framing Theory of Everyday Pro-environmental Behaviour

Gary Polhill and Nick Gotts

Abstract This chapter compares four implementations of (Lindenberg and Steg, J Soc Issues 63(1):117–137, 2007) Goal-Framing Theory of everyday pro-environmental behaviour. Two are from different versions of CEDSS (Community Energy Demand Social Simulator, versions 3.3 and 3.4); the other two are different versions of a completely different model that also draws on Goal-Framing Theory (Rangoni and Jager, Modeling social phenomena in spatial context. Lit Verlag, Zürich, Switzerland, 2013). We find that despite some similarities in the models, the implementations are different in a number of important ways, driven in part by the case studies to which they are applied, but also by areas where Goal-Framing Theory doesn't specify any mechanism. We anticipate that as more and more agent-based models draw on social theories, comparisons such as that herein will enable advances in both modelling and the social sciences.

Keywords Goal-framing theory • Agent-based model • Everyday pro-environmental behaviour

1 Introduction

One of the concerns expressed about agent-based modelling is that the algorithms they use to represent behaviour are too ad hoc (see for example [1]). Various approaches have been used to deal with this, including drawing on machine learning algorithms to develop decision-making algorithms (e.g. [2]), validating decision

G. Polhill (✉)
The James Hutton Institute, Aberdeen, AB15 8QH, UK
e-mail: gary.polhill@hutton.ac.uk

N. Gotts
Independent Researcher, Dunfermline, UK
e-mail: ngotts@gn.apc.org

© Springer International Publishing AG 2017
W. Jager et al. (eds.), *Advances in Social Simulation 2015*, Advances in Intelligent Systems and Computing 528, DOI 10.1007/978-3-319-47253-9_31

rules with stakeholders (e.g. [3]), using existing theories in economics (e.g. [4]), or other classical optimisation algorithms (e.g. mathematical programming—[5]; and genetic algorithms—[6]).

Another approach is drawing on existing theories in the social sciences to develop algorithms of human behaviour. Examples include Holtz's [7] model of practice theory [8], and Dubois et al. [9] use of the theory of planned behaviour [10]. This is where agent-based social simulation intersects with the project of Artificial Intelligence, and there has been some success deploying existing algorithms from that area that draw on psychological evidence on how people make decisions (e.g. [11], who use case-based reasoning [12]); while the widespread use of heuristics could be seen as having underpinnings in "folk-psychology").

Computer programs require precise specifications if they are to be successfully implemented. By contrast theories from the social sciences can be notoriously impenetrable and difficult to pin down, especially where there is a culture of rejecting positivism. Bourdieu [13], for example, is paraphrased by Sullivan [14] as asserting that "because the social world is complex, theories about it must be complicated, and must be expressed in complicated language". Though Lindenberg and Steg's [15] Goal-Framing Theory of everyday pro-environmental behaviour is not so extreme, there is still sufficient interpretability of it that implementing it need not result in the same computer program. This chapter explores four implementations of this theory in two models: two versions of CEDSS (Community Energy Demand Social Simulator—[16]) show how the model is refined to try and capture some more subtle features of the theory from one version (3.3) to the next (3.4); and two versions of a model of littering developed independently of CEDSS by different authors [17], but still applied to an everyday pro-environmental behaviour.

In the rest of this chapter, we briefly summarise Lindenberg and Steg's [15] Goal-Framing Theory, before describing each of the models. In a discussion, we examine the differences between them, and consider the implications for implementing social and psychological theories in agent-based models.

2   Goal-Framing Theory

Goal-Framing Theory [15, 18] proposes that goals "frame" how people perceive a choice situation, and so affect the choices they make. Three main clusters of goals guide decision-making. Hedonic goals concern immediate pleasure or comfort; gain goals focus on prosperity and financial security; while normative goals make people focus on "doing the right thing". Typically, multiple goals are active at any given moment but one is focal, the so-called goal-frame. The goal-frame influences what aspects of a situation are noticed, how they are assessed, what considerations come to mind when a decision is to be made.

Which goals predominate in making choices depends on the predominant values of the chooser, and also [18] on external cues that activate different values; for example, information about fashion or comfort would tend to activate hedonic values and thus goals of immediate enjoyment; prices or income expectations would tend to activate the "egoistic" values underlying gain goals; while reports of environmental problems might activate goals linked to normative (more specifically in this case, "biospheric") values.

Hedonic, egoistic and normative values, and the goals they promote, often conflict. Behavioural change can be encouraged by campaigns aimed to strengthen particular values (in the environmental case, biospheric values are of primary importance, but campaigns may also stress saving money, appealing to egoistic values). Another approach attempts to reduce goal conflicts, for example by subsidising pro-environmental alternatives or restricting the sale of damaging ones (reducing egoistic/biospheric conflict), or making such alternatives more enjoyable or promoting them as fashionable (reducing hedonic/biospheric conflict). However, over-reliance on targeting hedonic or gain goals may "crowd out" intrinsic pro-environmental motivations, reducing the likelihood of sustained pro-environmental actions [19].

3   Four Implementations of Goal-Framing Theory

This section discusses four implementations of Goal-Framing Theory in two versions each of two different models, CEDSS and a model of littering.

3.1   CEDSS

CEDSS (Community Energy Demand Social Simulator) is an empirically based model of direct domestic energy use for space- and water-heating, and household appliances, and the main outputs used in assessing the model are the amounts of energy (in the forms of electricity, gas and oil) used for these purposes by the set of households modelled. It is designed to model effects of policies and campaigns aimed at reducing domestic energy demand, in the context of economic scenarios for the period up to 2049 affecting energy prices and household income.

CEDSS agents (representing households) make decisions about the purchase of energy-using and energy-saving equipment, and influence such decisions made by other households in the model community; they have a dwelling, and a set of appliances, which will include a heating system. The dwelling's size, type and installed insulation affect the amount of energy required for space-heating. Households have three "value strength" or "goal-frame" parameters: "hedonic", "biospheric" and "egoistic", which are used to determine the basis on which their decisions are made: the "enjoy" goal-frame corresponds to hedonic values, the "sustain" goal-frame to biospheric values, and "gain" to egoistic ones.

Written in NetLogo [20], CEDSS originally formed part of the European Commission Framework 7 project GILDED (Governance, Infrastructure, Lifestyle Dynamics and Energy Demand—see http://gildedeu.hutton.ac.uk). It is focused on one of the GILDED study areas, Aberdeen and Aberdeenshire, and results of the quantitative survey of those areas formed the primary source of empirical data used to build and run CEDSS.

The model operates with a quarterly time-step during which it repeats the following:

- Some appliances break down, requiring replacement.
- New appliances may enter the market.
- Households do the following:

  – Determine their goal-frame. This is not a conscious choice—rather a stochastic outcome weighted by the relative strengths of values (hedonic, biospheric and egoistic).
  – Replace broken essential appliances.
  – Visit V households with whom they have a social-link, where V is a model parameter.
  – Depending on the goal-frame:

     Buy insulation (if they own the property).
     Update a wish-list of desired appliances.
     Buy new non-essential appliances.

  – Update social-links.
  – In version 3.4, implement external influences on value strength parameters.

An earlier version of CEDSS, CEDSS-3.3, is described in detail in Gotts et al. [16] and the associated downloadable "supporting material" (the code is however known to contain some errors, meaning the model's behaviour is not exactly as intended). The current chapter compares its *intended* behaviour with that of an extended version of the model, CEDSS-3.4; two of the three extensions were designed specifically to reflect aspects of goal-frame theory not included in the earlier version. The third is a first step in linking the modelled community to a wider cultural and informational context, in a way allowed for by the theory. These three modifications are:

1. The use of situational "triggers" with the potential to modify the goal-frame selection process carried out each time-step. The currently implemented types of trigger permitted are changes in the cost of energy, and changes in the level of savings currently held by the household. In CEDSS 3.4 runs where they are included, these "triggers" can cause the "enjoy" goal-frame to be replaced by the "gain" goal-frame if energy prices have risen or savings have fallen, and the reverse change to occur if energy prices have fallen or savings risen.
2. Goal-Framing Theory suggests that a person's actions can modify their sense of their identity and hence strengthen some values relative to others. Specifically, when people realise they have acted pro-environmentally, they may identify

themselves more as someone who values the environment, and because of that shift in identity, give more weight to biospheric values. CEDSS does not distinguish between identity and commitment to values, and its agents are households, but we can partially model such an effect by adding a rule that whenever a piece of equipment is bought while the goal-frame is "sustain" (and hence, is selected in a procedure that prioritises reducing energy-use), the strength of the "biospheric" value is increased, proportionally both to the cost of the item, and to a "biospheric boost" model parameter.

3. Third, CEDSS-3.4 incorporates a simple model of the effects of informational influences on goal-frame selection from beyond the household, and the community of which it is a part. CEDSS households influence each others' value-strengths via the mechanism described earlier. However, communities of the "neighbourhood" size represented in CEDSS are not self-contained, so it is desirable for CEDSS to include mechanisms whereby government policy and informational campaigns could influence domestic energy demand.

Goal-Framing Theory has been used to devise an "Integrated Framework for Encouraging Environmental Behaviour" [18], which proposes two routes to encouraging pro-environmental behaviour. First, by reducing goal conflict by decreasing the costs of pro-environmental choices; this was already implemented to a degree in CEDSS-3.3 by allowing governmental subsidies for pro-environmental purchases. Second, by strengthening normative goals, as represented by this addition to the model. The external communications influencing the community are embodied in a scenario-specific "external-influences file"—the file is thus fixed at the start of the run, on the reasonable assumption that the external influences can be treated as an exogenous input to intra-community dynamics.

At present, the entries in this file do not specify the source or content of communications, merely their effects. These do depend, however, on the current goal-frame of a household—representing the importance of the receptivity of the audience in any information or persuasion campaign. In principle, the external influences could operate for, and against, any of the three overarching values. The file consists of a time series each element of which specifies a time step, what the current goal-frame must be, the value-strength parameter to decrease, the value-strength parameter to increase, and an amount by which to make the adjustment.

CEDSS-3.4 can be run in a "default" configuration that omits all three of these modifications, so the parameter space of CEDSS-3.4 includes that of CEDSS-3.3.

3.2 Rangoni and Jager's Model of Littering

In Rangoni and Jager's [17] model of littering, Goal-Framing Theory is used to explore the motivations to drop litter in an urban street. The work draws on field experiments in Groningen by Keizer et al. [21], in which subjects had a leaflet attached to their bicycle while it was left in an alleyway that had to be removed before they could proceed on their way. The alleyway contained no rubbish bin, and the amount of litter in it was controlled. In a second intervention, they put a sign in the alleyway saying "no littering". People were significantly less likely to litter (throwing the leaflet on the ground, or hanging it on another bicycle) in the clean alleyway than in the alleyway with litter already in it; the presence of a sign exacerbating these tendencies.

The "norm" goal represents the desire to act appropriately (not littering), and the theory is that this goal is weakened in favour of hedonic (instant gratification) or gain (increasing and preserving personal resources) goals when people observe norms being violated. This weakening effect is larger when there is a sign explicitly prohibiting littering and evidence of norm violation. Difficult though it is to see throwing a piece of litter in the street as gratifying, as Keizer et al. [22] describe the situation, littering is seen as pursuing a hedonic goal because it avoids making effort.

Rangoni and Jager's [17] model therefore features just the hedonic and norm goals, and with an interest in investigating the interaction effect of norm following and norm violation, they use a reinforcement model to simulate the effect of the norm following/violation on an agent's norm activation level. This choice of reinforcement model is because, as they point out on p. 52: "The Goal Framing Theory does not explicitly describe the mechanism responsible for the activation [or] inhibition of the normative motive."

In the description of the reinforcement model on p. 53, agents i each have a norm activation level, N_i, a tolerance threshold, T_i, and a reinforcement amount, R_i. The norm activation level is updated thus (1): If the tolerance threshold exceeds the amount of littering in the alleyway (L), then the norm activation level is decremented by the reinforcement each time step (negative reinforcement); if it is less than the amount of littering in the alleyway, then the norm activation level is incremented by the reinforcement. A parameter S represents the effect of the "no littering" sign. ($S = 1$ when the sign is absent; $S > 1$ when present.)

$$N_i(t) = \begin{cases} N_i(t-1) - R_i S & T_i > L(t) \\ N_i(t-1) + R_i S & T_i < L(t) \end{cases} \tag{1}$$

The parameters of the model are described on pp. 54–55. These include population-level parameters for mean and standard deviation of hedonic and norm goal weight, and for the reinforcement parameter and tolerance threshold, as well as other environmental parameters. The link between the hedonic and norm goal weights and the norm activation level is not specified precisely. However, on p. 57, where the authors discuss an experiment with the model, they say that "the clean street reinforces the normative goal weight", and on p. 58 they say that at a critical point in the amount of litter, "a transition occurs as some agents' normative goal weight is suddenly weakened". Treating the "normative goal weight" as though it were the same variable as the "norm activation level" in the reinforcement process would not be inconsistent with either of these statements, but means that there is no apparent role for the hedonic goal weight. The latter seems to appear algorithmically only in a third version of the model, in which there is the option of throwing the litter in a bin. In this case, the goal weights are used to compute payoffs for each of the three options (take the litter away, throw the litter on the ground, throw the litter in the bin) based on the distance to the bin. The fact that there is no role for the hedonic goal weight in the "transition" experiment (as opposed to the "bin" experiment) is not problematic if, as in CEDSS, the sum of the goal weights for each agent is a constant.

4 Comparing Output

Experiments have been done using four versions of the CEDSS-3.4 model, which throw additional light on the differences found even between successive versions of the same model. These model versions were selected according to their performance in matching energy demand estimates from a GILDED survey of households in north-east Scotland [16]. As depicted in Fig. 1, the versions differ in whether they included a "biospheric boost" and/or situational triggers; the version without either is referred to here as B0-No, that with a biospheric boost only as B4-No, those including situational triggers as B0-Yes and B4-Yes. To improve its performance, B0-Yes differs in one other parameter from the other three, so results across the four are not strictly comparing the pure effect of the differences between versions. However, we can observe interesting difference in the *patterns* of results for the four

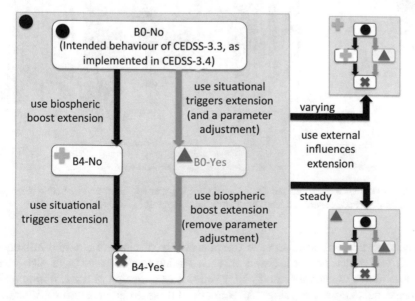

Fig. 1 Diagram showing the various versions of CEDSS used for runs in this section, and how they are related, together with coloured shapes showing how they are depicted in Fig. 2

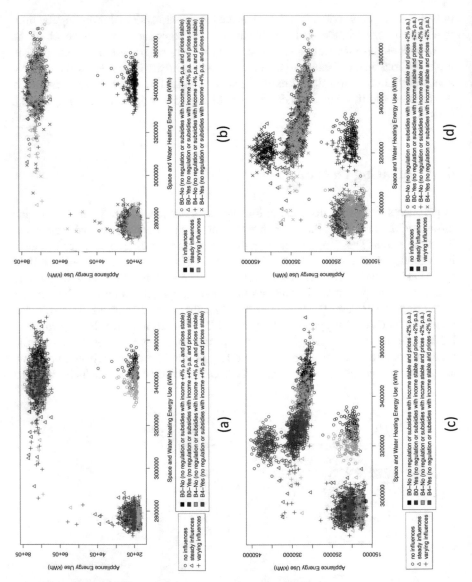

Fig. 2 CEDSS3.4 model: demand for domestic heating and appliance energy across a range of model versions and scenarios. Each point marks the outcome from one run of the model

models, particularly in relation to the presence or absence of external influences, which was varied across the runs of each model version (runs were also varied across a range of economic and policy futures; examination of the resulting multi-dimensional dataset is ongoing). An example of these different patterns is illustrated in Fig. 2.

The top two graphs (Fig. 2a, b) show the outcomes of simulation runs in the policy-economic family of scenarios: "no regulation or subsidies with income +4%pa and prices stable". The bottom two graphs ((c) and (d)) show "no regulation or subsidies with income stable and prices +2%pa." The left two graphs ((a) and (c)) show the model versions in colour (black = B0-No; red = B0-Yes; green = B4-No; blue = B4-Yes), and the external influences in shapes (circle = no influences; triangle = steady influences; + = varying influences). The right two graphs ((b) and (d)) show the same data but swapping colour (black = no influences; red = steady; green = varying) and shape (circle = B0-No; triangle = B0-Yes; + = B4-No; × = B4-Yes).

It is clear that the presence of external influences made a considerable difference in both families of scenarios, and across models, but effects within the different models are not the same. In the top scenario ((a) and (b)), all model versions appear in similar areas apart from the cluster of points on the bottom right. All points in this cluster have no influences (black colour in (b)), and all are apparently from model versions B0-No and B4-No (green and black in (a))—i.e. without situational triggers. B0-No with no influences (black circles in all graphs) lacks all three of the additions CEDSS-3.4 made to CEDSS-3.3.

In the bottom scenario ((c) and (d)), differences between model versions are considerably more marked. There are distinct clusters for the runs with and without situational triggers (black/green vs blue/red in (c)). On the right hand side (high heating energy use), when there are no influences (circles in the left graph (c); black in the right (d)), there is a more significant split on appliance energy use between the Yes/No runs (Yes at the very top; No at the bottom), and these two regions also seem to feature sub-regions with different concentrations of B0 and B4 model versions. In the middle on the right hand side is a "smear" with points covering all three influence scenarios, but again, with distinct, but overlapping regions covering each of the model version settings. This graph shows that the modifications to CEDSS have all had a visible effect, the complexities of which are a current focus of investigation. The tendency for very distinct clusters of results to appear is thought to result from the particular mechanisms CEDSS-3.4 uses in adjusting value strengths. Experimenting with modified approaches is on our agenda; this will produce still another dimension on which models based on Goal-Framing Theory can differ.

5 Discussion

A summary of the two versions of each of the two models is shown in Table 1. At the time of writing, there is no publicly available version of the implementation of the litter-picking model, hence we have only been able to explore the effect of changing implementations on output dynamics with respect to the CEDSS model. As is clear from the previous section, the extensions to CEDSS do change output dynamics. Comparison of the CEDSS and litter-picking models is therefore at the level of their

Table 1 Comparison of the four models' approaches to implementing features of goal-framing theory

Feature	CEDSS 3.3	CEDSS 3.4	Littering ("transition")	Littering ("bin")
Goals represented	Hedonic, gain, norm	Hedonic, gain, norm	Hedonic, norm	Hedonic, norm
Goal frame parameters	Weights ("value strength parameters") for each that sum to a constant	Weights ("value strength parameters") for each that sum to a constant	Numbers ("goal activation levels") in the range [0, 100]; can be effectively seen as summing to 100	Numbers ("goal activation levels") in the range [0, 100]; no constant sum
Influence on goal activation levels	Visiting neighbours; habit	Visiting neighbours; habit; "biospheric boost"; information campaigns	Norm following or violation by others; presence or absence of a sign	Norm following or violation by others; presence or absence of a sign; distance to bin
Goal frame choice	Weighted random choice	Weighted random choice with situational triggers	Threshold in relation to environmental variable	No explicit choice of goal frame
Effect of goal frame on decision	Chosen goal influences choice algorithm	Chosen goal influences choice algorithm	Selected goal determines whether or not agent litters	Goal activation levels used to compute "scores" for behaviours; maximum scoring behaviour chosen
Calibration	Survey and secondary data	Survey and secondary data	Field experiments	Field experiments

design, and it is this we turn to here. This discussion focuses on two versions of CEDSS: CEDSS-3.3 as implemented in CEDSS-3.4, and CEDSS-3.4 including all extensions to CEDSS-3.3 covered earlier.

What is interesting about all four models is that numerical representations underpin the parameters determining which goal will be dominant ("value strength parameters" in CEDSS; "goal activation levels" in the littering model), though there are differences in the use of numbers across the four versions. Use of numerical representation is arguably common in the formalisation process for many agent-based models, though it has been argued against (e.g. [23]). This commonality, however, should arguably be seen as a feature of modelling practice, rather than something stipulated by Goal-Framing Theory.

In Goal-Framing Theory, the choice of goal frame is driven by the values of the individual and situational cues in the environment. The models have adopted very different approaches to this, but this is driven mostly by the case studies, which, if nothing else, feature significantly different temporal scales. In CEDSS-3.3, the only environmental cues are related to households visiting each other, whereas in CEDSS-3.4, the introduction of situational triggers and information campaigns adds additional environmental cues. At a fairly abstract level, links could be seen between the litter in the street (norm following or violation by others) and the information households in CEDSS get when they visit each other; and between the "no littering" sign and the information campaigns. (In the latter case, however, it is not clear whether there is an "exacerbation" effect of information campaigns as there is for signs.) However, the mechanisms for implementing the effect on the choice of goal frame of the values of the individual and the environmental cues are different:

- In CEDSS, a weighted selection is made using the value strength parameters.
- In the "transition experiment" littering model, the goal frame chosen depends on the amount of litter in the environment crossing an agent-specific threshold.
- In the "bin experiment" littering model, there is no explicit selection of goal frame prior to choosing behaviour; rather, scores are computed for each behavioural option using the agent's goal frame parameters and appropriate environmental variables. The behaviour chosen is then the option with the highest score.

Adjusting the goal frame parameters is also different across the models. Comparing the normative influence:

- In CEDSS, when one household visits another, the values are adjusted by a fixed, agent-specific proportion of the difference between the two. The values corresponding to all three goal frames are adjusted in this way.
- In the littering model, the norm activation level is reinforced negatively or positively by an agent-specific constant depending on whether a threshold level of litter has been breached. However, there is no adjustment of the hedonic activation level.
- CEDSS-3.4 also adds an adjustment of the goal frame parameters when households make a pro-environmental purchase using the biospheric goal frame. This adds a proportion of the purchase price to the corresponding value of the household, with appropriate decrements in hedonic and egoistic values.

Here, although both models use heterogeneous parameters to model the adjustment, the algorithm for doing so is different. In particular, the adjustment in CEDSS is for pairs of agents at a time, whereas the adjustment in the littering model is stigmergic, responding to aggregate behaviour of all agents.

Another endogenous adjustment of goal frame parameters occurs only in CEDSS. Here, when a goal frame is selected, adjustments are made to the goal frame parameters to make the chosen goal frame more likely next time. This is done using a constant factor added to the goal frame parameter corresponding to the goal frame chosen, with associated decrement in the other two goal frame parameters to keep the sum of all three parameters constant.

Comparing the "exogenous" influence, there are also significant differences between the models:

- In CEDSS-3.3, there are no exogenous influences.
- In the littering models, the presence of a sign acts as a factor increasing the positive or negative reinforcement of the norm activation level.
- In CEDSS-3.4, the effects of the external influences are rule-based; the effects as parameterised are to increase biospherism by a constant, decreasing hedonism and/or egoism accordingly.

Both versions of the CEDSS model also share with the "transition experiment" version of the littering model the idea that a single dominant goal frame drives the behavioural choice; in CEDSS the link is more indirect in that the goal frame determines the method by which the eventual choice is made, whereas in the "transition experiment" littering model, the selected goal frame directly determines the behaviour. However, in the "bin experiment" version of the littering model, the goal activation levels are instead used to compute a score for each of the behavioural options, with the highest-scoring option being selected.

As noted in the above quotation of Rangoni and Jager [17], the lack of any detailed description of mechanism for parts of Goal-Framing Theory is, besides the differences in case study, an important contributing factor in the differences observed in the four models. Assertions such as that observing norm violation weakens normative goals are understandable in the context of a theory in environmental psychology, but, even if they do suggest a numerical representation of "goal strength" do not specify how much the strength of the goal is reduced by any given observation—the kind of information needed to develop a model. Insofar as methods of social research are constrained in their ability to ascertain these kinds of details empirically, developing proposed mechanisms will be left to modellers. However, posing questions about mechanisms through interaction with modellers may lead to the development of new research methods.

The fact that both models are calibrated against empirical data further suggests that statistical metrics showing calibration and validation of emergent outputs cannot be trusted to confirm that a proposed mechanism developed by modellers is somehow "correct". This is not just a problem for agent-based modelling, as traditional function-fitting methods essentially "hide" the underlying mechanism. The lower level of representation in agent-based modelling means that the issue becomes more apparent, however. Although Moss and Edmonds [3] recommend validating qualitatively at the micro-level to support quantitative validation at the macro, more general approaches to validating the model's ontology (including processes and algorithms or rules) are needed.

Thus far, the irregularly held model-to-model workshops [24, 25] and other model docking exercises (e.g. [26]) have tended to focus on reproducing simulation results by re-implementing or re-running models in other work. In the first model-to-model workshop, Cioffi-Revilla and Gotts [27] did some work showing how their models with ostensibly different application domains had sufficiently similar characteristics to identify the $TRAP^2$ class of models. To our knowledge, agent-based models have not previously been compared for their implementation of underpinning theory. This is an issue that will be less pronounced in other disciplines, where theories tend already to be expressed in formal languages; where agent-based models do draw on underpinning theory, this is often economic theory, which is itself formal. As more and more agent-based models draw on underpinning social theories, opportunities to conduct studies such as this can be expected to increase.

This study has shown that the deployment of Goal-Framing Theory in two different models has led to some significant differences in implementation details—even across versions of these models. Results from CEDSS have shown that these implementation details can also have notable differences in reported outcome. For those responsive to the potential for dialogue between the process of formalisation of theory, and the iterative refinement thereof, questions of how variant implementations are offer insights into where ambiguities in the theories lie, as well as exposing areas of modelling practice that could be challenged. An alternative view is that such variation exposes weaknesses in relying on social theory to ground social simulation and address concerns about "*ad hocery*". However, rejecting the formalisation of theories in the social sciences as an impossible project ignores opportunities to make significant gains in multiple disciplines, and Goal-Framing Theory has at least imposed some constraints on the ways in which decision-making on everyday pro-environmental behaviours have been represented.

Acknowledgements This work was funded by the Scottish Government Rural Affairs and the Environment Portfolio Strategic Research Theme 4 (Economic Adaptation), and the European Commission Seventh Framework Programme under grant agreement SSH-CT-2008-225383 (GILDED).

References

1. Agar, M.: My kingdom for a function: Modeling misadventures of the innumerate. J. Artif. Soc. Soc. Simulat. **6**(3), 8 (2003). http://jasss.soc.surrey.ac.uk/6/3/8.html
2. Sánchez-Maroño, N., Alonso-Betanzos, A., Fontenla-Romero, O., Brinquis-Núñez, C., Polhill, J.G., Craig, T., Dumitru, A., García-Mira R.: An agent-based model for simulation environmental behavior in an educational organization. Neural Process. Lett. **42**(1), 89–118 (2014)
3. Moss, S., Edmonds, B.: Sociology and simulation: statistical and qualitative cross-validation. Am. J. Soc. **110**(4), 1095–1131 (2005)
4. Brown, D.G., Robinson, D.T.: Effects of heterogeneity in residential preferences on an agent-based model of urban sprawl. Ecol. Soc. **11**(1), 46 (2006). http://www.ecologyandsociety.org/vol11/iss1/art46/

5. Schreinemachers, P., Berger, T.: An agent-based simulation model of human-environment interactions in agricultural systems. Environ. Model. Software **26**(7), 845–859 (2011)
6. Manson, S.M.: Agent-based modeling and genetic programming for modeling land change in the Southern Yucatán Peninsular Region of Mexico. Agr Ecosyst Environ **111**(1-4), 47–62 (2005)
7. Holtz, G.: Generating social practices. J. Artif. Soc. Soc. Simulat. **17**(1), 17 (2014). http://jasss.soc.surrey.ac.uk/17/1/17.html
8. Shove, E., Pantzar, M., Watson, M.: The Dynamics of Social Practice: Everyday Life and How It Changes. Sage, London (2012)
9. Dubois, E., Barreteau, O., Souchère, V.: An agent-based model to explore game setting effects on attitude change during a role playing game session. J. Artif. Soc. Soc. Simulat. **16**(1), 2 (2013). http://jasss.soc.surrey.ac.uk/16/1/2.html
10. Ajzen, I.: The theory of planned behavior. Organ. Behav. Hum. Decis. Process. **50**, 179–211 (1991)
11. Izquierdo, L.R., Gotts, N.M., Polhill, J.G.: Case-based reasoning, social dilemmas, and a new equilibrium concept. J. Artif. Soc. Soc. Simulat. **7**(3), 1 (2004). http://jasss.soc.surrey.ac.uk/7/3/1.html
12. Aamodt, A., Plaza, E.: Case-based reasoning: Foundational issues, methodological variations, and system approaches. AI Commun. **7**(1), 39–59 (1994)
13. Bourdieu, P.: In Other Words. Polity Press, Cambridge (1990)
14. Sullivan, A.: Bourdieu and education: how useful is Bourdieu's theory for researchers? Neth. J. Soc. Sci. **38**(2), 144–166 (2002)
15. Lindenberg, S., Steg, L.: Normative, gain and hedonic goal frames guiding environmental behavior. J. Soc. Issues **63**(1), 117–137 (2007)
16. Gotts, N.M., Polhill, J.G., Craig, T., Galan-Diaz, C.: Combining diverse data sources for CEDSS, an agent-based model of domestic energy demand. Struct. Dynam. **7**(1), (2014) https://escholarship.org/uc/item/62x9p0w4
17. Rangoni, R., Jager, W.: Can Naples ever be a litter free town? Simulating littering behavior using the Goal-Framing Theory. In: Koch, A., Mandl, P. (eds.) Modeling Social Phenomena in Spatial Context. Lit Verlag, Zürich, Switzerland (2013)
18. Steg, L., Bolderdijk, J.W., Keizer, K., Perlavicute, G.: An integrated framework for encouraging pro-environmental behaviour: the role of values, situational factors and goals. J. Environ. Psychol. **38**, 104–115 (2014)
19. Taufik, D., Bolderdijk, J.W., Steg, L.: Acting green elicits a literal "warm glow". Nat. Clim. Change **5**, 37–40 (2014)
20. Wilensky, U. NetLogo. http://ccl.northwestern.edu/netlogo/. Center for Connected Learning and Computer-Based Modeling, Northwestern University. Evanston, IL (1999)
21. Keizer, K., Lindenberg, S., Steg, L.: The reversal effect of prohibition signs. Group Process. Intergroup Relat. **14**(5), 681–688 (2011)
22. Keizer, K., Lindenberg, S., Steg, L.: The spreading of disorder. Science **322**(5908), 1681–1685 (2008)
23. Edmonds, B., Hales, D.: When and why does haggling occur? Some suggestions from a qualitative but computational simulation of negotiation. J. Artif. Soc. Soc. Simulat. **7**(2), 9 (2004). http://jasss.soc.surrey.ac.uk/7/2/9.html
24. Hales, D., Rouchier, J., Edmonds, B.: Model-to-model analysis. J. Artif. Soc. Soc. Simulat. **6**(4), 5 (2003). http://jasss.soc.surrey.ac.uk/6/4/5.html
25. Rouchier, J., Cioffi-Revilla, C., Polhill, J.G., Takadama, K.: Progress in model-2-model analysis. J. Artif. Soc. Soc. Simulat. **11**(2), 8 (2008). http://jasss.soc.surrey.ac.uk/11/2/8.html
26. Axtell, R., Axelrod, R., Epstein, J., Cohen, M.D.: Aligning simulation models: A case study and results. Comput. Math. Organ. Theor. **1**(1), 123–141 (1995)
27. Cioffi-Revilla, C., Gotts, N.: Comparative analysis of agent-based social simulations: GeoSim and FEARLUS models. J. Artif. Soc. Soc. Simulat. **6**(4), 10 (2003). http://jasss.soc.surrey.ac.uk/6/4/10.html

Lessons Learned Replicating the Analysis of Outputs from a Social Simulation of Biodiversity Incentivisation

Gary Polhill, Lorenzo Milazzo, Terry Dawson, Alessandro Gimona, and
Dawn Parker

Abstract This chapter reports on an exercise in replicating the analysis of outputs
from 20,000 runs of a social simulation of biodiversity incentivisation (FEARLUS-
SPOMM) as part of the MIRACLE project. Typically, replication refers to reconstructing the model used to generate the output from the description thereof, but
for larger-scale studies, the output analysis itself may be difficult to replicate even
when given the original output files. Tools for analysing simulation output data do
not facilitate keeping records of what can be a lengthy and complicated process.
We provide an outline design for a tool to address this issue, and make some
recommendations based on the experience with this exercise.

Keywords Social simulation outputs • Metadata • Analysis

1   Introduction

Replication of social simulation results has been highlighted as a significant issue
for the discipline for a number of years (e.g. [1]), and has even led to a short series of
workshops [2, 3]. The focus of replication work has thus far been on the model itself,
but as will be shown here, the analysis of the outputs of the model can potentially

G. Polhill (✉) • A. Gimona
The James Hutton Institute, Aberdeen, AB15 8QH, UK
e-mail: gary.polhill@hutton.ac.uk; alessandro.gimona@hutton.ac.uk

L. Milazzo
The James Hutton Institute, Aberdeen, AB15 8QH, UK

The University of Dundee, Dundee, UK
e-mail: lorenzo.milazzo@hutton.ac.uk

T. Dawson
The University of Dundee, Dundee, UK
e-mail: t.p.dawson@dundee.ac.uk

D. Parker
The University of Waterloo, Waterloo, ON, Canada
e-mail: dcparker@uwaterloo.ca

© Springer International Publishing AG 2017
W. Jager et al. (eds.), *Advances in Social Simulation 2015*, Advances in Intelligent
Systems and Computing 528, DOI 10.1007/978-3-319-47253-9_32

be just as complex, and no less difficult to replicate unless adequate records are kept. Schmolke et al. [4] TRACE protocol provides some guidance highlighting the need to keep a notebook of the analysis done; however, lessons can be learned from the replication exercise reported herein that provide more detailed guidance on the information that should be recorded, and on tools that could be used to support the process.

The output analysis replication in this paper concerns earlier work with FEARLUS-SPOMM, which is a coupled agent-based model of agricultural decision-making and species stochastic patch occupancy metacommunity model that has been used to explore incentivisation strategies to improve biodiversity [5, 6]. Belonging to the "typification" class of social simulations ("theoretical constructs intended to investigate some properties that apply to a wide range of empirical phenomena that share some common features"—[7]), recent work involved the analysis of the outputs from around 20,000 runs of the model using a number of techniques aimed at demonstrating nonlinearities in the relationship between incentivisation and biodiversity outcome.

Recording workflow data on the process used to create analysis can be challenging, and currently there are no codified standards as to how this should be done for ABMs. For FEARLUS-SPOMM, the methods used drew heavily on statistical techniques available as R packages that are as part of core R functionality. Although R allows transcripts of interactive terminal sessions to be saved, the work involved great deal of exploration of different ways of attempting to visualise and analyse the nonlinearities in the model results, not all of which were likely to be reported in the paper. Such logs are therefore not the best way to record the means by which the outputs were analysed, and hence the strategy used was to save each analysis or visualisation in a(n R) script. Since the output from the (Swarm) software that generated the output data being analysed used a mixture of text formats, some Perl scripts were also written to process that output into a CSV file for easy import into R. When the MIRACLE project [8] provided a context in which the replication of that analysis was necessary, an opportunity was created to test the viability of the above strategy.

In the rest of this chapter, we describe the information available for the output analysis replication process, how it was used to regenerate some of the figures in Polhill et al. [6] (and the information that was missing that would have facilitated this process), and we describe a prototype tool to support keeping the records needed to perform replication more easily, before making some recommendations in concluding remarks.

2 Replicating the Output Analysis

This section briefly summarises the experiments done in Polhill et al. [6], before describing the way in which the analysis was reconstructed from the information available.

2.1   Summary of the Analysis and Techniques Used

The original work conducted a series of simulations exploring the parameter space of the model, with a particular focus on the relationship between the amount of incentive to farmers offered by each mechanism and the resulting landscape-scale species richness after 200 iterations of the model. Four incentivisation mechanisms were explored covering the following two dimensions; each has as parameter an incentive amount offered:

- Providing incentives for specific activities aimed at improving biodiversity versus providing incentives for biodiversity outcomes.
- Providing incentives to farmers individually versus providing increased incentives to farmers when they and their neighbours delivered the activity or outcome the mechanism required (referred to as "clustered incentives").

To cover other circumstantial factors, the model explored two levels of aspiration for profit in the farming agents, two levels of input costs, and two stylised time series of the market prices for the goods produced on the farm. The combination of incentivisation mechanism, market, input costs and aspirations is referred to as a "scenario"; there are 32 of these.

Incentive values covering 1, 2, . . . , 10 were explored, with 20 runs per parameterisation. This makes 6400 runs. Differences between the incentivisation mechanisms mean that some naturally spend more money than others given the same set of circumstances. To correct for this, the 6400 runs were repeated twice further, once dividing the clustered incentive values by 2 (i.e. 0.5, 1, 1.5, . . . 5 instead of 1, . . . , 10), and once dividing these by 10 (i.e. 0.1, . . . , 1). This makes 19,200 runs. In a second phase of runs, the non-clustered incentivisation mechanisms were explored using incentive values 15, 20, 25, 30, 40, 50, 100, requiring a further 2240 runs.

Analysis of the outputs reported in the paper covered the following:

- Rejecting runs with levels of expenditure that are too high, or levels of bankruptcy that are too high. These runs were deemed unrealistic, and the rejection left 16,949 runs for use in subsequent stages.
- Testing for nonlinearity in each scenario in the relationship between incentive amount and landscape scale species richness in the last recorded time step of the model. Five tests were used, the details of which are not important here (see Polhill et al. [6] for details), but they involved building and comparing Generalised Additive Models (GAMs; [9]) and linear models of the relationship, as well as computing the Akaike Information Criterion [10].
- Using recursively partitioning classification trees [11] to see how scenario variables and expenditure combined to deliver landscape-scale species richness.

Archives of the raw simulation output files were kept. The goal of the exercise reported here is for another team member not involved in the original research to replicate the output analysis described above given these files.

2.2 Records Kept

The following details the available information for the reconstruction process:

- Diaries of research activities on analysing the outputs starting 19 October 2011 and running until 10 August 2012;
- Perl, shell and R scripts that were intended to act as documentation of these activities;
- A README file describing some of the Perl and R scripts, and associated output files.

The diaries were contained principally in two files covering consecutive time periods, but were also distributed in different versions across different computers used to perform the analysis. The same applied to the scripts, of which there are 74 R scripts 29 Perl scripts and 19 shell scripts. However, these scripts can be grouped as several of them are updated versions of others, fixing bugs or providing additional or variant functionality. As scripts essentially aimed at recording an analysis or visualisation that would otherwise be done interactively with the R console, and hence less "formal" computer software, version control software was not used to keep track of these updates. More importantly, several approaches to analysing and visualising the output were tried during the course of the work, most of which were "dead ends" or not included in the final paper for other reasons. Although the README file and diaries provided some guidance, they were not always sufficient to record precisely which version of which script was used to produce a particular result.

2.3 Reconstructing the Output Analysis

Figure 1 shows the process by which the various analysis and visualisation artefacts in the paper (shaded nodes) were produced. Each run of the model generates a number of files. The SPOMM half creates several CSV files storing data about species occupancy in various different ways. The FEARLUS half has a bespoke text output format for reports on the model, and another output format loosely based on ARC's Grid/ASCII format for reporting on spatial data in which multiple layers of spatial data for exactly the same region and division of space are saved in the same file.

The first task in analysing the output was to create a single CSV file containing salient information on each run. This involved processing the several thousand files from all the runs of FEARLUS-SPOMM using a Perl script that provided a summary of each run, one per line of the CSV file. There were six versions of this script on two different machines, four of which contained bugs fixed in later versions, and a fifth contained functionality used for a later set of runs. Although this should mean that it is trivial to deduce which version of the script was used to create

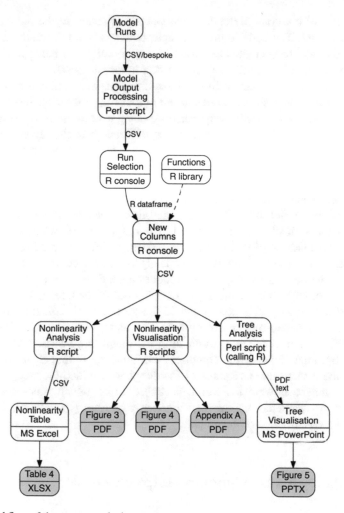

Fig. 1 Workflow of the output analysis process

the CSV file used for subsequent analysis (the one that has the fixed bugs but not the additional functionality), some of the R scripts associated with earlier analysis contained corrections to the data to address the bugs that were fixed later. Although the scripts used in the published analysis did not contain these corrections, it would have been more reassuring to have had a record of which version was used to create the CSV file used by these scripts, and indeed of which CSV file these scripts used.

As mentioned earlier, some runs reported in the resulting CSV file were eliminated from consideration because of overly high bankruptcy or expenditure rates. This was done using the R console, reading the file in, removing the offending runs from the data frame, and saving the result as a new CSV file. However, during the reconstruction process, it became clear that subsequent scripts assumed the existence of columns in the data that were not present in the output from the processing script. The code to produce these columns was eventually found in an R library that had been created because this functionality was being replicated in a number of the analysis scripts used during exploratory analysis.

The remaining scripts cover different visualisations and analyses reported on in the paper, and feature similar issues to the output file processing script discussed above. The analysis of nonlinearity was done using a script with five versions, and visualised using a script with four; however, differences in the layout and style of the visualisations in various figures in the Polhill et al. [6] paper meant that the scripts used for these were more definite. The five versions of the nonlinearity analysis scripts consist of an original version of the script, and four revisions of it that variously change default settings to those eventually used in the paper, and add some functionality not used. This also meant that it was possible to deduce which version of the script had been deployed. The output of this script (another CSV file) was subsequently loaded into a spreadsheet, which was used to generate the table used in the paper. The decision tree analysis was conducted with a Perl script that calls R while it is running. This produces a PDF containing graphs showing distributions of key model variables at each leaf node of the tree, and text describing the tree structure as part of the output to the terminal. The diagram used in the paper was constructed by hand from these files using presentation software, with the "notes" section of the slide containing the diagram consisting of text pasted from the terminal showing the command used and the output therefrom. Hence, although there were three versions of the script to produce the tree, the version used was known, as were the command-line options, which, since they include configuration parameters for the decision tree algorithm (rpart()), are critical for successful reconstruction of the output.

3   Designing Tools to Support Output Analysis

The missing information that would have critically helped in reproducing the output analysis described above is provenance-based—which script used which file(s) as input and generated which other file(s) as output, though other practices, discussed later, would also have facilitated this process. Identification of the key processes (events, actions) is one the crucial points that needs to be addressed in order to guarantee the replication (reproducibility) of any experiment or study. In the case of an in silico experiment, events and actions are associated with running applications, scripts and other software. At the most abstract level, the applications are making use of various forms of "container"; which in specific cases could be files (e.g. input/output files) or data structures (e.g. arrays or hash tables containing initialization or input data).

In general, an off-the-shelf Workflow Management System (e.g. Taverna-[12]) can be adopted to implement a (semi-)automated system to run the simulation, process and analyse the data. However, we propose a "light-weight" approach with

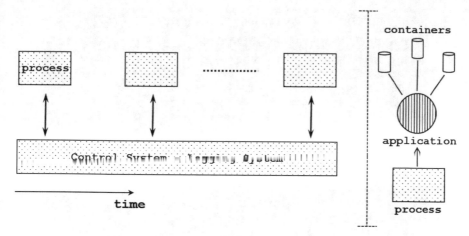

Fig. 2 Schematic diagram of the software system

a view to minimising the learning curve associated with a fully-fledged application, and decided to develop a bespoke application written in Python that also allows the replication of an entire study. Assuming a working practice in which scripts (e.g. in R) are written to cover each stage of the analysis process, the Python application is intended to "wrap" calls to the script, thereby keeping track of the required provenance.

The current version of the software consists of several components (Fig. 2) including: (a) a procedure for "forking" and monitoring processes, (b) a logging system (c) procedures to process different kinds of data (I/O, metadata), (d) an interface allowing the storage/retrieval of data into/from a relational database.

The logging system plays a crucial role in the system. A logging system has been implemented to perform bookkeeping, scheduling, and error handling, and includes capture of provenance metadata. The monitoring carried out by the logging system is closely related to the identification, organisation, and storage of the relevant information describing the simulation study. These key data are processed according to a set of pre-established metadata for social simulation outputs, which is the subject of on-going work.

4 Recommendations and Future Work

At least one of the stages in producing the final results could have been eliminated if the model had not produced output data in a bespoke text format. (Indeed, later versions of FEARLUS use XML rather than the bespoke text format for the runs reported here.) At last year's Social Simulation Conference in Barcelona, participants were asked to provide a record of their current practice in recording

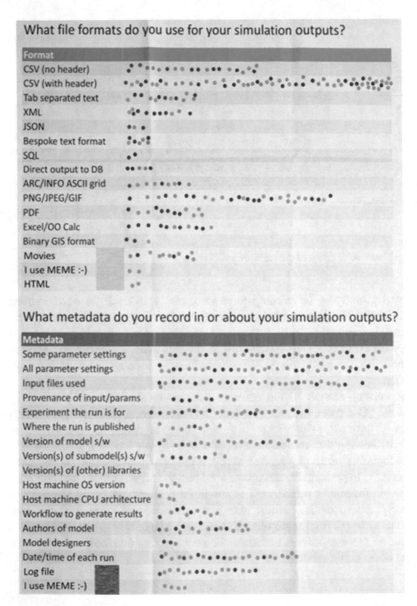

Fig. 3 Poster at last year's SSC capturing current practice. Entries added by participants are indicated with the *blue/purple rectangle*. There is no significance to the colours of the dots placed on the poster by participants

simulation outputs and associated metadata. The results (Fig. 3) show a strong preference for CSV formats (with or without a header line describing the data in the columns). Several people also use a tab-separated text format.

The reasons for this are obvious: CSV and tab-separated formats are almost universally readable by other software that might be used to do analysis of the outputs (e.g. R and spreadsheet software). Indeed, Repast Simphony and NetLogo (arguably the two most popularly-used agent-based modelling toolkits) use CSV format (or have an option to do so) in their parameter sweep and behaviour space (to use their respective nomenclature) batch run tools.

One important recommendation therefore is in encouraging developers of agent-based models to use output formats that are readily used by other software. CSV and tab-separated formats may not always be suitable for encoding outputs from social simulations, however. In particular, although CSV files can be used for raster spatial data using columns for x and y co-ordinates, vector-based representations may be less easy to encode this way, especially if the model makes important distinctions between point, line, area and volume spatial regions and the properties they have. Output data may also need to encode properties of different classes of object more generally. This suggests that broader support is needed for data formats such as XML or JSON that allow some encoding of the structure of the data. Capturing the relational structure of the output may also suggest adopting a practice of saving outputs in relational databases (though this was one of the least popular options in the sample). Tools such as sqlite3 offer options for providing such storage without the investment in configuration and management of larger RDBMS systems such as MySQL and PostgreSQL. R provides support both for reading XML files and for interacting with databases, but these tools are much less popularly used.

A second recommendation is that tools be provided that facilitate the recording of provenance metadata for the results of output analyses. Two important aspects of provenance would have helped with this reconstruction: (1) knowing which script was used to generate which output file, which input files were used, and which command-line options; (2) knowing which script is a revision of which other script or scripts. Although recommendations such as PROV [13] provide the basis on which such information can be structured, popularly used tools do not save such metadata. Log files of console sessions (both for the shell terminal and for R when this was used interactively) would have captured the information needed to perform the reconstruction, but they potentially contain a great deal of noise when attempting to find out specific information about how a particular visualisation or result was produced. Similarly, the diary contains information not relevant to the paper, since it documents numbers of other analyses that were not included. Worse, as free text, it would be much more challenging to provide an automated search facility that could theoretically extract the information associated with a specific query.

Using scripts to perform particular analysis and visualisation tasks does at least act as a record of that task, and one that can theoretically be reused for other tasks. Although there were several scripts, it took a only few hours to deduce which scripts were associated with which table or figure in the paper; something that could have taken a lot longer if searching through log files of console sessions or (worse) if no such log files were available at all. Nevertheless, the temptation to add functionality using command line options and switches poses an obstacle to reconstructing output analysis if it is not clear which options were used when generating particular files of interest. One approach to addressing the problem is to capture the entire workflow in a shell script; however, this only works if the analysis can be done on a single machine, and if certain stages in the process take a long time (and do not need replication once executed once), running the whole script each time a change is made to the analysis process is an inefficient use of computer resources. Dealing with the problem by making several copies of scripts effectively fixing combinations of options that would otherwise have been implemented in a single script would not necessarily facilitate output analysis reconstruction, and would compound problems with code maintenance and bug fixing during analysis. Although here an R library was created with a view to allowing commonly used functions to be reused from one script to the next, this is not necessarily recommended due to the overheads associated with maintaining R libraries (particularly in across R versions and operating systems).

For some researchers, there is a "flow" associated with data analysis that is interrupted if best-practice procedures of faithfully recording all activities are rigorously followed. Although in the work reconstructed here, best-practice has arguably been followed through keeping a diary of the work done, saving analyses in documented scripts, and in some cases saving terminal transcripts, these have not between them been sufficient to make the reconstruction process trivial. Some detective work has been necessary, and there have been cases where steps in the analysis have not been documented. For those willing to invest in learning how to use them, a tool such as AITIA's MEME [14] offers functionality that supports the analysis of batch runs of agent based models, and in this case, is licenced using a GNU GPL. The approach used in this paper has been to wrap scripts in another program that maintains provenance metadata.

Although facilities such as openabm.org are provided for archiving models, and journals are increasingly recommending the use of such archives, clearly there is an argument for recommending also archiving the processes by which the results were analysed and tables and diagrams in the paper generated.

Acknowledgements This work was funded by a number of agencies under the Digging into Data Challenge (Third Round) and by the Scottish Government Rural Affairs and the Environment Portfolio Strategic Research Theme 1 (Ecosystem Services). Computing facilities have been provided by Compute Canada and Sharcnet.

References

1. Edmonds, B., Hales, D.: Replication, replication and replication: some hard lessons from model alignment. J Artif Soc Soc Simulat 6(4), 11 (2003). http://jasss.soc.surrey.ac.uk/6/4/11.html
2. Hales, D., Rouchier, J., Edmonds, B.: Model-to-model analysis. J Artif Soc Soc Simulat 6(4), 5 (2003). http://jasss.soc.surrey.ac.uk/6/4/5.html
3. Roucher, J., Cioffi-Revilla, C., Polhill, J.G., Takadama, K.: Progress in model-to-model analysis. J. Artif. Soc. Soc. Simulat. 11(2), 8 (2008). http://jasss.soc.surrey.ac.uk/11/2/8.html

4. Schmolke, A., Thorbek, P., DeAngelis, D.L., Grimm, V.: Ecological models supporting environmental decision making: a strategy for the future. Trends Ecol. Evol. **25**(8), 479–486 (2010)
5. Gimona, A., Polhill, J.G.: Exploring robustness of biodiversity policy with a coupled meta-community and agent-based model. J. Land Use Sci. **6**(2-3), 175–193 (2011)
6. Polhill, J.G., Gimona, A., Gotts, N.M.: Nonlinearities in biodiversity incentive schemes: A study using an integrated age-based and metacommunity model. Environ. Model. Software **45**, 74–91 (2013)
7. Boero, R., Squazzoni, F.: Does empirical embeddedness matter? Methodological issues on agent-based models for analytical social science. J Artif Soc Soc Simulat **8**(4), 6 (2005). http://jasss.soc.surrey.ac.uk/8/4/6.html
8. Parker, D. C., Barton, NM., Dawson, T., Filatova, T., Jin, X., Lee, J.-S., Polhill, J. G., Robinson, K. and Voinov, A.: The MIRACLE project: a community library to archive and document analysis methods for output from agent-based models. Association of American Geographers 2015 Annual Meeting, April 21–25, 2015, Chicago, IL (2015)
9. Hastie, T.J., Tibshirani, R.J.: Generalized Additive Models. Chapman & Hall/CRC, Boca Raton, FL (1990)
10. Akaike, H.: A new look at the statistical model identification. IEEE Trans. Automat. Contr. **19**(6), 716–723 (1974)
11. Breiman, L., Friedman, K., Olshen, R.A., Stone, C.J.: Classification and Regression Trees. Chapman & Hall/CRC, Boca Raton, FL (1984)
12. Wolstencroft, K., Haines, R., Fellows, D., Williams, A., Withers, D., Owen, S., Soiland-Reyes, S., Dunlop, I., Nenadic, A., Fisher, P., Bhagat, J., Belhajjame, K., Bacall, F., Hardisty, A., Nieva de la Hidalga, A., Balcazar Vargas, M.P., Sufi, S., Goble, C.: The Taverna workflow suite: designing and executing workflows of Web Services on the desktop, web or in the cloud. Nucleic Acids Res. **41**(W1), W557–W561 (2013)
13. Moreau, L., Missier, P. (eds.): PROV-DM: The PROV Data Model. W3C Recommendation 30 April 2013. http://www.w3.org/TR/2013/REC-prov-dm-20130430/. (2013)
14. Ivanyi, M., Bocsi, R., Gulyas, L., Kozma, V., Legendi, R.: The Multi-Agent Simulation Suite. 22nd Conference on Artificial Intelligence (AAAI-07), Vancouver, British Colombia, 22–26 July 2007. (2007)

The Pursuit of Happiness: A Model of Group Formation

Andrea Scalco, Andrea Ceschi, and Riccardo Sartori

Abstract We developed an agent-based model with the aim of investigating the effect of the interaction among several virtual actors characterized by (1) a certain level of emotional intelligence and (2) an individual behavioral proneness to act positively or negatively within social interactions. The goal of each agent is to achieve a satisfactory internal state, which is consequential to the positive/negative effects derived by the incurred social interactions. As a result, when the simulation run, we observe the spontaneous emergence of groups. Moreover, it could be easily noted that the large majority of the defectors are incapable to join to any group, and the few groups that accept defectors are not able to maintain more than one of this kind of actors. Finally, we studied the ratios between virtual actors when stable configurations are reached.

Keywords Emotional intelligence • Agent-based model • Emergence • Group formation • Individual differences

1 Introduction

On the one hand, when we decide to look at the human social dimension, it could be easily observed that individuals are able to affect each other's thoughts and feelings during daily-life interactions: a field of study that was firstly understood by the extensive works by Watzlawick and colleagues (see above all [1]). On the other hand, the concept of emotional intelligence (EI), when compared to emotions, represents a higher psychological dimension that is associated with those individual differences related to the experience and the management of the emotional sphere [2, 7, 8]. Following these brief considerations, we can say that the effects on the individual emotional sphere that stem from the mutual influence during interactions are directly connected to people's ability to understand how emotions usually work [3].

A. Scalco (✉) • A. Ceschi • R. Sartori
Department of Human Sciences, University of Verona, Verona, Italy
e-mail: andrea.scalco@univr.it; andrea.ceschi@univr.it; riccardo.sartori@univr.it

© Springer International Publishing AG 2017
W. Jager et al. (eds.), *Advances in Social Simulation 2015*, Advances in Intelligent Systems and Computing 528, DOI 10.1007/978-3-319-47253-9_33

Now, it is interesting what has been reported in [4]: a high capability to understand other emotions does not necessarily coincide with a coherent and equivalent ability (or experience) to employ it (e.g. a manager can read dozens of books about how to motivate his staff, but the actual application of this knowledge is something completely different). In this way, Tatton [5], within an assessment center research, distinguished several categories of individuals based on their emotional knowledge and its actual application during interactions, among which:

- The *emotionally intelligent*: individuals who possess a high level of emotional intelligence and actively use it in an effective way to support other people.
- The *emotionally manipulative*: like the previous ones, this kind of people are endowed with a high level of EI, but, on the contrary, they use it with a "nefarious intent" [5, p.107] (e.g. lowering others' self-esteem to enhance themselves).
- The *emotionally intuitive*: these people are endowed with a lower level of emotional intelligence than the previous kinds of individuals; nonetheless, they are prone to interact positively with others.

These differentiations give us the inspiration to build a simulation in order to be able to observe how these kind of actors might interact and affect each other: in fact, as pointed out by Ceschi, Rubaltelli, and Sartori [6], social processes do not stem from isolated behaviors, but rather they are originated by the interactions among individuals over the time.

2 Aim

Following the previous considerations, the main aim of the simulation is to investigate the consequences of the interaction among the aforementioned kinds of actors (characterized by their behavioral tendency and level of EI), considering the effects of imaginary conversations exchanged among them. In particular, we were interested to observe the consequences on the long run of the defectors' behavior (i.e. the manipulative agents) compared to other actors.

3 Agents' Attributes and Simulation Model

The agents implemented inside the simulation are endowed with the following characteristics:

- A certain level of emotional intelligence (*EI*), which is implemented as a normally distributed characteristic over a continuum that ranges from 0.25 (low EI) to 0.99 (high EI). Given that the emotional intelligence is positively related to the ability to understand and to manage others' feelings [7, 8], this characteristic directly affects the intensity of the "emotional effect" of the conversation when agents interact among themselves.

- A behavioral tendency (*behave*), which indicates the value associated with the proneness of the agents to act positively (1) or negatively (−1) within the held conversations. This last one is independent by the agent's EI. Instead, it is related to the agent's nature (i.e. the kind of agent).
- An individual emotional state (*emo-state*), which ideally represents the internal state of the agent, and it is affected by the effects of the interactions. This latter could range within a negative–positive continuum.

When the simulation starts, each agent has the chance to move randomly: if it finds someone within the four adjacent cells (we considered Von Neumann neighborhood), it converses randomly with one of these. In this way, the emotional state of every agent involved into a conversation is altered consequentially to the effect that each agent arouses on the other one and vice versa. If the emotional state of the agent is positive, it indicates a satisfying condition: that is to say, the agent is "happy," and thus it has no need to move as long as *emo-state* remains positive. Otherwise, the agent persists to move and interact with other agents.

4 Results and Discussion

When the simulation run, it is possible to observe manifestly the emergence of several groups within the world. As groups come up the overall emotional state of agents grows, likewise group members strengthen themselves with continuous positive exchanges among them. It is also possible to notice the consequences of the defectors' behavior: in fact, when the simulation reaches a stable configuration (i.e. when every agent has a positive emotional state, and thus, no one needs to move anymore), the majority of defectors are left aside (Fig. 1): that is, they are not able to join a group. More interesting, although a group rarely accepts a strong defector, it is remarkable that the constituted groups are not able to maintain within them more than one of the manipulative agents.

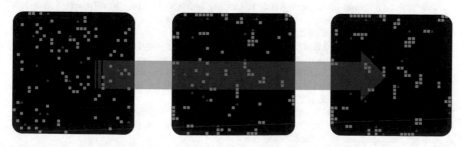

Fig. 1 A simulation example. Manipulative agents (i.e. defectors) are represented by a triangle, whereas emotionally intelligent ones by a square. When they start to interact with each other they tend to form autonomously group. Generally, only one defector can be accepted inside a group

Following these considerations, several repetitions of the simulation have been conducted in order to produce a spreadsheet on which record the number of emotional intelligent or intuitive agents that are necessary to contrast the negative effects of N manipulative actors (i.e. to generate a stable configuration) within a time limit of 1000 cycles. Succesively, we interpolated data to express in a formal way the relationships among virtual actors required to obtain stable groups formation. The earliest analyses conducted for this model show that the relationship between intelligent/manipulative agents can be expressed through a logarithmic curve ($R^2 = 0.96$): this indicates the fact that the effects of actors that are prone to interact in a positive manner are able to strengthen themselves when their number grows, contrasting efficiently the negative effects of defectors. Instead, the ratio between intuitive/manipulative agents in order to reach a stable configuration is best expressed by a linear interpolation ($R^2 = 0.96$): as expected, intuitive agents are less effective compared to emotional intelligent ones to contrast defectors' effects.

Starting from this, a further simulation has been recently developed, in order to gain more complexity: in fact, to get closer to reality, the simulation should take into account also the individual aptitude to cope with negative feelings through emotional regulation. This ability has been implemented as a reduction of the effects of negative interactions and, similarly to the emotional effects, it is positively associated with the level of emotional intelligence of the virtual actor. When this new simulation is run, it could be observed that the emerged groups are able to accept more than one defector, contrary to the previous model. This could provide support to the idea that the ability to cope with own emotions (an individual ability) could serve as a facilitator of social interactions and, consequently, to groups formation.

References

1. Watzlawick, P., Beavin, J.H., Jackson, D.D.: Pragmatics of Human Communication: A Study of Interactional Patterns, Pathologies, and Paradoxes. Norton, New York, NY (1967)
2. Scalco, A.: Trait emotional intelligence: modelling individual emotional differences in agent-based models. Trends Pract. Appl. Agents, Multi-Agents Syst. Sustain. PAAMS Collect. Springer International Publishing (2015)
3. Cunico, L., Sartori, R., Marognolli, O., Meneghini, A.M.: Developing empathy in nursing students: a cohort longitudinal study. J. Clin. Nurs. 21, 2016–2025 (2012)
4. Palmer, B.R., Stough, C., Harmer, R., Gignac, G.: The Genos Emotional Intelligence Inventory: A Measure Designed Specifically for Workplace Applications. Assessing Emotional Intelligence. Springer, New York, NY (2009)
5. Tatton, J.: Emotional intelligence or emotional negligence? Develop dimension international: white paper (2005). Cited in: Stough, C., Saklofske, D.H., Parker, J.D.A.: Assessing Emotional Intelligence. Springer, New York, NY (2009)
6. Ceschi, A., Rubaltelli, E., Sartori, R.: Designing a homo psychologicus more psychologicus: empirical results on value perception in support to a new theoretical organizational-economic agent based model. Adv. Intell. Syst. Comput. 290, 71–78 (2014)

7. Petrides, K.V.: Ability and Trait Emotional Intelligence. In: Chamorro-Premuzie, T., von Stumm, S., Furnham, A. (eds.) The Wiley-Blackwell Handbook of Individual Differences, pp. 656–678. Blackwell Publishing Ltd., Hoboken, NJ (2011)
8. Pérez-González, J.C., Petrides, K.V., Furnham, A.: Measuring Trait Emotional Intelligence. In: Schulze, R., Roberts, R.D. (eds.) Emotional Intelligence: An International Handbook, pp. 181–201. Hogrefe & Huber Publishers, Ashland, OH (2005)

The Social Learning Community-Modeling Social Change from the Bottom-Up

Geeske Scholz

Abstract Raising awareness through social learning is one important strategy to deal with today's challenges. Social learning is a complex and dynamic process, thus modeling may help to explore key dynamics. In this chapter I present a model of a Social Learning Community (SLC) which is not based on a network approach, but rather connects actors to different group settings where they interact. Within these social interaction rooms where actors meet and exchange they construct their own social reality, resulting in group phenomena. The social interaction contexts may span from such informal but long-lasting settings as a family to loose groups. Knowledge learned in one interaction setting can be brought into another one via the participating actors. The SLC is able to link social-psychological findings of group interaction (e.g., conformity) to social learning in wider social units.

Keywords Social learning • Societal learning • Convergent learning • Group interaction • Group learning • Agent-based model • Learning rooms

1 Introduction

Many pressing problems of today's world (e.g., overuse of resources or biodiversity loss) cannot be handled solely through technical improvements or the enforcement of rules. Rather a fundamental change of behavior is needed. For example, climate change is affected by emissions we produce in our daily lives, and which we could reduce if we changed our daily routines. To foster such behavioral change, raising awareness through social learning is one important strategy. Social learning can be defined as a process that produces a change in understanding in the individuals involved, which occurred through social interactions and becomes situated within wider social units [1]. Furthermore, convergent learning (referring to an increase in shared understanding) should be observable [2, 3].

G. Scholz (✉)
Institute of Environmental Systems Research, Osnabrück University, Barbarastraße 12, Osnabrück, 49076, Germany
e-mail: gescholz@uos.de

© Springer International Publishing AG 2017   373
W. Jager et al. (eds.), *Advances in Social Simulation 2015*, Advances in Intelligent Systems and Computing 528, DOI 10.1007/978-3-319-47253-9_34

Social learning is a complex and dynamic process, where modeling may help to explore some key dynamics. While there is a growing body of literature on opinion dynamics models (e.g., [4–7]), models that analyze awareness shifts and consider more complex types of group interaction are still rare. For the simulation of social learning, modeling social contagion is not sufficient, because social contagion is not sufficient to produce behavioral change. To go beyond modeling social contagion, one needs to include cognitive reasoning like the possibility of knowledge comparison and knowledge about other agents [8]. Furthermore, to simulate communication and interpretation during interaction, one needs to model the interpretation of messages, memory, and deliberation processes [9]. Thus, a model to address awareness raising through social learning should be able to simulate individual learning, group learning, and social learning. It should address relational aspects and cognitive knowledge. Furthermore, it should span over several scales to be able to really address awareness shifts and paradigm changes on societal level. Finally, social processes and influences (such as conformity) important during group interaction should be considered, because groups (of various sizes) are the cells where social learning processes start. In this chapter I present a social learning community which connects actors to different group settings where they interact. Within these social interaction "rooms" where actors meet and exchange they construct a social reality, resulting in group phenomena (e.g., conformity). This approach differs from models using a network structure to connect agents to each other (thereby focusing on one-to-one interaction).

The rest of this manuscript is structured as follows: in the next section I describe a conceptual model capable of simulating a Social Learning Community (SLC). This model builds upon and integrates a model of learning and group discussions from Scholz et al. [2]. In the last section I present some concluding remarks on the benefits of this approach.

2 A Virtual Social Learning Community (SLC)

2.1 Assumptions to Start With

In the SLC, Actors jointly construct a social reality in group interactions. This is an emergent phenomena, influencing Actors participating in the group interaction. To conceptualize such settings, I build upon the IAD framework from Ostrom [10], using the term "Action Situation." An Action Situation is referring to a "structured social interaction context" [10]. Furthermore, the two working hypotheses for the SLC are:

1. Every community is made up of diverse social interaction settings (Action Situations), and
2. Actors take part in different Action Situations.

2.2   Model Structure

The overall structure of the SLC is quite simple: Actors (having relational and cognitive knowledge) participate in various Action Situations. These may span from such informal but long-lasting and influential settings as a family to loose groups, e.g., a student learning group. In all of these Action Situations the participating Actors interact, constructing their own contextual social reality. Knowledge learned in an Action Situations is saved within the Actor, and may be brought into the interaction process of another Action Situation in which this Actor is involved. Figure 1 gives an overview of the main structure.

2.3   The Action Situations

To model the Action Situations and Actors, I build upon CollAct, which is an agent-based model of group discussion, learning, and consensus finding [2]. CollAct simulates how Actors discuss about certain topics at stake, how they may come to a consensus, and learn from each other. Thereby, relational influences are taken into account by implementing roles for all Actors. Roles influence speech probability, and also the probability of learning from each other. This implementation allows to model conformity, which is of crucial importance when looking at group interaction [11]. Note that roles in the SLC are context specific, i.e., they may differ from Action Situation to Action Situation and also in between Actors. Furthermore, they may shift during the simulation.

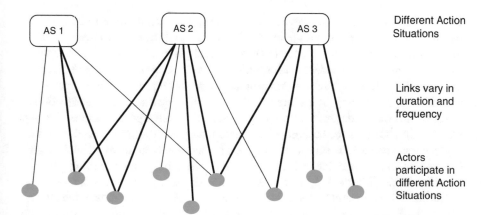

Fig. 1 Actors (*green circles*) participating in diverse action situations

2.4 The Actors

Actors also build upon CollAct [2]. Actors possess relational and cognitive knowledge, thus they know what they are talking about and whom they are talking to. They use this knowledge to evaluate incoming messages. For example, if Actor A is having a very positive opinion of Actor B, this leads to a higher probability of A learning from B. Learning can take place in both relational and cognitive knowledge, thus if A disagrees on B's statements a couple of times this may lead to A having a minor opinion of B.

Social influences are also implemented in the Actors in the form of cognitive biases leading to conformity [2].

2.5 The Linkages

Linkages represent an affiliation of an Actor in a specific Action Situation. After a linkage between an Actor and an Action Situation is established, Actors become participants within an Action Situation. Linkages can vary in their duration (e.g., membership in a committee constituted for a year versus family membership) and their frequency (how often are certain Action Situations visited). Thus, linkages are dynamic and can change during the simulation.

3 Concluding Remarks and Outlook

The SLC is still work in progress. Nevertheless, it is already able to link social-psychological findings of group interaction (e.g., conformity) to social learning in wider social units, such as, for example, paradigm shifts. Models that analyze awareness shifts and consider more complex types of group interaction are still rare. In the SLC, cognitive and relational knowledge and group processes are considered, as well as the grouping of Actors in various social settings (Action Situations). Through this model design it is possible to simulate convergent learning at group level, leading towards the building of a shared understanding, and to analyze under which parameter settings such a learning result may spread out in wider social units. Convergent and divergent learning may serve different purposes regarding possible transitions. Van Mierlo [12] found that while divergent learning may open up possibilities for innovation interpretation, convergent learning may foster the closing of a niche. Furthermore, she accounts for different process conditions fostering the two types of learning [12]. Different process conditions can be simulated through the choice of parameter settings.

Simulation experiments will have different foci, the main being the emergence of a new paradigm, such as, awareness to the impact of our daily actions on climate change. An exploration of factors influencing awareness raising might eventually help to design better environmental communication strategies.

References

1. Reed, M.S., Evely, A.C., Cundill, G., Fazey, I., Glass, J., Laing, A., et al.: What is social learning? Ecol. Soc. **15**(4), 10 (2010). http://www.ecologyandsociety.org/vol15/iss4/resp1/
2. Scholz, G., Dewulf, A., Pahl-Wostl, C.: An analytical framework of social learning facilitated by participatory methods. Syst. Pract Action Res. (2014). doi: 10.1007/s11213-013-9310-z
3. Scholz, G., Pahl Wostl, C., Dewulf, A.: An agent-based model of consensus building. Proc. Soc. Simulat. Conf. (2014)
4. Deffuant, G., Amblard, F., Weisbuch, G., Faure, T.: How can extremism prevail? A study based on the relative agreement interaction model. J. Artif. Soc. Soc. Simulat. **5**(4) (2002)
5. Hegselmann, R., Krause, U.: Opinion dynamics driven by various ways of averaging. Comput. Econ. **25**(4), 381–405 (2005)
6. Lorenz, J.: Continuous opinion dynamics under bounded confidence: a survey. Int. J. Mod. Phys. C **18**(12), 1819–1838 (2007)
7. Meadows, M., Cliff, D.: Reexamining the relative agreement model of opinion dynamics. J. Artif. Soc. Soc. Simulat. **15**(4) (2012). http://jasss.soc.surrey.ac.uk/15/4/4.html
8. Conte, R., Paolucci, M.: Intelligent social learning. J. Artif. Soc. Soc. Simulat. **4**(1), U61–U82 (2001). http://jasss.soc.surrey.ac.uk/4/1/3.html
9. Troitzsch, K.G.: Simulating communication and interpretation as a means of interaction in human social systems. Simulation **88**(1), 7–17 (2012). 10.1177/0037549710386515
10. Ostrom, E.: Understanding Institutional Diversity. Princeton University Press, Princeton (2009)
11. Baron, R.S., Kerr, N.L.: Group process, group decision, group action, 2nd ed., p. 271. Open University Press, (2003)
12. Van Mierlo, B.: Convergent and divergent learning in photovoltaic pilot projects and subsequent niche development. Sustain. Sci. Pract. Pol. **8**(2), 4–18 (2012)

Geeske Scholz is lecturer at the Institute of Environmental Systems Research at Osnabrück University. She holds a PhD in Applied Systems Science. Her research interests are social learning and social change, and how modeling, specifically computer simulations, can help us to understand and facilitate these phenomena. She is a member of the organizing team of ESSA@work.

Understanding and Predicting Compliance with Safety Regulations at an Airline Ground Service Organization

Alexei Sharpanskykh and Rob Haest

Abstract Failures to comply with safety regulations are currently a major issue at many airline ground service organizations across the world. To address this issue, approaches based on an external regulation of the employees' behavior have been proposed. Unfortunately, an externally imposed control is often not internalized by employees and has a short term effect on their performance. To achieve a long-term effect, employees need to be internally motivated to adhere to regulations. To understand the role of motivation for compliance in ground service organizations, a formal agent-based model is proposed in this chapter based on theories from Social Science. The model incorporates cognitive, social, and organizational aspects. The model was simulated and partially validated by a case study performed at a real airline ground service organization. The model was able to reproduce behavioral patterns related to compliance of the platform employees in this study.

Keywords Compliance • Agent-based model • Motivation • Cognitive models • Social contagion

1 Introduction

Nowadays commercial air transport is one of the safest modes of transportation. However, the ever increasing amount of traffic and introduction of new increasingly autonomous systems pose a difficult challenge of ensuring safety targets set by regulatory organizations. According to statistics [1], most of the safety occurrences happen not during the flight, but on the ground, e.g., during aircraft ground handling operations and aircraft maintenance operations. Decreasing the number of ground safety occurrences has a high priority in many airlines in different countries. To achieve this aim some airlines use Ramp Line Operations Safety Assessments (LOSA) [1], a monitoring tool for measuring and identifying the adherence to safety regulations on the platform. Monitoring and recording of violations of the

A. Sharpanskykh (✉) • R. Haest
Faculty of Aerospace Engineering, Delft University of Technology, Kluyverweg 1, Delft, 2629 HS, The Netherlands
e-mail: o.a.sharpanskykh@tudelft.nl; r.c.s.haest@tudelft.nl

© Springer International Publishing AG 2017
W. Jager et al. (eds.), *Advances in Social Simulation 2015*, Advances in Intelligent Systems and Computing 528, DOI 10.1007/978-3-319-47253-9_35

safety regulations using ramp LOSA is done on a regular basis by experienced platform employees. It was expected that collecting information about violations and presenting this information employees should make them aware of their undesirable behavior. Unfortunately, the introduction of ramp LOSA in the ground service organization under study did not result in a decrease of the number of ground safety occurrences.

An approach to address this issue using the STAMP framework was proposed in [2]. To resolve the problem of compliance of the platform employees with safety regulations, this approach appeals to top-down external control of the employees by management. The approach proceeds by identifying and mitigating deficiencies of the organizational control structure. However, as pointed in [3] such an externally imposed control is often not internalized by agents and has a short term effect on the performance of employees. To achieve a long term effect, employees need to be internally motivated to adhere to regulations [3]. Therefore, the main focus of this paper is on modeling and analysis of motivation of the platform employees to comply with safety regulations. Specifically, the following research questions were formulated:

- *Which and how cognitive, social and organizational factors influence the motivation of an employee to comply with safety regulations?*
- *Can a motivation model be developed to predict deviations from safety regulations in a real airline ground service organization?*

To answer these questions an agent-based motivation model was developed based on several theories from social science combined in an integrated framework. The model was applied in the context of a specific task of the aircraft arrival procedure—Foreign Object Damage (FOD) check. Foreign object is any object that should not be located near aircraft as it can damage aircraft or injure personnel. According to Boeing [4], the improper execution of FOD checks costs airlines and airports millions of dollars every year. Especially, the effects of FOD on maintenance costs, predominantly engine damage repairs, can be significant. Nevertheless, the ramp LOSA statistics showed that FOD checks are often not performed by platform employees.

The proposed model elaborates the motivation and decision making of the platform employees whether or not to perform the FOD check. During this elaboration individual cognitive, social and organizational factors are taken into account. Furthermore, the model includes individual and social learning of agents representing the employees and addresses two modes of reasoning of agents—explicit rational and implicit automatic (habits). To initialize the model, an extensive 1 year study was performed at a real ground service organization. Data were gathered by observation, questionnaires and interviews with employees playing different roles in the organization. The collected data were separated in two data sets. The first set contained data on the organizational context (i.e., formal organizational structures and processes, norms and regulations) and on local processes and characteristics of the organizational agents. This dataset was used for the model initialization. The second set contained data describing global organizational or systemic properties (such ramp LOSA statistics), which were used for the model validation. A part of the model validation results is discussed in the paper. The model was able to capture behavioral patterns of the platform employees and reflect the ramp LOSA statistics concerning the FOD check compliance in the real ground service organization.

The paper is organized as follows. In the following Sect. 2 the theoretical basis of the model is described. In Sect. 3 the proposed agent-based model is provided. Main results of the simulation study are discussed in Sect. 4. The paper ends with conclusions and discussions.

2   Theoretical Background

The theoretical basis of the model comprises several theories from social science described below. These theories address universal human needs, the way how humans reason about their needs and make choices to act based on this reasoning. All the theories used for the model development have a good empirical support.

Self-determination theory [3] is a theory of human motivation, which addresses people's universal, innate psychological needs and tendencies for growth and fulfillment. Specifically, the theory postulates three types of basic needs:

- *the need for competence* concerns the people's inherent desire to be effective in dealing with the environment;
- *the need for relatedness* concerns the universal disposition to interact with, be connected to, and experience caring for other people;
- *the need for autonomy* concerns people's universal urge to be causal agents, to experience volition.

In line with other motivation theories [5], in addition to the needs listed above *the need for safety* was added, which is particularly relevant for the ground service organization, in which physical injuries are not uncommon.

Based on needs individual goals can be defined. Higher level individual goals may be refined in goal hierarchies as described in [6]. A goal is state, which the individual desires to achieve or maintain. To achieve or maintain his or her goals, an individual considers different behavioral options (actions or plans). One of the theories that explain why individuals choose one option over another is *the Expectancy Theory of Motivation by Vroom* [5]. Advantages of the Expectancy Theory include: (a) it can be formalized; (b) it allows incorporating the organizational context; (c) it has received good empirical support. According to the theory, when an individual evaluates alternative possibilities to act, he or she explicitly or implicitly makes estimations for the following factors: *expectancy, instrumentality*, and *valence*.

Expectancy refers to the individual's belief about the likelihood that a particular act will be followed by a particular outcome (called a first-level outcome). Its value varies between 0 and 1.

Instrumentality is a belief concerning the likelihood of a first level outcome resulting into a particular second level outcome; its value varies between -1 and $+1$.

Instrumentality takes negative values when a second-level outcome has a negative correlation with a first-level outcome. A second level outcome represents a desired (or avoided) state of affairs that is reflected in the agent's goals.

Valence refers to the strength of the individual's desire for an outcome or state of affairs; it is also an indication of the priority of goals.

Values of expectancies, instrumentalities and valences may change over time, in particular due to individual and social learning. *The motivational force* of an individual i to choose option to act k is calculated as:

$$F_{k,i}(t) = \sum_{l=1}^{n} E_{kl,i}(t) \sum_{h=1}^{m} V_{h,i}(t) I_{klh,i}(t) \tag{1}$$

Here $E_{kl,i}(t)$ is the strength of the expectancy that option i will be followed by outcome j; $V_{h,i}(t)$ is the valence of the second level outcome (a goal) h; $I_{klh,i}(t)$ is perceived instrumentality of outcome l for the attainment of outcome h for option k.

The Vroom's theory describes the process of rational decision making. However, repetitive actions such as occur during aircraft handling may over time become automatic, i.e., a habit. *The dual process theory* [7] distinguishes System 1 and System 2 thinking. While System 2 is used for rational, rule-based and analytic thinking, System 1 is associated with unconscious, implicit and automatic reasoning. Depending on the dynamics of environmental changes, an individual switches between the systems. Both systems are used in the model and the case study considered in the paper.

In the following Sect. 3 it is demonstrated how the theories from this section were integrated in an agent-based model.

3 The Agent-Based Model

To develop the model, the steps of the generic methodology for modeling of agent organizations from [8] were used. Because of the space limitations, only selected steps of the methodology will be elaborated in this paper. A full formal description of the model is provided in [9].

3.1 Identification of Organizational Roles

The following roles were identified: Platform Employee, Team Leader, and Sector Manager. Platform Employee role and Team Leader role form a composite role Team. For each Team 5 Platform Employee role instances and one Team Leader role instance are specified. There is one Sector Manager role instance in the model.

3.2 Identification of Interactions Between Roles and with the Environment

All roles are able to interact with each other. Platform Employee role and Team Leader role are able to observe the tasks to be accomplished (i.e., aircraft to be handled), outcomes of own actions (task is finished successfully, a safety occurrence happened), task execution by the other platform employees. Furthermore, Team Leader and Sector Manager roles are able to detect violations of procedures and to provide reprimands to Platform Employee role. Reprimands can be observed by all roles.

3.3 Identification of Tasks and Workflows

Each Team role handles one aircraft at a time. Aircraft to be handled are organized in a FIFO cue.

Specific agents are allocated to the role instances, which in contrast to the roles and role instances may have internal cognitive states and dynamics, which are described in the following steps.

3.4 Identification of Characteristics of Agents

Platform employee agents have three characteristics: *risk aversion* (reflected in parameters of the expectancy theory model), *openness to new experience* and *expressiveness in communication*. All the characteristics range in the interval [0,1].

3.5 Identification of Goals and Needs of Agents

In accordance with the self-determination theory and the field study a number of goals and their subgoals were identified (Table 1).

3.6 Identification of Beliefs of Agents

All the expectancies, instrumentalities and valences of the Platform Employee agents are represented by their beliefs.

Table 1 The goals and states of the decision making model provided in Fig. 1

Goals	States
G1 Achieve a high level of competence	S1 Action saves time
G1.1 Achieve highest time efficiency	S2 Action costs additional time
G1.2 Prevent aircraft, equipment, and/or infrastructural damage	S3 Action results in aircraft, equipment, or infrastructural damage
G2 Achieve a high level of occupational safety	S4 Action prevents aircraft, equipment, or infrastructural damage
G2.1 Prevent personal injury	S5 Action results in personal injury
G3 Maintain sense of belonging and attachment to colleagues	S6 Action prevents personal injury
	S7 Action is in alignment with the team member norms
G3.1 Maintain high team acceptance	S8 Action is not in alignment with the team norms
G3.2 Maintain high management acceptance	
G4 Achieve a high control over own behavior and goals	S9 Action is in alignment with the team leader norms
G4.1 Achieve a high level of freedom in the execution of tasks	S10 Action is not in alignment with team leader norms
G4.2 Achieve high psychological ownership of rules	S11 Action is in alignment with sector management norms
	S12 Action is not in alignment with sector management norms
	S13 Reprimand received from team member
	S14 Reprimand received from team leader
	S15 Reprimand received from sector manager

3.7 Specification of Decision Making of Agents

Decision making by the Platform Employee agents whether or not to perform FOD
check was modeled by using the Vroom's expectancy theory (Fig. 1). To initialize
the expectancies, instrumentalities and valences of the model for each agent three
classes were introduced: Low, Medium and High. This was done to address the
issue of uncertainty in these parameters and individual variations of the platform
employees. Most of the numerical scales of these parameters were divided equally
among the classes. For a few scales the division between the classes was adjusted
depending on the interpretation of the corresponding parameters in the context of
the case study. To determine the values (i.e., specific classes) of the parameters,
an ethnographic study was performed in the real ground service organization
by observation of organizational practices, interviews and questionnaires with
employees. Furthermore, secondary sources were used such as safety reports, safety
statistics and reports on previous operational studies at the organization.

The expectancy theory model was used for System 2 reasoning. When the same
operations were routinely executed by a Platform Employee agent, the agent's
System 2 reasoning was gradually shifting to System 1 reasoning—a habit had been
formed. This shift was modeled by the dynamics of agent's i openness parameter α_i:

$$\alpha_i (t + \Delta t) = \alpha_i(t) + \zeta \left(\alpha_i^{\min} - \alpha_i(t)\right) \Delta t \qquad (2)$$

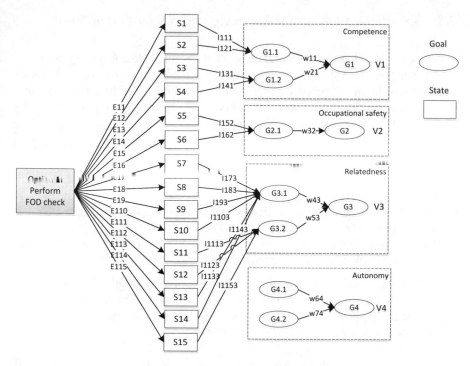

Fig. 1 Decision making model of an Platform Employee agent for performing FOD check based on the expectancy theory with expectancies (E), instrumentalities (I), states (S), and valences (V). For readability purposes the time parameter and agent indexes were omitted

where $\alpha_i^{\min}$ is the minimum perceptive openness of agent i (set to 0.1 in the simulation), ζ is the rate of transition from System 2 to System 1. It depends on the execution frequency of the operation by the agent, as well as on the agent's personal characteristics. In the simulation $\zeta = 0.015$, meaning that it takes around 2 months to form a new habit.

When procedural rules change, an agent needs to adapt to a new situation and reconsider options by switching from System 1 to System 2: the agent i's openness is set to its initial value $\alpha_i(0)$ and the process of the new habit formation starts again.

A similar expectancy theory model was created for option 2—"Not to perform FOD check." Note that for option 2 there are several factors other than for option 1 that play a role in the decision making.

In the simulation, every time when an agent considers explicitly (System 2) or implicitly (System 1) whether or not to perform FOD check, motivation forces $F_{1,i}$ and $F_{2,i}$ for both options are calculated by (1). Then, the agent performs FOD check with probability $(F_{\max}+F_{1,i})/(2F_{\max}+F_{1,i}+F_{2,i})$. The normalization with $F_{\max}$ is used to compensate for the negative values of the instrumentalities.

3.8 Specification of Agent Learning and Social Interaction

Two types of learning were modeled: individual and social learning of agents.

An agent learns individually by observing a feedback from the environment to its action. In the decision making model from Fig. 1 the individual learning was realized by updating values of expectancies (E) based on the following observations:

- An agent observes whether or not a reprimand from other agents is provided, when the agent does not comply with regulations (update of E113, E114, E115).
- After the successful execution of a task an agent observes how much time it took and how it influenced the total execution time of the operation (update of E11, E12). The task durations were determined based on operational data from the organization under investigation.
- When an agent does not perform a FOD check, a safety occurrence could occur. The agent is able to observe such occurrences (update of E13, E14).

Furthermore, the Platform Employee agents are able to observe the execution of operations by other agents in their teams and to learn from these agents by verbal communication. Social learning is modeled as the process of social contagion [10]. By this process expectancies $E_{kl,i}(t)$ were updated as:

$$E_{kl,i}(t + \Delta t) = E_{kl,i}(t) + \delta_{kl,i}(t)\Delta t \qquad (3)$$

Here

$$\delta_{kl,i}(t) = \sum_{j \in T} \gamma_{j,i}(t) \left(E_{kl,j}(t) - E_{kl,i}(t)\right) / \sum_{j \in T} \gamma_{j,i}(t)$$

is the amount of change of the agent i's state; T is the set of the agents in the team. A weight $\gamma_{j,i} \in [0,1]$ is the degree of influence of agent j on agent i defined as:

$$\gamma_{j,i}(t) = \alpha_i(t)\varepsilon_j(t)\beta_{ji} \qquad (4)$$

$\alpha_i(t)$ and $\varepsilon_j(t)$ are the agent characteristics—the openness of information recipient agent i and the expressiveness of information provider agent j, and $\beta \in [0,1]$ is the strength of the information channel between the two agents.

The communication style in the teams of platform employees is direct, informal, and of a high frequency. For direct communication between agents $\beta_{ji}=1$.

Sector managers communicate with the platform employees directly during observation tours and dedicated meetings. Indirect communication with the management occurs by using messages on information screens, posters, memos etc. $(\beta_{ji}=0.3)$.

The expectancy values of the agents are updated once per simulated day, taking into account the frequency of interactions during that day.

3.9 Identification of Shared Beliefs, Norms and Values of (Groups of) Agents

By field observations and interviews a team norm was identified. The norm applies to situations in which a team arrives too late at an aircraft stand while the aircraft is waiting for the docking process. To save time, the FOD check is omitted and the arrival procedure starts directly. Field data revealed that employees who execute the check in the described situation get a social reprehension from other team members. This influences the achievement of goal G3.1, which is driven by the alignment of the decision option with the team norms and team leader norms.

Expectancy of agent i $E_{17,i}$ (t) that decision option 1 is in alignment with the team norms depends on the degree of similarity between decision option 1 and the most relevant team norm (if one exists, otherwise $E_{17,i}$ $(t)=0$) and the likelihood that agent i is familiar with this norm. The value of corresponding I_{173} is determined as the product of the perceived importance of the norm in the team and the degree of connectedness of agent i in the team: $\sum_{j \in T, \, i \neq j} \left(\gamma_{ij} + \gamma_{ji} \right) / 2 \left(|T| - 1 \right)$, where T is the set of agents in the team.

Intuitively, to have a large positive contribution to team relatedness goal G3.1, the agent i's decision option should be in line with an important team norm and agent i should be well connected in the team, i.e., be its prototypical member. This way of reasoning is inspired by prototypicality theories [11].

The Team Leader agent's and Sector Manager agent's norms are in line with the organizational regulations.

3.10 Specification of the Environmental Dynamics

The number of aircraft to be handled by the teams of platform employees was modeled according to the actual operational statistics.

4 Simulation Study

In this section first a simulation setup is described (Sect. 4.1), then simulation results are discussed in Sect. 4.2.

4.1 Simulation Setup

Several scenarios based on the developed model have been simulated. In this paper we discuss only a scenario, which occurred in reality and which could be validated by empirical data from ramp LOSA.

In this scenario three periods of the organizational operation are considered:

- the first period with a limited managerial control over the execution of the platform operations and limited safety information provision;
- the second period (8 weeks) with a high managerial control after many safety occurrences happened in the first period;
- the third period in which the release of managerial control occurs over time (a linear transition from High control to Low control mode, Table 2).

The parameters for these periods determined by using the field study data are provided in Table 2.

A team consists of five agents, a team leader and four platform employees. In the simulation the agents in the teams communicated with each other in random order.

In line with the empirical findings, two types of agents in the teams were modeled: more expressive agents with $\varepsilon_i \in$ [0.5, 0.9] and less expressive agents with $\varepsilon_i \in$ [0.1, 0.5]. Each agent can be of either type with an equal probability. The openness of an agent α_i was assigned a wide range [0.1, 0.9] to represent the diversity of agents. In each simulation run the agents' parameters were randomly instantiated from the uniformly distributed ranges introduced above.

In the simulation every time step corresponds to a decision moment for executing or not executing the FOD check. One simulation day is divided in three shifts (morning, afternoon/evening and night shift). During normal operations, on average, the arrival procedure is executed three times each shift. The simulated time period was 200 working days; $\Delta t = 1/3$ indicating that on average 3 decisions are made by each Platform Employee agent per day.

The parameters of the expectancy theory model were initialized based on the field study data. A complete specification thereof is provided in [9].

Table 2 The parameters of the managerial control modes used in the simulation

Parameter	Low control	High control
Probability of reprimand from Sector Manager agent	0.05	0.11
Probability of reprimand from Team Leader agent	0.05	0.33
$\varepsilon_{SM}(t)$ in direct communication	0.06	0.6
$\varepsilon_{SM}(t)$ in indirect communication	0.03	0.3

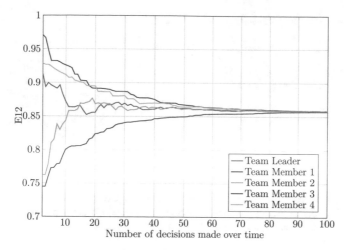

Fig. 2 Dynamics of expectancy $E_{12}(t)$—belief about the likelihood that the execution of the FOD check would cost additional time—for all agents in a team

4.2 Simulation Results

In the following main simulation results obtained for the scenario under consideration are described.

Due to frequent observation and communication of the agents in a team, their expectancies related to the information exchanged tend to converge over time (Fig. 2). This is a well-known effect in the social contagion literature [10, 12].

Simulation experiments with a varying composition of teams indicated that the individual characteristics of Team Leader agent play a crucial role in the dynamics of social contagion and largely determine expectancies in a team.

In the first period, after the initialization phase, most of the agents have a relatively constant motivational force for both decision options (Fig. 3). The motivational forces to perform FOD checks are low as the organization neither sufficiently controls the execution of operations nor creates a sufficient awareness about the importance of FOD checks. Some agents in the team even prefer not to perform the check. By the end of the phase the agents function in System 1 mode of reasoning.

In the beginning of the second period (indicated by the dotted vertical line in Fig. 3) the organization introduces more frequent managerial control and reprimands. To adapt to the new circumstances, the agents switch to System 2 mode of reasoning. Such a change results in an increased motivation to perform FOD checks and a decreased motivation not to do so of all agents in all teams. The differences in motivation are explained by differences in the individual characteristics of the agents. However, when after 8 weeks the control and information provision was gradually removed, the agents start returning gradually to their previous state. This form of motivated behavior is known in the literature as externally regulated

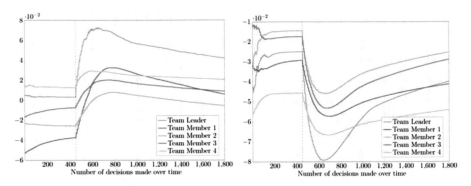

Fig. 3 The motivational forces of five agents in a team to perform FOD check (*left*) and to not perform FOD check (*right*) in the three periods of the scenario under consideration. The *dotted vertical line* indicates the beginning of second period (increased control)

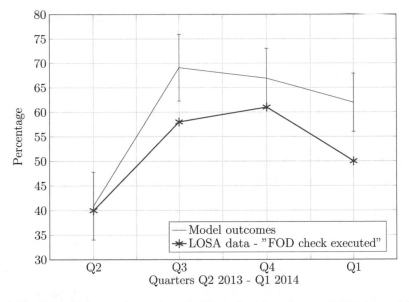

Fig. 4 The FOD check execution statistics (in %) obtained from the ramp LOSA data and from the simulated model

behavior [3]. Such a form of motivation is sustained by continuous presence of reprimands and rewards. These behavioral patterns were also observed in the ground service organization under study.

The model simulation outcomes were also compared to the ramp LOSA statistics of the FOD checks execution in the ground service organization (Fig. 4). The model was able to capture the trends in real ramp LOSA data. The Student's two-sample T-test performed on real and simulated data supported the null-hypothesis at the significance level 5 %.

5 Discussion and Conclusions

In this chapter a formal agent-based motivation model is presented, based on an integrated theoretical basis from social science. All the theories used for the model development were extensively validated by experiments with human subjects. In the study presented in the paper a good agreement is demonstrated between the simulated data obtained using the proposed model and the real data from the ground service organization under study.

In addition to the study described in the chapter, to improve organizational compliance, a model analysis study was performed, including sensitivity analysis. Based on this study improvement options were provided to the organization. The outcomes of this study will be presented elsewhere.

To the best knowledge of the authors, it is a first attempt to approach the problem of compliance in airline ground service organizations by a model-driven simulation study of the employees' motivation. Previous studies on safety occurrences at airline ground service organizations (e.g., [13]) had mostly focused on statistical data analysis and informal identification of possible causes of these occurrences.

In the chapter the compliance of employees to a specific task—FOD check—was investigated. However, the same modeling approach could be used to study other tasks too. Moreover, in a discussion with another ground service organization from a different national culture, findings from this study were recognized as relevant for this organization too. However, to determine the generality of the proposed model-driven approach more detailed studies at other organizations are required.

The proposed formal agent-based modeling and simulation approach has a potential to become a useful decision support tool for managers at airline ground service organizations. It allows investigating consequences of diverse organizational changes for the motivation of employees and their behavioral choices.

Acknowledgement We are grateful to Gert Jan van Hilten and Patrick Pronk for practical support of this research. Furthermore, we would like to thank David Passenier, Dr. Robert Jan de Boer, and Dr. Nicoletta Dimitrova for stimulating discussions in the course of this research.

References

1. de Boer, R.J., Koncak, B., Habekotté, R., van Hilten, G.J.: Introduction of ramp-LOSA at KLM Ground Services. In: Human Factors of Systems and Technology, Human Factors and Ergonomics Society Annual Meeting. Shaker Publishing, Maastricht (2011).
2. de Boer, R.J., de Jong, S.: Application of STAMP to facilitate interventions to improve platform safety. In: Proceedings of the 3rd STAMP Workshop, Boston, (2014)
3. Deci, E.L., Vansteenkiste, M.: Self-determination theory and basic need satisfaction: understanding human development in positive psychology. Ricerche Psicholog. **27**, 17–34 (2004)
4. Boeing.: Foreign object debris and damage prevention. (1998). http://www.boeing.com/commercial/aeromagazine/aero_01/textonly/s01txt.html

5. Pinder, C.C.: Work motivation in organizational behavior. Prentice-Hall, Upper Saddle River, NJ (1998)
6. Popova, V., Sharpanskykh, A.: Formal modelling of organisational goals based on performance indicators. Data Knowl. Eng. **70**(4), 335–364 (2011)
7. Kahneman, D.: Thinking, fast and slow. Farrar, Straus and Giroux, New York (2011)
8. Sharpanskykh, A., Stroeve, S.: An agent-based approach for structured modeling. Analysis and improvement of safety culture. Comput. Math. Organ. Theor. **17**, 77–117 (2011)
9. Sharpanskykh, A., Haest, R.: Understanding and predicting compliance with safety regulations at an airline ground service organization by agent-based modelling technical report. TU Delft. (2015). (http://homepage.tudelft.nl/j11q3/papers/tech_report_172014.pdf)
10. Hegselmann, R., Krause, U.: Opinion dynamics and bounded confidence: models, analysis and simulation. J. Artif. Soc. Soc. Simulat. **5**(3) (2002)
11. Monti, A., Soda, G.: Perceived organizational identification and prototypicality as origins of knowledge exchange networks. In: Brass, D.J., Giuseppe (JOE) Labianca, Mehra, A., Halgin, D.S., Borgatti S.P. (eds.) Contemporary Perspectives on Organizational Social Networks (Research in the Sociology of Organizations, Volume 40) Emerald Group Publishing Limited, pp. 357–379. (2014)
12. Axelrod, R.: The dissemination of culture: a model with local convergence and global polarization. J. Conflict Resolut. **4**, 2023–2226 (1997)
13. Balk, A.: Safety of ground handling (Research Report No. NLR-CR-2007-961). Accessed from EASA http://easa.europa.eu/essi/ecast/wpcontent/uploads/2011/08/NLR-CR-2007-961.pdf: National Aerospace Laboratory. (2008).

Opinions on Contested Infrastructures Over Time: A Longitudinal, Empirically Based Simulation

Annalisa Stefanelli

Abstract Using an agent-based model, we investigate the opinions of individuals on contested infrastructures over time. The model employs a psychological theory and a longitudinal study design. The longitudinal empirical data served as implementation and validation for the model. From the emerging results, we gained insights into the mechanisms of opinion adaptation; and discuss the relevance of opinion disaggregation, and the utilization of psychological theories and empirical data.

Keywords Agent-based simulation • Empirical data • Opinion dynamics • Contested infrastructures • Social judgment theory • Longitudinal

1 Introduction

Opinion dynamics models are useful for investigating how people exchange and adapt their opinions [1, 2]. The algorithms behind many such models employ arbitrary values and formulae to describe the dynamics of changes in opinion. Some scholars employed a more psychological approach, basing their models on theories from social psychology, such as the social judgment theory (SJT) by Sherif and Hovland [3–5]. This theory postulates that new information is compared with the individual's own anchor (or point of view) as soon as the message is perceived. The persuasiveness of this message depends on how the receiver evaluates its position on three possible latitudes [6]:

1. Latitude of rejection (information is unreasonable or objectionable);
2. Latitude of acceptance (information is acceptable or worthy of consideration);
3. Latitude of non-commitment (information is neither acceptable nor objectionable).

A. Stefanelli (✉)
Institute of Environmental Decision (IED), ETH Zürich, Zürich, Switzerland
e-mail: annalisa.stefanelli@usys.ethz.ch

© Springer International Publishing AG 2017
W. Jager et al. (eds.), *Advances in Social Simulation 2015*, Advances in Intelligent Systems and Computing 528, DOI 10.1007/978-3-319-47253-9_36

The ways in which a person judges new information describe the cognitive structure of their attitude or opinion. Adaptations of personal attitudes or opinions occur in a second step, where the person's opinion moves away from (if the new information is in the latitude of rejection) or towards (if the new information is in the latitude of acceptance) the new information. These two mechanisms of adaptation are explained by contrast and assimilation effects.

However, in order to obtain more representative agent profiles, real data need to be incorporated into the model. In the present case, the empirical data are taken from the first wave of a longitudinal online questionnaire. The aim of our agent-based model (ABM) is to investigate changes in opinion over time on the specific topic of a contested infrastructure in Switzerland (i.e., the construction of deep geological repositories for nuclear waste). Our ABM is therefore an opinion dynamics model, applying the SJT and using longitudinal empirical data.

2   Methods and Agent-Based Model

In our ABM, we simulate the dynamics of the opinions of German-speaking Swiss citizens over time with respect to the contested infrastructure of deep geological repositories for nuclear waste. Data collected via online questionnaires conducted in January 2014 (first wave, $N = 1328$) and January 2015 (second wave, $N = 844$) served as input data for the agents' profiles as well as for the choice of the social network (i.e., a 2-dimensional space neighborhood with eight links).

Respondents' opinions were compiled as disaggregated arguments [7, 8] of different categories (i.e., risk-, benefit-, and process-oriented). This disaggregation is important because it represents more realistically the interactions between individuals on a specific topic [8]. Ten arguments were rated in the online questionnaire according to their valence (from 1 *"absolutely against"* to 7 *"absolutely in favor"*) and importance (from 1 *"not important at all"* to 7 *"very important"*). To preserve the facility of inspection, the model incorporated three arguments per agent, one for each category. Each argument assumed a normalized value A_x [−1, 1] and was computed as the product of valence V_x [−1, 1] and importance I_x [0, 1] for each argument ($A_x = V_x \times I_x$).

Furthermore, information about the latitudes of the arguments described in the SJT was collected. Participants were asked to report if they evaluate the argument as being acceptable, objectionable or neither acceptable nor objectionable to be used in a discussion about the topic of deep geological repositories. Therefore, each agent profile included empirical information about the location of the arguments on the individual's opinion structure. The latitudes were held constant in the model, as over all participants in the sample no significant change occurred among the two waves.

One time step t in the simulation represented one interaction between all randomly chosen pairs of agents. The one-directional interaction is depicted in Fig. 1. The interacting agent T_a compared the argument values of agent T_b;

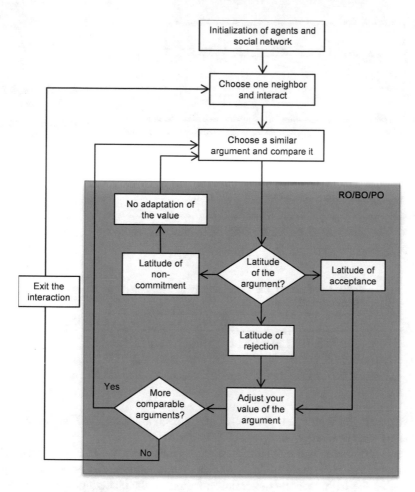

Fig. 1 Flowchart of the interaction between two agents. Each argument (*RO* risk-oriented, *BO* benefit-oriented, *PO* process-oriented) is sequentially compared in each time step. The interaction is represented in the *gray* field

depending on the value of the argument and its location along the social judgment continuum (see Appendix, Tables A1 and A2), the interacting agent T_a moved away from or towards the argument value of agent T_b.

3 Results

Our model was able to successfully represent the agent profiles (i.e., value for each argument; positions of the arguments on the latitudes) from the empirical data. Furthermore, the dynamics of interactions based on the theoretical assumptions of the SJT occurred in the predicted directions, showing face validity for our model [9].

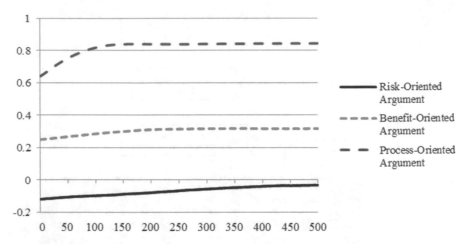

Fig. 2 Changes in values of the three arguments ($t=500$)

Table 1 Values of the three argument types in wave 1 and 2 (empirical validation)

	Wave 2 at $t=1$	Wave 1 (t needed)
Process-oriented argument	0.652	0.652 (6)
Risk-oriented argument	−0.106	−0.106 (49)
Benefit-oriented argument	0.286	0.286 (100)

We observe different patterns of change for each argument. In Fig. 2, the means of the arguments are depicted over 500 time steps.

These differences in the dynamics of change are directly related to the different positions of the arguments on the social judgment continuum. Sensitivity analysis shows that the latitude is the determining factor for the changing dynamics of the argument values. Moreover, the mean values for individual agents show pluriformity in the long run for both benefit-oriented and risk-oriented arguments (see Appendix, Figs. A1 and A2). In contrast, the process-oriented argument showed uniformity towards positive values (see Appendix, Fig. A3).

The second phase of the empirical investigation served as empirical validation [9] for our model. The values for every argument of the second wave occur in the simulation runs implemented using the data from the first wave. An interesting observation is that those values are reached after different time steps for every argument (see Table 1), demonstrating that the dynamics of how the changes occur are crucial. As mentioned above, the key factors in our model are the positions of the arguments along the social judgment continuum. The findings indicate that individuals revise their opinions (if at all) depending on how they evaluate new information, and that they do so at different rates.

4 Discussion

Our model simulates the dynamics of individual opinions on a contested infrastructure in Switzerland. The input and validation data are taken from an empirical, longitudinal online questionnaire. The opinions are disaggregated into arguments that are located on a social judgment continuum. The interactions and adaptations are based on the assumptions of SJT.

Our results show different patterns of adaptation for each argument. These differences are directly linked to the positions of the arguments on the social judgment continuum. Furthermore, individual differences are relevant to how people exchange information on a specific topic, because every agent follows their own adaptation dynamic, according to their initial profile and the interactions in which they are involved.

The model was validated with a second wave of longitudinal data, which showed that the model is robust and confirmed the relevance of the social judgment continuum and its role in the dynamics of adaptation.

The combination of empirical data and psychological assumptions together with the simulation model provides insights into the dynamics provides insights into the dynamics of how individuals interact and exchange information, and how this process relates to changes in opinion.

The model offers possibilities for extension and application to more complex interactions (e.g., the inclusion of more arguments, experimental manipulation). Moreover, it provides additional information that is relevant to empirical research on opinion dynamics, in which rates of change depend on the arguments and their positions on the social judgment continuum.

A.1 Appendix

Table A.1 Equations used when the argument x is in the latitude of acceptance

Argument x in the latitude of acceptance		A_{xb}	
		$+$	$-$
A_{xa}	$+$	$A_{xa(t+1)} = A_{xa(t)} + (A_{xb}^2 \times s)$	$A_{xa(t+1)} = A_{xa(t)} - (A_{xb}^2 \times s)$
	$-$	$A_{xa(t+1)} = A_{xa(t)} + (A_{xb}^2 \times s)$	$A_{xa(t+1)} = A_{xa(t)} - (A_{xb}^2 \times s)$

Note. A_{xa} = value of argument x of agent T_a; A_{xb} = value of argument x of agent T_b; t = time step in the model; s = speed of change constant ($s = 0.01$)

Table A.2 Equation used when the argument x is in the latitude of rejection

Argument x in the latitude of rejection		A_{xb}	
		+	−
A_{xa}	+	$A_{xa(t+1)} = A_{xa(t)} - (A_{xb}^2 \times s)$	$A_{xa(t+1)} = A_{xa(t)} + (A_{xb}^2 \times s)$
	−	$A_{xa(t+1)} = A_{xa(t)} - (A_{xb}^2 \times s)$	$A_{xa(t+1)} = A_{xa(t)} + (A_{xb}^2 \times s)$

Note. A_{xa} = value of argument x of agent T_a; A_{xb} = value of argument x of agent T_b; t = time step in the model; s = speed of change constant ($s = 0.01$)

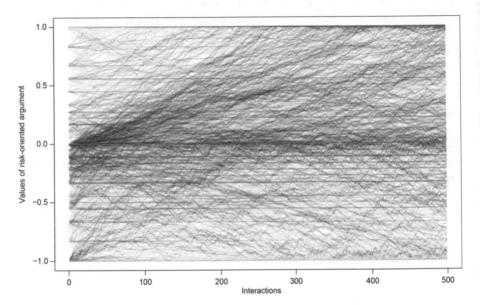

Fig. A.1 Values of the risk-oriented argument over 500 interactions. Each curve represents an agent in the model. *Red* = argument in the latitude of rejection; *Yellow* = argument in the latitude of acceptance; *Orange* (parallel curves) = argument in the latitude of non-commitment

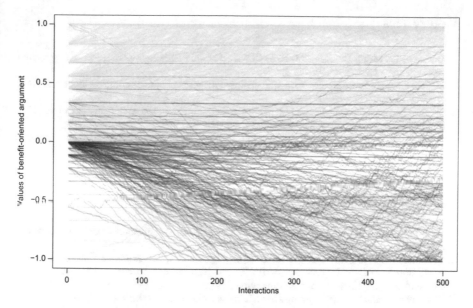

Fig. A.2 Values of the benefit-oriented argument over 500 interactions. Each curve represents an agent in the model. *Red* = argument in the latitude of rejection; *Yellow* = argument in the latitude of acceptance; *Orange* (parallel curves) = argument in the latitude of non-commitment

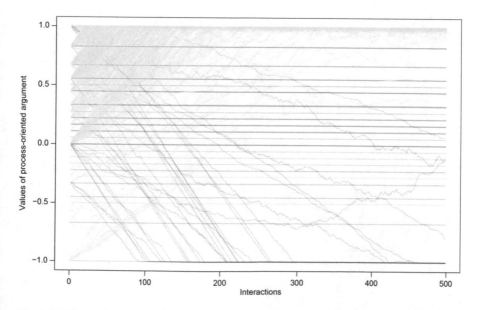

Fig. A.3 Values of the process-oriented argument over 500 interactions. Each curve represents an agent in the model. *Red* = argument in the latitude of rejection; *Yellow* = argument in the latitude of acceptance; *Orange* (parallel curves) = argument in the latitude of non-commitment

References

1. Deffuant, G., Neau, D., Amblard, F., Weisbuch, G.: Mixing beliefs among interacting agents. Adv. Complex Syst. **3**, 87–98 (2000)
2. Deffuant, G.: Comparing extremism propagation patterns in continuous opinion models. J. Artif. Soc. Soc. Simulat. **9**, (2006)
3. Sherif, M., Hovland, C.I.: Social Judgment: Assimilation and Contrast Effects in Communication and Attitude Change. Yale University Press, New Haven, CT (1961)
4. Huet, S., Deffuant, G., Jager, W.: A rejection mechanism in 2D bounded confidence provides more conformity. Adv. Complex Syst. **11**, 529–549 (2008)
5. Jager, W., Amblard, F.: Uniformity, bipolarization and pluriformity captured as generic stylized behavior with an agent-based simulation model of attitude change. Comput. Math. Organ. Theory **10**, 295–303 (2004)
6. O'Keefe, D.J.: Social Judgment Theory. In: O'Keefe, D.J. (ed.) Persuasion: Theory and Research. Sage, Newbury Park, CA (1990)
7. Hegselmann, R., Krause, U.: Opinion dynamics and bounded confidence models, analysis, and simulation. J. Artif. Soc. Soc. Simulat. **5**, (2002)
8. Stefanelli, A., Seidl, R.: Moderate and polarized opinions. Using empirical data for an agent-based simulation. In: Proceedings of the Social Simulation Conference 2014 of the 10th Conference of the European Social Simulation Association, pp. 61–64. Barcelona (2014)
9. Klügl, F.: A validation methodology for agent-based simulations. In: Proceedings of the 2008 ACM symposium on applied computing (SAC), pp. 39–43. ACM Press, New York (2008)

Road Repair Sequencing for Disaster Victim Evacuation

Kumiko Tadano, Yoshiharu Maeno, and Laura Carnevali

Abstract Disaster victim evacuation is one of the most urgent disaster relief efforts in saving lives after a disaster strikes a populated area. In urban areas, candidate routes to take for evacuation are basically determined based on static hazard maps depending on the types of disasters (e.g., routes which are not in areas at risk of landslide disaster in case of a flood, or tsunami or liquefaction in case of an earthquake). However, when a large-scale disaster occurs, unpredictable simultaneous road closures on a road network may be caused by various factors such as broken roads, traffic accidents, building collapse and outage of traffic lights by electricity failure. Since it takes time to repair roads and required resources for repairing activities are usually limited, it is necessary to determine a plan to sequence roads repairs. The plan on repairing damaged roads affects decisively how quickly and reliably evacuation is completed. To determine the optimal plan, we use stochastic time Petri nets to sequence one road repair after another for multiple evacuation origin–destination pairs with different speeds on different routes.

Keywords Disaster • Evacuation • Road repair • Sequencing • Stochastic time Petri nets

1 Introduction

Unpredictable disasters range widely from such natural disasters as an earthquake and an eruption of a volcano to such man-made disasters as an industrial plant accident and a terrorist attack. Disaster victim evacuation is one of the most urgent efforts in saving lives after a disaster strikes a populated area. Generally, evacuation means the movement of people away from the threat of an impending disaster or the

K. Tadano (✉) • Y. Maeno
Knowledge Discovery Research Laboratories – NEC Corporation, Kawasaki, Japan
e-mail: k-tadano@bq.jp.nec.com; y-maeno@aj.jp.nec.com

L. Carnevali
Department of Information Engineering, University of Florence, Florence, Italy
e-mail: laura.carnevali@unifi.it

© Springer International Publishing AG 2017   401
W. Jager et al. (eds.), *Advances in Social Simulation 2015*, Advances in Intelligent
Systems and Computing 528, DOI 10.1007/978-3-319-47253-9_37

ensuing deadly destruction. For example, in 2011, almost 400,000 evacuees were caused by the 3.11 Great East Japan Earthquake [1], the subsequent tsunami, and the nuclear power plant accidents a week later.

Previous studies analyze the pros and cons of commonly applied practices in disaster relief efforts, and propose mathematical models in operations research, optimization, and decision theory to improve them. The disaster relief efforts are classified in three categories. The first category is coordination between organizations, warehouses, and transportation in resolving gaps between abruptly increasing demands and instable supplies. A network theoretical study is proposed in investigating collaboration, coordination, and facilitation among government agencies, nonprofit and profit organizations in response to the 9/11 terrorist attack in 2001 [2]. The second category is preventive evacuation and inventory prepositioning for mitigating possible future disasters. A stochastic programming model is proposed in making an inventory plan [3]. The inventory plan describes how humanitarian supplies are positioned in a network of cooperative warehouses in preparation for such predictable disasters as a hurricane, and how they are distributed urgently in spite of possible traffic congestion that results from disaster victim evacuation. The third category is disaster victim evacuation and traffic management to recover from the damage. An integrated approach is proposed for road construction, contraflow setup, and resource planning to move victims in different urgency statuses to destinations [4, 5].

The road network is critical for evacuation. In urban areas, candidate routes to take for evacuation are basically determined based on static hazard maps depending on types of disasters (e.g., routes which are not in areas at risk of landslide disaster in case of a flood, or tsunami or liquefaction in case of an earthquake). However, when a large-scale disaster occurs, unpredictable simultaneous road closures on a road network may be caused by various factors such as broken roads, traffic accidents, building collapse, and outage of traffic lights by electricity failure. Since it takes time to repair roads and required resources for repairing activity are usually limited, it is necessary to determine a plan to sequence roads repairs. The plan on repairing damaged roads affects decisively how quickly and reliably evacuation is completed. It affects even more if complicated evacuation routes are planned between many origin–destination pairs. Road repairs are conducted by road repair resources. A road repair resource is a project unit consisting of operators, a work force, heavy machinery equipment, and construction materials. And, it moves on undamaged roads along with evacuees to a damaged road. If it moves along a long detour, the repair is by no means immediate. Nor is road repair resource mobilized intensively in a large-scale disaster. Sequencing one road repair after another is essential for these reasons.

In this study, we leverage a probabilistic model to sequence multiple road repairs for multiple evacuation origin–destination pairs. The model is formalized through stochastic Time Petri Nets (sTPN) [6] which allow the concurrent firing of multiple generally distributed transitions. sTPN features are suitable for capturing various aspects of concurrent actions of both evacuation and road repairs.

This paper is organized as follows. Section 2 describes the design concepts of the proposed method. Section 3 defines the problem of road repair sequencing. A model for road repair sequencing is proposed in Sect. 4. An example is shown in Sect. 5. Section 6 gives our conclusions.

2 Design Concepts

In this Section, we describe the concept of road repair sequencing. Usually, the time to complete the evacuation and its variability differ depending on the sequence (order) of multiple road repairs. The larger the number of damaged roads and origin–destination pairs of evacuees are, the more difficult it is to identify which sequence minimizes the time to complete the evacuation. This study is motivated by the literature [4, 5]. The integrated approach proposed in [4, 5] only considers one road repair and does not consider sequencing multiple road repairs. In addition, if there are multiple damaged roads, the assignment of repair resources and the time to move to a new geographical point for next road repair from the current point are also needed to be taken into account. After a disaster, repair resources are often limited and are not always positioned in ideal places for repairing the damaged roads. Hence, simultaneous repairing of all damaged roads is unrealistic, and we need to prioritize (and partly give up) road repairs and to consider the required time to move to a suitable place for the next road repair after the current road repair finishes. We incorporate the road repair sequencing and the limitation of the number and the travel time of repair resources into our model to sequence multiple road repairs for multiple evacuation origin–destination pairs.

In this study, we make the following assumptions. First, each evacuee goes to its destination (e.g., a shelter) from its origin (i.e., the start point) along a given fixed route (e.g., the safest travelable route) which is determined by the state of road repairs. If one road repair is completed, the fixed route can be changed to a new different fixed route. Note that if the evacuee is not on the new fixed route after completion of the road repair, the evacuee goes along the former (same) fixed route. Second, each evacuee has its own fixed destination. Third, since the uncertainty on the state of roads increases after a disaster, the travel time for each road in fixed routes is not deterministic but represented by a probability distribution.

3 Problem Description

In this section, we present the general problem definition of road repair sequencing. As a boundary condition, the problem includes multiple source–destination pairs, multiple road failures, and multiple road repair resources on a road network. The objective function is the time to complete the evacuation. The solution of the problem is the derived optimal plan, which represents the repair prioritization, and optionally possible updates of evacuation routes on a road network. The repair sequencing is the combination of repair sequence and repair resource movement.

4 Road Repair Sequencing

The evacuation and road repairs are modeled as a stochastic Time Petri Net (sTPN) [6]. Then the sTPN is analyzed through the approach of stochastic state classes [7] to evaluate given candidate plans of road repair sequences. sTPN are a special class of Petri nets. The Petri nets are a mathematical model for the description of a distributed system. A Petri net consists of places, transitions, arcs, and tokens. The road network and the movements of evacuees and repair resources are represented in the form of a sTPN.

4.1 Stochastic Time Petri Net

We model the evacuation and road repair using sTPNs [6]. An sTPN is a tuple $\langle P, T, A^-, A^+, A^\cdot, m_0, EFT, LFT, \mathcal{F}, C, E, L \rangle$, where: P is a set of places, T is a set of transitions, $A^- \subseteq P \times T$ is a set of precondition arcs, $A^+ \subseteq T \times P$ is a set of postcondition arcs, $A^\cdot \subseteq P \times T$ is a set of inhibitor arcs, m_0 is the initial marking associating each place with a nonnegative number of tokens, $EFT : T \to Q_0^+$ and $LFT : T \to Q_0^+ \cup \{\infty\}$ associate each transition with a static earliest firing time and a (possibly infinite) static latest firing time, respectively $(EFT(t) \leq LFT(t), \forall t \in T)$, $\mathcal{F} : T \to F_t^s$ associates each transition $t \in T$ with a static Cumulative Distribution Function (CDF) F_t^s supported over $[ETF(t), LFT(t)]$. We assume that F_t^s is absolutely continuous over its support and that there exists a Probability Density Function (PDF) f_t such that $F_t^s = \int_0^x f_t(y)dy$. $C : T \to \mathbb{R}^+$ associates each transition with a weight used to resolve the random switch between concurrent transitions with the same firing time. A transition t is called *immediate* (IMM) if $[ETF(t), LFT(t)] = [0, 0]$ and *timed* otherwise; a timed transition t is called *exponential* (EXP) if $F_t^s(x) = 1 - e^{\lambda x}$ over $[0, \infty]$ for some rate $\lambda \in \mathbb{R}_0^+$ and *general* (GEN) otherwise; a GEN transition t is called *deterministic* (DET) if $EFT(t) = LFT(t) \geq 0$ and *distributed* otherwise. $E : T \to \{true, false\}^{N^P}$ associates each transition with an *enabling function* that, in turn, associates each marking with a boolean value. $L : T \to \mathcal{P}(P)^{N^P}$ associates each transition with a *flush function* that, in turn, associates each marking with a set of places. For space limitations, we refer the reader to [6] for a complete discussion on syntax and semantics of sTPNs.

4.2 Road Repair Sequencing Model

To evaluate given candidates of road repair sequences, the evacuation and road repairs are modeled as an sTPN. We call the sTPN a *road repair sequencing model*. The road repair sequencing model consists of two kinds of sub models: (1) *evacuation models* and (2) *repair models*. The evacuation model represents the state of an evacuee, i.e., the current position of the evacuee in the road network. The repair model represents the state of road repair, i.e., the pgress of a given candidate sequence of repairs of damaged roads. Different states are represented by different markings (i.e., distribution of tokens). The evacuation models and the repair models interact with each other through enabling functions in the models depending on the markings. As time passes, evacuees become closer to destinations and road repairs become near completion. Details of each model are described in the following.

Evacuation Model

The evaluation model represents the condition of a road network and the movement of an evacuee along its given fixed route. The evaluation model consists of the following sTPN parts.

- A geographical location is represented by a place of an sTPN. There are multiple places in an evacuation models including its origin, destination, and intermediate locations for an evacuee.
- A position of an evacuee is represented by a token in a place of an sTPN. The token moves to different places through the firing of an enabled transition which is connected to the place having the token by an input arc.
- A road between two locations is represented by a timed transition and by its input and output arcs. The arcs connect the transition with two places representing the two locations. The direction of arcs is along a given fixed route. The roads which are not included in the fixed route for the evacuee are omitted. The damaged road is not travelable until the repair of the road is completed. The transition representing the damaged road is disabled by the repair model through enabling functions determined by markings, until the road repair activity finishes. A probability distribution is assigned to a transition based on the feature of the travel time of the road.
- The destination of the evacuee is represented by a special place which is reachable from its origin. The place does not have an input arc connected to a transition, i.e., the evacuee does not move once the evacuee reaches its destination.

Repair Model

The repair model represents the given sequence (procedure) of repairing of damaged roads. It includes road repair activities and movements of repair resources if necessary. The repair model consists of the following sTPN parts.

- Each state of progress of the sequence of repairs of the damaged roads is represented by a place of an sTPN. There are multiple places in a repair model, each of which represents the movement of a repair resource or a repair activity of a damaged road.
- The current state of repair progress is represented by a token in one of the above places.
- A repair activity is represented by a place with a token, a timed transition, input and output arcs, and enabling functions. An enabling function disables a transition representing one of damaged roads in the evacuation model. The arcs connect the transition to two places representing the repair of a damaged road and the movement of a repair resource. The direction of arcs is determined in accordance with a road repair sequence to be evaluated. A probability distribution is assigned to the transition based on the feature of the road repair activity (e.g., human reliability). Once a repair activity is completed, a token moves to its next place. The disabled transition corresponding to the repaired damaged road by the activity in the evacuation model is enabled. This might cause the change of the fixed route to take (from the old route to the new one).
- A movement of a repair resource is represented by a timed transition, input arcs and output arcs of a sTPN. A probability distribution is assigned to the transition based on the feature of the travel time.

5 Example Problem

To evaluate the feasibility of the proposed model shown in the previous section, this section demonstrates an example of sequencing repairs of damaged roads. A simple example of the road repair sequencing model is evaluated through transient analysis based on the Sirio framework [8] using the Oris Tool [9], which supports the derivation of the transient probability of reachable markings within a given time bound. This example includes two source–destination pairs for two groups of evacuees E1 and E2, seven geographical locations in the road network, two road failures f1 and f2, and a road repair resource. In this example, the groups E1 and E2 have different moving speed and priority in moving on the road. E2 represents evacuees with special needs in movement such as people with wheel chairs, family with small children and old people. E1 represents adults without special needs. We assume the speed of movement of E1 is faster than E2. To reduce the completion time for evacuation of all evacuees, E2 should take the shortest travelable route and E1 should take a route different from the E1's route so that they can avoid congestion. However, lengthy detour might make the completion time of evacuation of E1 unacceptably long. To determine the optimal repair plan with the shortest completion time of evacuation of all evacuees with different routes and speeds, we use stochastic time Petri nets to sequence one road repair after another for multiple evacuation origin–destination pairs. Since there are two road failures f1 and f2, there are five possible road repair sequences: SQ1, SQ2, SQ3, SQ4, and SQ5 as follows.

- SQ1: no road is repaired
- SQ2: only f2 is repaired
- SQ3: only f1 is repaired
- SQ4: f1 is repaired, then f2 is repaired
  SQ5: f2 is repaired, then f1 is repaired

In this example, we focus on the sequences including multiple road repairs, i.e., SQ4 and SQ5.

In addition to the assumptions described in Sect. 2, we made the following assumptions for this example problem. Roads in each evacuation model are one-way. The two groups of evacuees E1 and E2 have different priority for evacuation. As already mentioned in the previous paragraph, E2 has higher priority than E1. E1 moves faster than E2 when E1 and E2 do not share the same road. On the other hand, when E1 and E2 share the same road, the moving speed of E1 is reduced to the speed of E2. This is a reasonable assumption because a walking person needs to slow down if someone with slower moving speed is in front of the person. To represent this interaction between E1 and E2, in the evacuation model for E1, probability distributions assigned to transitions for roads which are not shared with E2 are exponentially distributed with $\lambda = 2$, while probability distributions assigned to transitions for roads which are shared with E2 are exponentially distributed with $\lambda = 1$. In the repair model for E2, all the probability distributions are exponentially distributed with $\lambda = 1$. To show the general feature of the problem we use exponentially distributed transitions, although the Oris Tool we use in this study supports the analysis of multiple concurrently enabled generally distributed transitions depending on the characteristics of movement of evacuees. In general, the transition rates can be affected by various factors such as the width of a road, the flatness of road surface, and the number of the evacuees.

The road network and the given fixed routes for evacuees E1 and E2 are shown in Figs. 1 and 2, respectively. The fixed routes are updated by the completion of road repairs. There are four possible states of the road network in terms of road repairs: (1) f1 and f2 are damaged (initial state), (2) f1 is repaired, (3) f2 is repaired, and (4) f1 and f2 are repaired. We assume that the initial position of the repair resource is "4" in Figs. 1 and 2, and it may move from "4" to "5," and from "5" to "4." In order to repair f2 on the road connected to "4," the repair resource need to be in "4." On the other hand, to repair f1 on the road connected to "5," the repair resource needs to be in "5." Hence, in SQ4, the repair resource is in "4" when their road repairs complete. In SQ5, the repair resource is in "5" when their road repairs complete.

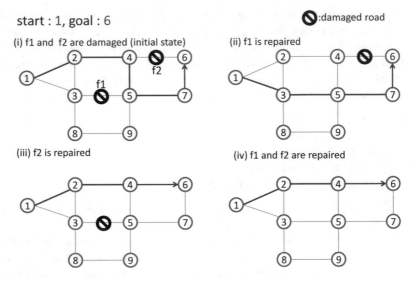

Fig. 1 Candidate road repair sequences SQ4 ((i) → (ii) → (iv)) and SQ5 ((i) → (iii) → (iv)) and the corresponding fixed route for evacuee E1

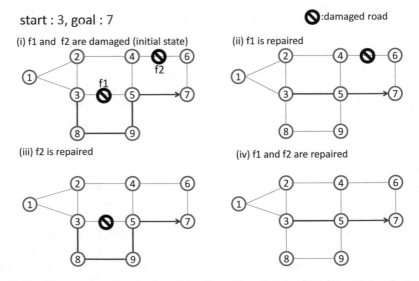

Fig. 2 Candidate road repair sequences SQ4 ((i) → (ii) → (iv)) and SQ5 ((i) → (iii) → (iv)), and corresponding fixed route for evacuee E2

Figure 3 shows the road repair sequencing model, consists of two evacuation models for E1 and E2, and one shared repair model. By changing the initial marking in the repair model, different candidates of road repair sequences can be evaluated. Each candidate of road repair sequences has its own initial marking (distribution of tokens). The initial positions of tokens in the repair model are given as follow.

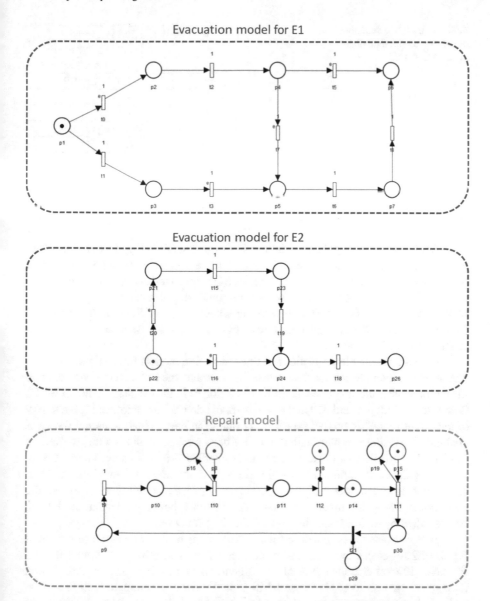

Fig. 3 The road repair sequencing model. Timed transitions are drawn as white rectangular boxes

- SQ4: p9, p8, p15, and p29 have a token
- SQ5: p14, p15, p8, and p18 have a token

Note that the numbers of tokens in p29, p18, p14, and p9 are changed to determine start and end points of road repair sequences in the initial marking. For SQ4, p29 and p9 have a token, but p14 and p18 have no token. For SQ5, p14 and p18 have a token, but p29 and p9 have no token.

Table 1 Enabling functions
of the evacuation models for
(a) E1 and (b) E2

(a) Transition	Enabling function
t0	P16! = 1 && P19! = 1
t1	p8! = 1 && P19! = 1
t3	p8! = 1 && P19! = 1
t7	p19! = 1
t5	p15! = 1
(b) Transition	Enabling function
t16	p8! = 1
t20	p16! = 1

The interactions between the evacuation
models and the repair model are defined
through the enabling functions

In SQ4, after repairing the road failure f1 (shown in (ii) in Fig. 1), two roads ("3" to "5" and "5" to "7") are shared with E2. This makes E1's moving speed slower but the route for E2 is short. In contrast, in SQ5, no roads are shared with E1 and E2. So E1 can move faster than SQ4, but the route for E2 becomes longer than that of SQ4. Hence, it is not trivial to determine whether SQ4 or SQ5 should be chosen from the perspective of completion time of evacuation of all evacuees with different characteristic of movement.

Table 1 shows enabling functions of the evacuation models for (a) E1 and (b) E2. The interactions between the evacuation models and the repair model are defined through the enabling functions. A timed transition t0 in the evacuation model for E1 is enabled if both p16 and p19 in the repair model do not have a token. If p8 and p19 in the repair model do not have a token, t1 and t3 in the evacuation model for E1 is enabled. If p19 in the repair model does not have a token, then t7 in the evacuation model E1 is enabled. In the same way, t5 is enabled if p15 does not have a token. Regarding the evacuation model for E2, if p8 does not have a token, t16 is enabled. t20 is enabled if p16 does not have a token. To improve readability we represent the interaction among sub models by enabling functions, but an equivalent model can be described by inhibitor arcs instead of enabling functions.

Figure 4 shows the completion time distributions of the evacuation of evacuees E1 and E2 for each candidate road repair sequence, respectively. Although E1 and E2 have different characteristics of movement, E1 and E2 have similar completion time of evacuation in SQ4. the variability of the time to complete the evacuation of E1 is large in comparison with that of E2. Taking into account the difference in sensitivity to the change of road repair sequences might help to determine the most suitable road repair sequences for evacuees' requirements.

Figure 5 shows the completion time distribution in the worst case (i.e., complete time distribution for the group of evacuees with longer completion time is selected). From the point of view of the worst case analysis, SQ4 is better than SQ5. Note that the results may vary if the priority assigned to each group of evacuees is different from the assumptions of this example.

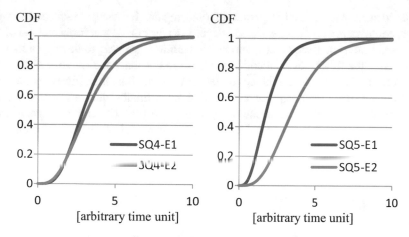

Fig. 4 Completion time distribution of evacuation in case of repair sequences SQ4 (*left*) and SQ5 (*right*)

Fig. 5 Completion time distribution of evacuation in the worst case

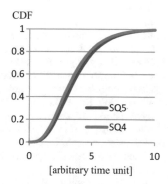

From the results, the evacuation time varies depending on the sequence of road repairs, even if the final state of the road network is the same. Hence it is important to carefully select the sequence of road repairs by considering the specific needs of the evacuees.

6 Conclusions

We present a model for road repair sequencing. The model allows us to identify the sequence of repairs of damaged roads to minimize the time to complete the evacuation. An sTPN is introduced to sequence multiple road repairs for multiple evacuation origin–destination pairs. From the result of the example problem, the evacuation time varies depending on the sequence of road repair, even if the final state is the same. Hence it is important to carefully select the sequence of road repair by considering specific needs of the evacuees.

To identify the optimal sequence of road repairs for many evacuees in large-scale, complex road networks, we will develop an automatic synthesis method of the proposed road repair sequencing model. In addition, we plan to develop a method to improve the efficiency of identification of the optimal road repair sequence for evacuees in order to reduce the computation time, such as the efficient generation of candidate road repair sequences. Based on the optimal sequences of road repairs obtained by the road repair sequencing model, we plan to carry out agent-based simulation of evacuation with a detailed behavior model of evacuees for efficient and realistic risk scenario planning for evacuation.

References

1. Cabinet Office Team in Charge of Assisting the Lives of Disaster Victims. Evacuees of the number of evacuees. http://www.cao.go.jp/shien/en/2-count/annex1-2.pdf. Accessed 23 July 2014
2. Schweinberger, M., Petrescu-Prahova, M., Vu, D.Q.: Disaster response on September 11, 2001 through the lens of statistical network analysis. Soc. Networks **37**, 42–55 (2014)
3. Davis, L.B., Samanlioglu, F., Qu, X., Root, S.: Inventory planning and coordination in disaster relief efforts. Int. J. Prod. Econ. **141**, 561–573 (2013)
4. Wang, J.W., Wang, H.F., Zhang, W.J., Ip, W.H., Furuta, K.: Evacuation planning based on the contraflow technique with consideration of evacuation priorities and traffic setup time. IEEE Trans. Intell. Transport. Syst. **14**, 480–485 (2013)
5. Wang, J.W., Ip, W.H., Zhang, W.J.: An integrated road construction and resource planning approach to the evacuation of victims from single source to multiple destinations'. IEEE Trans. Intell. Transport. Syst. **11**, 277–289 (2010)
6. Vicario, E., Sassoli, L., Carnevali, L.: Using stochastic state classes in quantitative evaluation of dense-time reactive systems. IEEE Transactions on Software Engineering **35**(5), 703–719 (2009)
7. Horváth, A., Paolieri, M., Ridi, L., Vicario, E.: Transient analysis of non-Markovian models using stochastic state classes. Perform. Eval. **69**(7–8), 315–335 (2012)
8. Carnevali, L., Ridi, L., Vicario, E.: A framework for simulation and symbolic state space analysis of non-Markovian models. Proc. SAFECOMP. 409–422 (2011)
9. Bucci, G., Carnevali, L., Ridi, L., Vicario, E.: Oris: a tool for modeling, and evaluation of real-time systems. Int. J. Software Tools Technol. Transfer **12**(5), 391–403 (2010)

Using Empirical Data for Designing, Calibrating and Validating Simulation Models

Klaus G. Troitzsch

Abstract Many simulation models just model stylised facts, and as such they are interesting and often helpful. But simulation models can be seen as an implementation of theory in a computer, and this is why at least an empirical validation should be aimed at. And if a simulation model is to be validated in a concrete empirical setting, it should be initialised with empirical data in order that one can test whether the model behaves the same way as the target system. The paper discusses two models, their empirical background and validation results, distinguishing between retrospective/predictive validity and structural validity.

Keywords Simulation • Validation • Calibration • Gender segregation • Extortion racket • Mafia • Norm orientation

1 Introduction

1.1 The Role of Simulation in the Research Process

Many simulation models just model stylised facts, and as such they are interesting and often helpful. Even if recently the frequency of approaches to base agent-based models on empirical research has considerably grown [11] there is still a large proportion of agent-based models without empirical background, 'one of the most frequently reported criticisms in [Waldherr's and Wijerman's] survey … that agent-based models are too abstract and too far from reality' [15, 3.8].[1] But if one

---

[1] A preliminary analysis of 213 papers in the Journal of Artificial Societies and Social Simulation (JASSS) since 2011 showed that at most 19.2 % compare their quantitative simulation results to quantitative empirical data. Another 17.4 % discuss the necessity and/or possibilities of such a comparison but do not perform it, usually for lack of a sufficient dataset. A more detailed analysis of JASSS papers with respect to the issue of quantitative validation is under preparation.

K.G. Troitzsch (✉)
Universität Koblenz-Landau, Project Global Dynamics of Extortion Racket Systems (GLODERS), 56070 Koblenz, Germany
e-mail: kgt@uni-koblenz.de

© Springer International Publishing AG 2017
W. Jager et al. (eds.), *Advances in Social Simulation 2015*, Advances in Intelligent Systems and Computing 528, DOI 10.1007/978-3-319-47253-9_38

413

understands simulation models as an implementation of theory in a computer [7], one has to conclude that an empirical validation should at least be aimed at [17].

Simulation, and particularly agent-based simulation, means deducing macro structures from micro specifications with the help of computer programs where it is impossible or extremely difficult to do such a deduction with classical mathematics. This is why simulation results have the same status as the outcome of a mathematical deduction. A simulation is 'a thought experiment which is carried out with the help of a machine, but without any direct interface to the target system'. [12, p. 46] But a 'simulation experiment' is not an experiment in the same sense that we use the word in 'an experiment in empirical research'. Simulation yields inference, 'transforms knowledge of the world already gained' [4, p. 409] whereas experiments in empirical research discover the real world. Although simulations and experiments have something in common, they 'still differ in the epistemic aspect of interest here'. [4, p. 409]

Simulation has—strictly speaking—nothing to do with the real world. It generates virtual worlds or—in the case of computational social science—artificial societies. But if a simulation model is to be validated in a concrete empirical setting, it should be initialised with empirical data in order that one can test whether the model behaves the same way as the target system. More often than not important features of the real world, and particularly of social systems and their components, cannot be observed, let alone measured. In these cases simulation can help to find the values of those unobservable parameters which generate the observable features of the real system.

1.2 Types of Validity and Validation

With Zeigler [19, p. 5] we should distinguish between three types of validity and three different stages of model validation (and development):

- replicative validity: the model matches data already acquired from the real system (retrodiction),
- predictive validity: the model matches data before data are acquired from the real system,
- structural validity: the model 'not only reproduces the observed real system behaviour, but truly reflects the way in which the real system operates to produce this behaviour'.

For the two simpler types of validity the paper will discuss what a match between empirical and simulated data means: in agent-based models we will have to content ourselves with a match in probability, i.e. with empirical frequency distributions which match the theoretical distributions which the simulation models yield. This is shown in much more depth in two detailed examples, whereas the question of structural validity can only be tackled with the help of individual empirical data which usually are scarce. What they could look like will be discussed with respect to the two examples.

The paper discusses two models, their empirical backgrounds and validation results, distinguishing between retrospective/predictive validity and structural validity. It tries to give at least preliminary answers to the following research questions:

- Can agent-based models provide the relation between explanatory variables and parameters [on the one hand] and the outcome [on the other hand] in an explicit form?
- How does such an explicit-form relation help to validate an agent-based model?
- Can structural validity of an agent-based model be achieved at all (when even the real actors often cannot tell how they 'operate to produce [their] behaviour')?

2 The Models and the Empirical Data Behind Them

2.1 Overcoming Gender Segregation in German High Schools

In her PhD thesis, Rita Wirrer [18] collected data about male and female children and teachers in about 150 high schools (*'Gymnasien'* in German) in the federal state of Rhineland-Palatinate from 1950 to 1990. In a conference paper [9] a first analysis of the time-dependent frequency distribution of these data was published. A recent re-analysis of Wirrer's data yields the result in Fig. 1.

The original simulation run [9],[6, p. 117] was based on a few simple assumptions:

1. that all teachers leaving their jobs are replaced by men and women with equal overall probability, following the equal opportunities principle already laid down in article 3 line 2 of the German Basic Law of 1949 (although ordinary legislation procrastinated the equal opportunities for women for many years, as the simulation results will also show),
2. that men stay in their jobs twice as long as women, which is an approximation to the historical data, and

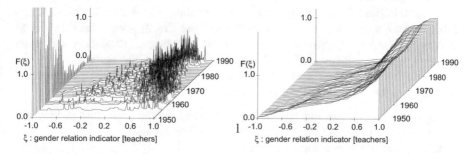

Fig. 1 Time-dependent distributions of gender relations among teachers in about 150 high schools in Rhineland-Palatinate 1950–1990. *Left*: frequency density functions, *right*: cumulative frequency functions

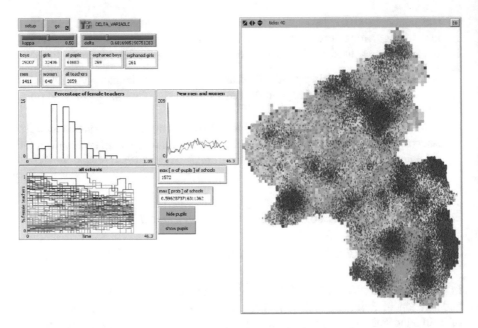

Fig. 2 Interface of the co-education model. The side length of each patch corresponds to approximately 2.2 km; *green patches* represent regions with low population density. Map background: © GeoBasis-DE / BKG 2015, http://www.bkg.bund.de. The *top plot* shows the distribution of the percentage of female teachers at the end of the simulation, the *bottom plot* shows the history of this percentage for each school (*red*: girls' schools, *blue*: boys' schools, *green*: co-educational schools)

3. that new women are assigned to an individual school with a probability depending on the percentage of women among its teachers, at the same time making sure that at all times men and women have the same overall probability of replacing retired teachers.

The simulation was initialised with a gender distribution close to the empirical distribution of 1950.

The re-analysis in Fig. 1 and the new model use a scale ranging from −1 to +1, −1 indicating 100 % females and +1 indicating 100 % males. The NetLogo [16] simulation model which we present here (see Fig. 2) goes a little farther as it is initialised with approximate geographical coordinates of schools and exact empirical gender relations of staff and pupils found in the statistics of 1950. Additionally the teachers and the pupils are placed in plausible geographical positions. Moreover two assumptions are added:

4. the numbers of male and female pupils increase at the same rate as they did according to the accumulated data of the empirical statistics,
5. the number of teachers increases accordingly.

The patches where the school agents are located generate teacher agents according to the historical records, and the school agents generate the pupil agents and place them around their own patches. During the simulation, the pupils, the schools and the teachers are updated according to the following specifications:

- Pupil agents generate a small number of new pupil agents, each with a probability of 0.011 per period, and these become females with probability 5/7 and males with probability 2/7. This makes sure that the overall population of pupils of both sexes increases at the same rate as it did historically. The new pupil agent is assigned to the same school if this school is either co-educational or if this school is gender segregative and the gender of the new pupil allows this assignment. Otherwise the new pupil is assigned to an appropriate school in the neighbourhood. If there is no appropriate school in the neighbourhood such a pupil agent is assigned to no school at all, but has to wait until a gender segregative school in its neighbourhood is turned into a co-educational school. That pupil agents generate new pupil agents is, of course, a simplification, but it seemed inappropriate to represent parent agents for this model. The distribution of pupil gender relations is not discussed in this paper.

- School agents first count how many teachers are about to retire, calculate the gender relation among their teachers and calculate the probability that the next retiring teacher is replaced by a woman and that the next newly employed teacher is a woman (for the latter case see below). This probability is calculated as follows: $P(W|\xi_s) = v(t)\delta \exp(\kappa \xi_s)$, where ξ_s is the difference between the numbers of men and women among the staff of school s divided by their sum, $\xi_s = (m_s - w_s)/(m_s + w_s)$; δ and κ are two free parameters whose values can perhaps be calibrated when one compares the historical data to the simulation results of several runs with varying δ and κ, and v is a parameter which makes sure that for $\delta = 1$ at all times men and women have the same overall probability of being employed ($v = t_r/(2w_r)$ where t_r is the number of all teachers in all schools currently to be employed) and w_r is the number of women in all schools to be currently employed as calculated with the value of v of the previous period, such that v is periodically adjusted. The parameter κ describes the strength of the dependence of $P(W|\xi)$ on ξ whereas δ describes the equal opportunities policy of the ministry in charge of employing teachers ($\delta = 1$: equal opportunities, $\delta < 1$: men are preferred, $\delta > 1$: women are preferred). We start with a simulation keeping $\delta = 1$ and $\kappa = 0.5$ constant (which replicates the example in [6]).

  Besides a school agent does some bookkeeping, counting the unserved pupil agents in its neighbourhood which it could not accept and counting their pupils and staff for output to an analysis file.

- Teacher agents become mainly active when they retire; in this case they are replaced with a newly initialised teacher whose gender is calculated from the probability which its school has calculated during its own update. Except for the teachers generated during the initialisation (their age is a normal variable with mean 40 and standard deviation 7—which seems plausible but reliable information about the age structure of teachers in 1950 is not available), teacher agents are employed at age 30, and the time they spend active is a normal random variable with standard deviation 5 and mean 30 for men and 15 for

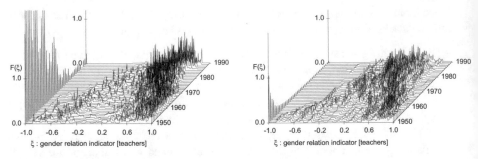

Fig. 3 Time-dependent distributions of gender relations among teachers (*left*: historical data as in Fig. 1, *right*: simulation result) in high schools in Rhineland-Palatinate 1950–1990. Frequency density functions

women—these distributions approximate the historical data satisfactorily. Moreover and with a probability of 0.0025 per period a teacher agent generates a new teacher (whose gender depends on the probability calculated by the school agent) to take account of the fact that the number of teachers increased in Rhineland-Palatinate between 1950 and 1990 (although at a much smaller rate than the number of pupils increased).

With this model the time-dependent gender relations of Rhineland-Palatinate high schools can be reconstructed in an even more satisfactory manner than with the original model. Figure 3 shows the historical and the simulated distributions.

Apart from the fact that schools with only female teachers prevailed for a longer time and with higher frequencies than the simulation could replicate the two graphs in Fig. 3 look surprisingly similar. But a closer inspection of the data is necessary as the visual inspection can be misleading. Hence for each school year the distributions of the two gender relation indicators of one simulation run were compared to the respective historical distributions, using the Kolmogorov–Smirnov test for independent samples. The result can be seen in Fig. 4 (left diagram). The indicator for the teachers remains unsignificant ($\alpha > 0.05$) for the first two decades, the one for the pupils remains insignificant for 12 out of the first 13 years.

The simulation run reported here generated the approximate historical numbers of boys, girls and teachers for the last school year considered, but not the historical gender relation of teachers in that year: The historical numbers are 1615 male and 438 female teachers, but the simulation yields 1420 and 665, respectively—which is a clear indicator that the equal opportunities assumption of the model was violated by the historical data or, to put it the other way round, that the actual chances of female teachers to be employed were worse than they should have been if the ministry of education had diligently observed the equal opportunities principle. This is in line with the findings of [8], and it shows that the model is capable of showing deviations between what is and what ought to be—as assumption 1 is a normative, not an empirical or stylised-fact assumption. Another observation is that at the end of the simulation run only 1 % of boys and girls are left unserved, which

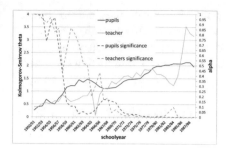 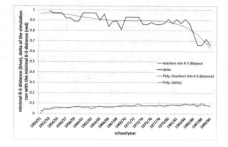

Fig. 4 *Left*: time-dependent Kolmogorov–Smirnov test statistic for the comparison between historical and simulated distributions of gender relation indicators, sample run with $\delta = 1$ and $\kappa = 0.5$, *right*: δ of the run matching the empirical data best for each year (random κ)

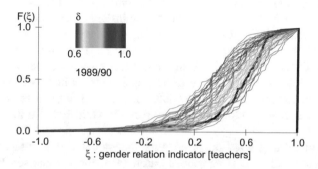

Fig. 5 Distribution of gender relations among teachers at the end of the observed historical period (*black*) and at the end of 60 simulations with varying κ (not shown) and δ (*colours*). Cumulative density functions

is certainly not a reason to open new schools, and only a weak reason to convert gender segregative schools to co-educational schools (but this reason was given in the interviews documented in [8, 18]).

It is an interesting question what the result will be when the historical data are compared to the results of a large number of simulation runs (and varied κ and δ), hence we varied κ randomly between 0 and 1 and δ between 0.6 and 1.0, thus generating 60 different parameterisations, and calculated the Kolmogorov–Smirnov distances between the cumulated frequency distribution function (CDF) of the empirical data and each of the 60 CDFs generated in the simulated simulation runs. Figure 5[2] shows the CDFs at the end of each of these runs with colours representing the δ values used and the empirical CDF in black, making clear that the empirical CDF is within the range of simulation-generated CDFs. Different κ does not seem to make a difference (variance reduction in a bivariate polynomial

---

[2]An animated version of this figure which shows this distribution for each year can be downloaded from http://userpages.uni-koblenz.de/~kgt/GR/CalValSim.ppsx, slide 12; this file also contains all the other figures of this paper in colour and high resolution.

regression up to the second order is far below 1 %), hence they are not marked in Fig. 5, whereas δ alone reduces about two thirds (68.9 %) of the variance of the Kolmogorov–Smirnov distance between empirical and simulated data (κ and δ together 68.5 %, such that the role of κ can be neglected), minimising the latter at 0.22 for $\delta = 0.79$—which means that during the four observed decades the equal opportunities assumption was far from fulfilled.

Wirrer [18] yields a number of hints at decision mechanisms reported by the historical decision makers, and although these are sometimes anecdotal they can be used to check structural validity. The simulation model calculates probabilities with which a teacher is randomly employed and allocated to a school of a certain characteristic. Although the historical decisions were, of course, not exactly random, they must have contained some randomness (or at least features which lend themselves to be modelled stochastically): The employing administration never had a full choice of the best adequate candidate (for instance, fulfilling all requirements to teach a certain subject at school), and the candidates were not always willing to be sent to a school in a region which they did not like—which means that the deterministically made optimal decision could not always be realised and instead a second or third best decision had to be implemented which depended on more or less random circumstances. This conclusion can lead to the assessment that the model is, although still simplified, at least partially validated also in the structural sense.

Another interesting question is whether the model can be used to calibrate what one could call $\delta(t)$. If one looks carefully at the comparison between the empirical distribution and the simulated distributions for varying δ and different schoolyears one could try and use the δ of the best matching simulation run for a certain schoolyear as an estimation of the historical propensity to abide by the equal opportunities principle. The result is shown in Fig. 4 (right diagram) and insinuates that in the first decade this principle was followed at level between 0.89 and 0.97, whereas in the second and third decade it was still observed at a level between 0.81 and 0.96, but dropped to a level of 0.66 in the middle of the 1980s. The reason for this steep decrease remains an open question.

2.2 Extortion Racket Systems

Extortion racket systems were the target systems of the GLODERS project ('Global dynamics of extortion racket systems') which analysed criminal groups, their victims and the society around them both with simulation models and with qualitative and quantitative empirical analysis. The story behind this model is the conflict between extortion racket groups (like the Mafia, the 'Ndrangheta', the Camorra, but also motorcycle gangs such as Hells Angels, Bandidos, or even local independent criminal gangs) which 'sell' defence to shops, restaurants, bars, etc., against their own (or other groups') raids when money—protection money or *pizzo* in Italian—or equivalent assets are ceded to the criminals. This happens in a societal context where customers of shops, restaurants, bars, etc., might abstain from buying (or continue to buy) from victims who pay this protection money, or where the police force fights (or collaborates with) the criminals, and where the behaviour of all persons and groups is governed by law and social norms which more often than not are in deep conflict.

One of GLODERS' simulation models[3] lends itself for an analysis similar to the one in the previous section, as its results, too, can be compared to empirical and historical data, although here these data are even more scarce and less reliable than in the co-education case. What we have is qualitative and anecdotal knowledge about the behaviour and actions of the actors involved in extortion rackets and incomplete quantitative data about attempted and completed extortion. The quantitative data collected by the Palermo group of the GLODERS project contains 629 extortion cases which became known to the police in Calabria and Sicily mainly between 2005 and 2011. As a matter of course, these data do not include extortions which remained secret between extorter and victim, and we do not even have estimates about the dark figure of crime, i.e. the estimated number of unreported and undetected cases of extortion, but this is an unsolvable problem. The main ingredients of the NetLogo model of extortion racket systems and their main features are the following (for more details see [10, 13, 14]):

- Extorter agents approach agents representing shop owners and other entrepreneurs, demanding extortion money and menacing punishment in case this money is not paid. This can be successful or not, depending on the victim's reaction.
- Shop agents either pay the requested money or decide to denounce the extorter to the police (or rarely do both).
- Consumer agents decide whether they do their shopping only with shops which abstain from paying extortion money or do not care whether shops pay or denounce.
- Police agents try to prosecute extorters, bring them to investigation custody and see to it that they are convicted for a longer term in jail.
- An agent representing the state collects assets from convicted extorters and redistributes part of these assets to shop agents which fell victim to extortion to compensate them for their losses. This is in line with Italian legislation.

In a simple version of the NetLogo model, all agents have fixed probabilities to make their respective decisions when such decisions are due, whereas in a more sophisticated version all agents keep a long-term memory of past experience with extortions they observed or suffered from and positive or negative sanctions on extortion-related behaviour. The contents of their memories are used to calculate the salience of several norms, different for the four agent types. Shop agents, for instance, know about norms related to paying extortions and to denouncing,

---

[3]It can be found at http://www.gloders.eu/components/com_jwiki/mediawiki/images/d/d5/ NOERS.zip.

but for both actions two conflicting norms exist in their society: not to pay, as
extortion is illegal, and to pay, as extortion has a long tradition (as it actually has in
different regions in Southern Italy), to denounce, as this is legally bidden, and not
to denounce, as denunciation would violate a long tradition. Consumer agents know
about a norm not to buy from shops which pay extortion, as this is, for instance,
recommended by an NGO called 'addio pizzo' in Italy, and about a norm to pay
from pizzo payers as well, as paying pizzo has a long tradition. Extorters know
the legal norm to denounce their accomplices, but perhaps the social norm not to
betray—the 'omertà'—might be stronger. And finally the police know their legal
norm to prosecute wherever they get to know about extortion, but perhaps they, too,
might want to keep themselves safe and to tolerate traditional criminal behaviour.
The salience calculation follows a complicated formula first defined in [3, Text S1]:

$$\sigma = \alpha \left(\beta + \frac{C-V}{C+V}w_c + \frac{O_c - O_v}{O_c + O_v}w_o + \frac{\max(0, (O_v + V) - P - S)}{O_v + V}w_{npv} \right.$$

$$\left. + \frac{Pw_p + Sw_s}{\max(P+S, O_v + V)} + \frac{E_c - E_v}{E_c + E_v}w_e \right) \tag{1}$$

where

- C and V are counters for events when the agent itself complied with the respective
  norm or violated it, respectively,
- O_c and O_v are counters for events when an agent observed another agent which
  complied with the respective norm or violated it, respectively,
- P and S are events when an action was punished or sanctioned, respectively,
- E_o and E_v are events when an agent explicitly invoked the respective norm
  because of a compliance or a violation, respectively,
- w_c, w_o, w_{npv}, w_p, w_s and w_e are the weights for the six factors ('norm
  cues') derived from [5] and defined in [3] (see also [2]), namely *own norm
  compliance/violation*, *observed norm compliance/violation*, *non-punished
  violation*, *punishment*, *sanction* and *explicit norm invocation* where these factors
  are calculated from the counters listed above and
- α and β have to be chosen dependent on the weights w_c, w_o, w_{npv}, w_p, w_s and w_e
  in a way that $0 \leq \sigma \leq 1$.

Whenever a decision is to be made which action is to be taken the saliences of all
related norms are updated, and the norm with the higher salience is applied for this
decision. Hence the distribution of norm saliences changes over time.

The simulation reports every extortion with several attributes similar to those
reported in the Sicily and Calabria database with more than 630 cases, mainly
between 2005 and 2011, such that every simulation run can be compared to the
database (which can be split into extortions that occurred in specific regions or
that were executed by specific families). Figure 6 (left) shows a scattergram which

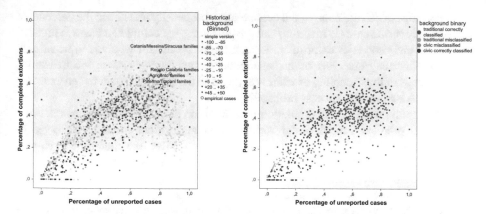

Fig. 6 Scattergram of the percentage of completed cases against the percentage of undenounced cases. Simulation results of both versions of the model and historical extortions executed by four groups of Cosa Nostra and 'Ndrangheta' families (*left*) and results of the normative version with classification of the background input parameter as a function of the percentage of completed cases and the percentage of undenounced cases (*right*)

represents 1000 runs of the simple stochastic version of the model (grey dots), 1000 runs of the sophisticated model (coloured dots) and four groups of Mafia families which are differentiated according to their regions of activity. Each dot represents either one run with several hundred attempted or completed, denounced and undenounced extortions or one of the four groups of Mafia families, the coordinates marking the percentages of unreported cases and of completed cases (these are the only two variables for which the database provides a sufficient amount of non-missing data). For the sophisticated version the dots are marked in different colours according to the initialisation of the respective simulation run. When the simulation starts, all agents' memories are filled with a number of extortion-related experiences and observations—in the case of violet, blue and green dots these are experiences and observations which one would make in a traditional society, in the case of yellow and red dots these are the ones which one would make in a civic society.

From Fig. 6 two kinds of conclusions can be drawn which both relate to the research questions raised in section 1.2:

- The relation between the percentage of unreported cases and the percentage of completed cases is more or less the same for both versions of the model. These two variables show a positive correlation when their values are both below 0.5— and this is the case for simulated societies with a 'civic' background (and for the simple version, not shown in Fig. 6 (right), for a high propensity to denounce extortions), and they show a slightly negative correlation when the opposite holds (and the variance of the percentage of completed cases is high). In artificial societies which are neutral, the denunciation propensity and hence the percentage of unreported cases is medium, and the percentage of completed cases is highest, as in civic societies extortions rare and rarely successful whereas in traditional societies they are most often undenounced, and the success of the extortions depends of the capability of victims to pay, and the competition between extorters might decrease the success probability.

– The comparison of the simulation runs with the empirical cases shows that the model predicts a much wider variety of outcomes than could be observed in Southern Italy. The empirical cases are most similar to simulation runs with a highly traditional background (which one finds in Southern Italy) whereas simulation runs with a civic background have no empirical correlates, at least not in Southern Italy. Criminal statistics from other countries analysed during the GLODERS project in fact show that extortion is rare, for instance, in Germany such that they do not even form a special category in the statistical records, hence no reasonable statistics can be calculated which would lend themselves to be compared to simulation results. In this respect, this model is a typical example of a 'case 1 model' as 'one or more relevant behaviours are relatively rare' such that only few 'behavioural assumptions for agents can be derived from surveys' [11, p. 10] (see the discussion in [14]).

– Both scattergrams in Fig. 6 can in principle be converted into an explicit form in the sense of the first research question in subsection 1.2, combining the two output variables with the explanatory variable with the input parameter, but with the restriction that it is mainly the sign and much less the absolute value of the background variable which influences the output variable. Hence here we restrict ourselves to discriminance analysis with the sign of the background variable ($\mathrm{sgn}(b)$) dependent on the percentage of unreported cases (x) and the percentage of completed cases (y) and find that only in one out of six runs the sign of the background is misclassified. The closed form is

$$\mathrm{sgn}(b) = \begin{cases} 1 \text{ if } 0.3436 - 0.2077x - y > 0.135 \\ -1 \text{ if } 0.3436 - 0.2077x - y < 0.135 \end{cases} \tag{2}$$

Thus, answering the second question in Sect. 1.2, one could validate the model for the full range of the historical background variable if one had, for instance, in a future Eurobarometer or European Values Study, sufficiently many data about the exposure of the interviewees to extortion and similar crimes and data about the salience of traditional and more modern, civic or liberal norms. Unfortunately such a dataset does not yet exist, but the model can provide information to design an appropriate survey. But even this would not guarantee that real actors would credibly unbosom how they 'operate to produce [their] behaviour'.

3 Conclusion

The two models and their target systems discussed in this paper do not have much in common with respect to the theoretical assumptions behind them and the empirical scenarios from which they were derived. But from all methodological point of view they have—notwithstanding some differences—several features in common:

- In both cases a preliminary version of the respective model was drafted without considering all empirical data. In the co-education case the first model only tried to replicate the distributions of the gender relations over the period for which data were available as these quantitative data were available prior to the results of qualitative case analyses, whereas in the extortion racket case the quantitative data were made available only after several different versions of the simulation models had been designed according to the stylised facts and anecdotal material which allowed for some qualitative understanding (in the sense of [1]).
- In both cases, extended versions of the respective models were designed, taking into account the additional material which had been collected alongside the model design. In the co-education case this was the inclusion of the development of the sheer increase in numbers of boys and particularly girls attending high schools during the 1960s, 1970s and 1980s which gave rise to a decrease in the number of schools, taking into account that in less densely populated regions it would have been much too costly to open separate schools for girls in the same provincial town where a boys' school already existed, thus modelling the change in teacher gender relations as not only dependent on some first principles but also on the change of the available labour force. In the extortion racket case this was mainly the idea that not only the behaviour of the public (shop owners and consumers), but also the behaviour of police and extorters is not purely random but governed by norms which are being learnt during the communication among all four (and perhaps even other) groups.

In the end, both simple cases showed that a satisfactory replicative validity could be achieved when comparing their simulation results to the (unfortunately still scarce) empirical data. If one compares the results of the more sophisticated versions of the two cases, one finds that the replicative validity improves—slightly in the extortion racket case, considerably in the co-education case—which for the latter case is not really a surprise as part of the process is governed by the exogenous data about the development of the numbers of teachers and of male and female pupils. If these had had to be estimated endogenously this might have been more questionable.

To conclude, a detailed analysis of two quite different target systems with different access to empirical data has shown the difficulties of thorough validation of simulation models against observational data. At the same time it has shown that such a validation is indeed possible and that it opens new perspectives on empirical data to be retrieved, answering research questions which without a simulation model—in the sense of a theory implemented in computer code—would never have been asked.

Acknowledgements The research leading to the results of Sect. 2.1 received funding from Deutsche Forschungsgemeinschaft between 1991 and 1995 under grant agreement no. KR 960/5-1 and -2 ('Einführung und Auswirkung der Koedukation. Eine Untersuchung an ausgewählten Gymnasien des Landes Rheinland-Pfalz'/'The introduction of co-education. A study of educational history at selected grammar schools in the German State of Rhineland-Palatinate').

The research leading to the results of Sect. 2.2 has received funding from the European Union Seventh Framework Programme (FP7/2007–2013) since 2012 under grant agreement no. 315874 (http://www.gloders.eu, 'Global dynamics of extortion racket systems').

Comments of two anonymous reviewers are gratefully appreciated.

References

1. Abel, T.: The operation called "Verstehen". Am. J. Sociol. **54**, 211–218 (1948/1949)
2. Andrighetto, G., Castelfranchi, C.: Norm compliance: the prescriptive power of normative actions. Paradigmi [6] **2**, 120–135 (2013)
3. Andrighetto, G., Brandts, J., Conte, R., Sabater-Mir, J., Solaz, H., Villatoro, D.: Punish and voice: punishment enhances cooperation when combined with norm-signalling. PLoS ONE **8**(6), e64941 (2013)
4. Beisbart, C., Norton, J.D.: Why Monte Carlo simulations are inferences and not experiments. Int. Stud. Philos. Sci. 403–422 (2013). http://dx.doi.org/10.1080/02698595.2012.748497
5. Cialdini, R.B., Reno, R.R., Kallgren, C.A.: A focus theory of normative conduct: recycling the concept of norms to reduce littering in public places. J. Pers. Soc. Psychol. **58**, 1015–1026 (1990)
6. Gilbert, N., Troitzsch, K.G.: Simulation for the Social Scientist, 2nd edn. Open University Press, Maidenhead, New York (2005)
7. Ihrig, M., Troitzsch, K.G.: An extended research framework for the simulation era. In: Diaz, R., Longo, F. (eds.) Emerging M&S Applications in Industry and Academia Symposium and the Modeling and Humanities Symposium 2013, 2013 Spring Simulation Multiconference. Simulation Series, vol. 45.5, pp. 99–106. Curran Associates Inc., Red Hook, NY (2013). http://dl.acm.org/citation.cfm?id=2499751.2499763
8. Kraul, M., Wirrer, R.: Koedukation gegen Lehrerinnen? Die Berufschancen von Lehrerinnen an Gymnasien des Landes Rheinland-Pfalz. Die Deutsche Schule **88**(3), 313–327 (1996)
9. Kraul, M., Troitzsch, K.G., Wirrer, R.: Lehrerinnen und Lehrer an Gymnasien: Empirische Ergebnisse aus Rheinland-Pfalz und Resultate einer Simulationsstudie. In: Sahner, H., Schwendtner, S. (eds.) Kongreß der Deutschen Soziologie Halle an der Saale 1995. Kongreßband II: Berichte aus den Sektionen und Arbeitsgruppen. pp. 334–340. Westdeutscher Verlag, Opladen (1995). http://www.ssoar.info/ssoar/handle/document/17224
10. Nardin, L.G., Andrighetto, G., Conte, R., Szekely, A., Anzola, D., Elsenbroich, C., Lotzmann, U., Neumann, M., Punzo, V., Troitzsch, K.G.: Simulating the dynamics of extortion racket systems: a Sicilian Mafia case study. J. Auton. Agent Multi Agent Syst. 30(6), 1117–1147 (2016). doi:10.1007/s10458-016-9330-z
11. Smajgl, A., Barreteau, O. (eds.): Empirical Agent-Based Modelling — Challenges and Solutions. Volume 1, The Characterisation and Parameterisation of Empirical Agent-Based Models. Springer, New York (2014)

12. Troitzsch, K.G.: Social simulation – origins, prospects, purposes. In: Conte, R., Hegselmann, R., Terna, P. (eds.) Simulating Social Phenomena. Lecture Notes in Economics and Mathematical Systems, vol. 456, pp. 41–54. Springer, Berlin (1997)
13. Troitzsch, K.G.: Distribution effects of extortion racket systems. In: Amblard, F., Miguel, F.J., Blanchet, A., Gaudou, B. (eds.) Advances in Artificial Economics. Lecture Notes in Economics and Mathematical Systems, vol. 676, pp. 181–193. Springer, Berlin (2015)
14. Troitzsch, K.G.: Extortion racket systems as targets for agent-based simulation models. comparing competing simulation models and empirical data. Adv. Complex Syst. **18** (2015, accepted for publication). http://www.worldscientific.com/doi/pdf/10.1142/S0219525915500149
15. Waldherr, A., Wijermans, N.: Communicating social simulation models to sceptical minds. J. Artif. Soc. Soc. Simul. (2013). http://jasss.soc.surrey.ac.uk/16/4/13.html
16. Wilensky, U.: NetLogo (1999). http://ccl.northwestern.edu/netlogo
17. Windrum, P., Fagiolo, G., Moneta, A.: Empirical validation of agent-based models: alternatives and prospects. J. Artif. Soc. Soc. Simul. (2007). http://jasss.soc.surrey.ac.uk/10/2/8.html
18. Wirrer, R.: Koedukation im Rückblick. Die Entwicklung der rheinland-pfälzischen Gymnasien vor dem Hintergrund pädagogischer und bildungspolitischer Kontroversen. Blaue Eule, Essen (1997)
19. Zeigler, B.P.: Theory of Modelling and Simulation. Krieger, Malabar (1985). Reprint, first published in 1976, Wiley, New York, NY

A Methodology for Simulating Synthetic Populations for the Analysis of Socio-technical Infrastructures

Koen H. van Dam, Gonzalo Bustos-Turu, and Nilay Shah

Abstract Modelling socio-technical systems in which a population of heterogeneous agents generates demand for infrastructure services requires a synthetic population of agents consistent with aggregate characteristics and distributions. A synthetic population can be created by generating individual agents with properties and rules based on a scenario definition. Simulation results fine-tune this process by comparing system level behaviour with external data, after which the emergent behaviour can be used for analysis and optimisation of planning and operation. An example of electricity demand profiles is used to illustrate the approach.

Keywords Agent-based model • Socio-technical system • Synthetic population

1 Introduction

To analyse and understand the operation of socio-technical systems, in which physical systems interact with social networks, simulation models need to include the behaviour of the actors in the model [1]. For example, the emergent behaviour of a group of actors, modelled as agents, could generate the demand for services provided by the socio-technical system. Individual characteristics of agents are necessary to include heterogeneity in the simulation model while remaining consistent with aggregate/average values for the population [2, 3]. In this chapter a methodology to generate a synthetic population given certain land-use and general population characteristics is proposed. The approach followed uses geographical information system (GIS) data as input which is enriched with land-use data (e.g. population density, floor space of offices) combined with statistics on the population (e.g. employment rates, car ownership). A case study on electricity consumption is used as an example of the proposed methodology and some illustrative results are presented for an area in West London.

K.H. van Dam (✉) • G. Bustos-Turu • N. Shah
Centre for Process Systems Engineering, Imperial College London, London, UK
e-mail: k.van-dam@imperial.ac.uk; gb1612@imperial.ac.uk; n.shah@imperial.ac.uk

© Springer International Publishing AG 2017
W. Jager et al. (eds.), *Advances in Social Simulation 2015*, Advances in Intelligent Systems and Computing 528, DOI 10.1007/978-3-319-47253-9_39

2 Methodology

The core idea is that a synthetic population can be generated during the initialisation of a simulation model using a few basic properties for the area considered, key characteristics for the population and distributions of the activities they engage in which together form the scenario definition. An "agent factory" then generates a population of agents to make up the synthetic population, after which *system level behaviour* can be *analysed* or used for *validation* of the generation step. Figure 1 shows a schematic for this workflow from scenario definition to output analysis. The population data and general rules are characterised by probability distributions (e.g. uniform or normal), while the individual data and individual rules at the agent level represent a specific instance (e.g. a chosen departure time). The input of the agent factory includes spatial (i.e. built environment) data, socio-demographic data and technical parameters which can be provided through input files (e.g. shape files, structured documents, or as variable values in the model definition). Examples of typical inputs are given in Table 1.

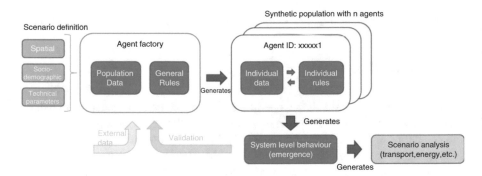

Fig. 1 Methodology of generating a synthetic population based on a scenario definition

Table 1 Examples of scenario definition (with some sub-properties shown in brackets)

Input category	General	Scenario specific
Built environment (spatial data)	Roads (type, speed), buildings or areas (land-use, number of floors, total footprint)	Electricity distribution network (capacity and layout)
Socio-demographic	Population density, household size, car ownership level, employment rate, activity schedule (activity, departure, deviation, occurrence)	Electric vehicle type market share, PEV adoption level
Technical parameters	Peak and base electricity demand per household, average speed etc.	Charging infrastructure access level, charging power rates

From this input data, the agent factory determines how many agents to create and where their activity locations are. The number of agents living in a certain location depends on the population density. Agents may be created in different groups (e.g. workers or non-workers) again taking socio-demographic data into account. The total floor space, based on footprint and number of floors, combined with the land-use (e.g. leisure or commercial) of a building or neighbourhood affects the probability that an agent chooses it as a destination for its activities.

In this work, the general activity schedule AS_k for each group k of agents is defined with a list of 4-tuples:

$$AS_k = \{(ACT_j, MDT_j, SD_j, PD_j)\} \tag{1}$$

For each activity ACT_j, the departure times are modelled as a stochastic variable following a normal distribution with a mean departure time MDT_j representing the peak hour of that period and considering a standard deviation SD_j to account for variability in the departure time among agents. A departure probability PD_j is included in the model to simulate that not every agent in that group undertakes all the activities.

Three types of analysis can be done then by making changes to the scenario input:

- Changes in land use or spatial planning can be explored by adjusting the spatial configuration input files. This can be used to explore the impact of different masterplans and proposed developments on the same population.
- Different population characteristics (i.e. socio-demographic data) can be provided to see how they impact the way people use the available city infrastructures. This way one could experiment how changes in the population living in an area could be reflected on the demand for infrastructure services.
- The user can experiment with the impact of different behavioural rules. This could be used to test the consequence of incentives or policies that change how people travel, work and engage in activities, by changing the occurrence and the timing.

In next section an example is given following this methodology, including case-specific data and behavioural rules for the agents.

3 Case Study: Simulating the Electricity Consumption in an Urban Neighbourhood, Including Electric Vehicles

The example application used in this paper is the charging of plug-in electric vehicles (PEVs) in an urban area, following the description in [4]. Driving reduces the state of charge (SOC) of the battery which is recharged when the car is plugged into a charging unit. Simulation of the spatial and temporal distribution of the

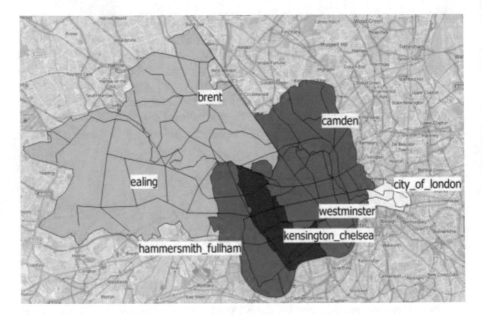

Fig. 2 City layout used for the simulation. *Colours* represent the density for each borough

heterogeneous PEV owner activities generates PEV journeys and individual demand for charging, giving detailed insights about where and when there is the potential to recharge the PEV batteries. To assess the impact of this "mobile" load, local distribution network conditions (including local demand for electricity from "static" loads leading to constraints on the distribution network) have to be determined. The spatially and temporally explicit static load is also simulated using the area's occupancy that is estimated using the same population and their activities and transport demand, leading to electricity consumption profiles in the different areas of the simulated city.

For this example, the number of PEVs simulated for each area is calculated based on the number of cars (based on the population size, household size and car ownership levels per area[1]) and level of PEV diffusion. Finally, the population for each area is based on the density[1] and footprint area[2] (see Fig. 2). A combination of general and scenario specific data is thus used (see Table 1) to set up the scenario.

During the initialisation of the model the agents are created based on the number of PEVs in each area. Then, each agent is linked to a home, office and typical place for shopping and leisure activities (based on the land-use) and its charging infrastructure access level is defined. Next, this agent is linked with a PEV with an initial SOC. Finally, the activity schedule for each agent is created based on the

---

[1]Extracted from Office for National Statistics data (http://neighbourhood.statistics.gov.uk/).

[2]From OrdnanceSurvey MasterMap data (https://www.ordnancesurvey.co.uk/).

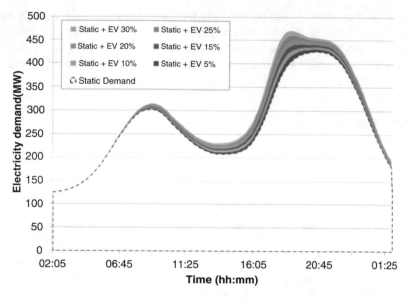

Fig. 3 Output of simulated electricity demand for 5–30 % PEV adoption

general activity schedule of the agent's group (see Eq. 1). If desired this process can be deterministic to enable replication of scenarios or stochastic to explore parameter space.

Once the population is created the simulation is run for a working day and the residential and PEV electricity demands are generated. Different scenarios of PEV adoption (from 5 to 30 %) are considered to analyse the impact of PEV on the distribution network in different areas of the city (see Fig. 3). Although the simulation generates individual agent data, the outputs can be aggregated spatially, or according to different group attributes of agents (workers/non-workers etc.).

4 Conclusions

Social simulation is essential for realistic analysis of urban infrastructure systems in which human activities drive demand [5]. If these demands are not part of the simulation but provided as input (e.g. average values obtained from surveys) it means the model cannot provide decision support for the impact of incentives and behavioural change, or explore how such demands may change in the future. By including the actors and their behaviour, the bottom-up generation of demands means feedback loops can be explored by seeing what the response of the population is on infrastructure changes. This way the social, the physical or both systems can be varied to study the impact on overall system behaviour. For the case presented this means that one can to experiment with detailed socio-economic data, looking at the impact of social factors on PEV adoption levels or smart charging operation. Separating generation of the synthetic population from the rest of the simulation means this model component is reusable and flexible in multiple case studies, but also makes it transparent how the agents are generated based on selected input data.

References

1. Van Dam, K.H., Nikolic, I., Lukszo, Z.: Agent-Based Modelling of Socio-Technical Systems. Springer, New York (2013). ISBN 978-94-007-4932-0
2. Harland, K., Heppenstall, A., Smith, D., Birkin, M.: Creating realistic synthetic populations at varying spatial scales: a comparative critique of population synthesis techniques. J. Artif. Soc. Soc. Simulat. 15(1), 1 (2012)
3. Moeckel, R., Spiekermann, K., Wegener, M.: Creating a synthetic population. Proceedings of International Conference on Computers in Urban Planning and Urban Management (CUPUM), Sendai, Japan (2003)
4. Bustos-Turu, G., van Dam, K.H., Acha, S., Shah, N.: Estimating plug-in electric vehicle demand flexibility through an agent-based simulation model. 5th IEEE PES Innovative Smart Grid Technologies (ISGT), Istanbul, Turkey (2014) Accessed 12–15 Oct 2014
5. Sovacool, B.K., Ryan, S.E., Stern, P.C., Janda, K., Rochlin, G., Spreng, D., Pasqualetti, M.J., Wilhite, H., Lutzenhiser, L.: Integrating social science in energy research. Energ. Res. Soc. Sci. 6, 95–99 (2015)

Modeling the Individual Process of Career Choice

Mandy A.E. van der Gaag and Pieter van den Berg

Abstract Making a suitable career choice is a difficult task. Every year, many adolescents prematurely end their studies, commonly citing "having made the wrong choice" as the main reason. This is a problem, both for the adolescents making these choices, and for society, which bears at least part of the cost of higher education. A thorough understanding of how adolescents make these career choices is essential to identifying the factors responsible for why the wrong choices are often made. Identity development theory emphasizes the role of exploration in career choice, but neglects many of the micro-level processes likely to play an important role. Similarly, traditional decision theory often focuses on optimization of choice, thereby neglecting the cognitive mechanisms that may explain deviations from optimal choice. Here, we present a novel computational approach to modeling long-term decision making. We combine elements of the macro-level theory on identity development with a firm rooting in micro-level cognitive processes. Specifically, we model decision making as an iterative process in which individuals can explore new options or more deeply investigate options that are already under consideration. The output of our model allows us to analyze how the quality of decisions depends on various factors, such as aspiration levels, the tendency to explore new options, and the ability to judge the fit of an option with one's interests and capabilities. We present some preliminary results that already show our approach can lead to surprising conclusions, encouraging further development of this model in the future.

Keywords Intra-individual computational modeling • Information processing models • Decision making • Career choice • Identity development

M.A.E. van der Gaag (✉)
Faculty of Behavioral and Social Sciences, University of Groningen, Grote Kruisstraat 2/1, Groningen, 9712 TS, The Netherlands
e-mail: mandyvandergaag@gmail.com

P. van den Berg
Lab of Socioecology and Social Evolution, KU Leuven, Naamsestraat 59, Leuven, 3000, Belgium

© Springer International Publishing AG 2017
W. Jager et al. (eds.), *Advances in Social Simulation 2015*, Advances in Intelligent Systems and Computing 528, DOI 10.1007/978-3-319-47253-9_40

1 Introduction

Choosing the right career is by no means a straightforward process for the vast majority of adolescents in Western cultures. At a relatively young age (typically between 16 and 20), students in secondary school are faced with making the important decision of choosing a major in higher education. For example, in the Netherlands, prospective students have to commit to a specialization even before entering university, choosing between more than a thousand relatively narrowly defined subjects.

For adolescents, making an important life choice out of so many options can be a daring task. This is not only difficult because of the sheer number of options, each with many facets, but also because adolescents are still very much in the process of identity development. Because of this, many adolescents do not have a clear idea of what their preferences and interests actually are; this makes the evaluation of options all the more difficult. Making such an important decision is further complicated by an imbalance in adolescent brain development: as limbic structures develop more quickly than prefrontal structures, rational cognitive control is limited, and emotional motivations are more likely to drive decisions [1].

A recent study among Dutch students [2] found that the most important reason for dropping out of higher education was having made a wrong education choice. Dropping out is a common phenomenon in Europe; 20–55 % of university students do not complete their education [3]. Needless to say, this is a problem that not only frustrates adolescents, but also comes at a significant cost to society.

1.1 Identity Development

Getting a clear idea of one's own interests is crucial in making a fitting career choice. Early identity development theorists Erikson [4] and Marcia [5, 6] posited that ideas on "who you are and what you want" develop mainly in adolescence through a process of exploration. Exploration is a broad behavioral construct: it can be defined as any kind of behavior aimed at eliciting information (be it cognitive, emotional or social in nature) about the self or the environment in order to make a decision about an important life choice [7]. Different types of exploration have been distinguished [8]; an important distinction is between "exploration in breadth" (globally investigating multiple options) and "exploration in depth" (investing time and energy to gain more information on a particular option). Germeijs and Verschueren [9] found that both types of exploration are important for developing suitable career commitments.

Although identity development theory is relevant for describing macro-level variables relevant to making a career choice, research in this field offers little knowledge on what happens to individuals on a micro-level [10]. Consequently, this framework does not provide a clear notion of what the basic mechanisms of career exploration are, and how individual differences in these mechanisms may affect the quality of choices made.

1.2 Information Processing Models

To be able to work towards policies to help adolescents make more suitable career choices, it is vital that we understand this decision making process in more detail. Decision science has a long tradition of modeling micro-level choice processes. Traditionally, the study of decision making has been dominated by classic expected utility theory. In this framework, decision making is presumed to be a rational process of optimization between available options, given a function determining the desirability (utility) of each option, based on various characteristics. Although framing decision making as a process of sampling and subsequent optimization may appear intuitively appealing, there is mounting evidence that very few human decision making processes can be adequately modeled in this way [11]. In fact, there is a growing movement of grounding models of decision making in basic nonlinear cognitive processes (e.g., Decision Field Theory [12]; Query Theory [13]), rather than assuming a "black box" psychology that is an optimization machine. These information processing models are currently rapidly gaining ground (indeed, even causing a paradigm shift; [11]).

1.3 Current Study

Here, we introduce a novel approach to modeling decision making processes that combines macro level identity development theory with micro-level information processing models. In contrast to existing models on career choice, our approach allows us to study the effect of key factors within the process of exploration that may differ between individuals (such as aspiration level or clarity of preferences). In addition, by explicitly modeling the dynamics of decision making, we can gain insights in the process of decision making, and how different processes are related to different outcomes. Although our model is currently still under development, we have already produced some interesting preliminary results. For example, we observe conditions where it is always more beneficial to explore in breadth, than to exploit options in depth.

2 The Model

We are developing an event-based simulation model of individual career choice processes in $C++$. The general assumptions of our model are partly grounded in information processing theory. Specifically, we model decision making as an iterative process proceeding for a number of time steps (following Query Theory; [13]), and assume that time is limited (following Decision Field Theory; [12]). In line with personal identity development literature, we assume that there are two ways to investigate options: exploration (in which new options are sampled), and exploitation (in which options of particular interest are investigated in more depth). We further assume that individuals judge options by their perceived fit with their interests and capabilities; only options that are associated with a high perceived fit are exploited and eventually chosen.

We assume that the focal individual has a set of options (S) under consideration (where the size of S is limited to a maximum N; see Table 1 for an overview of model parameters and variables). In each time step, the individual explores a new option with probability m; this may lead to the addition of the newly explored option to S. With the complementary probability, she randomly exploits one of the options that is already in S. In the very first time step, the individual does not yet have any options under consideration, and can therefore only engage in exploration. In any time step, an option may be chosen. The model runs for a maximum of T time steps; if no option is chosen before time runs out, the individual is forced to choose the option in S with the highest perceived fit.

2.1 Exploration

Exploration is modeled as the random sampling of an option from a pool of potential options. We assume that each of the potential options has an "objective fit" (x_o), drawn from a standard normal distribution. This is meant to reflect that some options are more suitable to the focal individual than others, and that options that fit very well or very poorly are rarer than options with an intermediate fit (other distributions can also be considered). We further assume that individuals are not able to directly perceive the objective fit of an option. Rather, their perceived fit (x_p) of an option is subject to some error, such that

$$x_p = x_o + \varepsilon \tag{1}$$

where ε is drawn from a normal distribution with mean 0 and standard deviation a. The parameter a determines how accurately the individual is able to judge the fit of an option, which captures differences in level of identity development (i.e., how clearly defined own interests and preferences are). A newly explored option will be included in S if there are fewer than N options under consideration. Alternatively, the newly explored option may replace the option in S with the lowest perceived fit, if it has a higher perceived fit. Otherwise, the newly explored option is discarded.

Table 1 Parameters and variables of the model

Parameter	Description
T	The number of time steps available for exploring/exploiting options before a decision has to be made
N	The maximum number of options the individual can have under consideration at any point in time
m	The probability with which the individual explores a new option in any time step. With the complementary probability, the individual exploits an option already under consideration
a	The standard deviation of the normal distribution from which perception errors are drawn (the mean of this distribution is 0). The perception error determines the distance of the perceived value of the fit of an option from the objective fit. With increasing a, the individual is less accurate in her assessment of the fit of an option
t_1	The first aspiration level. If the perceived fit of an option exceeds this number, the individual takes this option under consideration
t_2	The second aspiration level. If the perceived fit of an option exceeds this number, the individual chooses this option
ρ	The recency factor. This number determines the weight past experiences relative to the current experience with an option. If smaller than one, the recency factor leads individuals to discount the past
Variable	Description
S	The set of options under consideration
x_o	The objective fit of an option
x_p	The perceived fit of an option
r	Number of times an option has been exploited

2.2 Exploitation and Choice

When exploitation occurs, one of the options in S is selected for further investigation. In this case, the individual randomly draws an option from the options in S that have a perceived fit that exceeds their first aspiration level (t_1). This first aspiration level is meant to reflect the idea that individuals will only exploit options that they are at least moderately interested in. Through exploitation (be it hands-on experience with the option, discussing the option with friends, further reading, or otherwise), the individual may update her perceived fit of the option so that it eventually comes closer to the objective fit. Exploitation occurs in a similar fashion as exploration, but past experience is taken into account when updating the perceived fit of the option. Specifically, the updated perceived fit ($x_p{}'$) depends on the previous perceived fit (x_p) as follows:

$$x'_p = \frac{\rho r x_p + x_o + \varepsilon}{\rho r + 1} \tag{2}$$

where ε denotes an error term drawn from a normal distribution with mean 0 and standard deviation a (as in Eq. 1), ρ represents a recency factor (ensuring that a new experience is weighed more heavily than experiences in the past), and r denotes the number of times the option has already been evaluated in the past (this ensures that the influences of new experiences diminishes as the total experience with an option increases). Over time, repeated exploitation will lead x_p to approach x_o. If the perceived fit of any of the options in S exceeds a second aspiration level t_2, the individuals decides for this option. If time T has run out before an option has exceeded t_2, the option in S with the highest perceived fit is chosen.

2.3 Simulation Setup

There may be variation between individuals in the number of options that they are able to consider at the same time (N), the time and effort they invest in the decision making process (T), the accuracy with which they are able to judge the fit of an option (a), their tendency to explore new options relative to their tendency to exploit options already under consideration (m), the emphasis they place on recent experiences with an option, relative to experiences further in the past (ρ), and their aspirations levels, both for whether they are willing to consider an option at all (t_1), and for their final choice (t_2). With this in mind, we have run preliminary simulations exploring a relatively wide range of parameter settings for a (number of parameter settings $[n] = 51$), m ($n = 51$), t_1 ($n = 4$), and t_2 ($n = 4$). For now, we have kept three parameters constant: T (100), N (3), and ρ (0.5). For each of the in total 41,616 parameter combinations, we have run 1000 replicate simulations (a total of 41,616,000 simulations).

3 Preliminary Results

Figure 1 shows a single simulation run of the model. Although this specific run may of course not necessarily be illustrative of the overall patterns, it does give an intuition for how our way of modeling long-term decision making can lead to patterns that would not be observed with more classical optimization-based approaches. For example, if individuals have trouble accurately assessing the fit of an option (i.e., they have a relatively high value of a), they may choose an option that is actually below their aspiration level for making a final choice (t_2), even if they may have been likely to explore a better option before time runs out. In Fig. 1 for

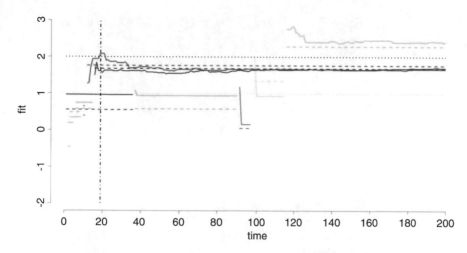

Fig. 1 A single simulation run of a career choice process over time. Each pair of *solid and dashed lines* with matching colors represents an option in *S*. *Solid lines* represent the perceived fit of options (x_p), *dashed lines* represent their objective fit (x_o). The *horizontal grey dotted line* represents the first aspiration level (t_1); the *horizontal black dotted line* represents the second aspiration level (t_2). The *vertical dot dashed line* represents the moment a decision is made (after which the career choice process ends, but we show this to illustrate the dynamics of the model). Options with a perceived fit below t_1 are not exploited; all *solid lines* below t_1 are unchanging (note that this may occur even if the objective fit does exceed t_1; see the *lime-colored* option around $t=100$). Options with a perceived fit above t_1 are exploited (their perceived fit changes over time, and tends to approach the objective fit). When the perceived fit of an option exceeds t_2, that option is chosen. Parameter values for this simulation run are as follows: $T = 200$; $N = 3$; $m = 0.1$; $a = 0.5$; $t_1 = 1.0$; $t_2 = 2.0$; $\rho = 0.9$

example, the purple option is chosen, even though the objective fit of that option is below t_2. For illustrative purposes, the dynamics of the simulation after the moment of choice are also shown (even though the choice cannot change after this point). After the moment of choice, the perceived fit of the purple option drops below t_2. Also, at a later point, a much better option than the purple option is explored (the turquoise option). This illustrates that if an aspiration level is relatively low, especially in combination with a low accuracy of estimating options, this can lead to relatively poor choices.

Figure 2 shows an overview of simulation outcomes across a wide range of parameter combinations. Perhaps not surprisingly, individuals tend to make poor choices if they have a very low aspiration level for their final choice (t_2). However, if this aspiration level is very high, their choices are not necessarily good either. This is probably because with too high aspiration levels, combined with a low tendency to explore (m), individuals may "get stuck" in investigating options that will not cross their choice threshold, until time runs out and they are forced to choose the best of inferior options.

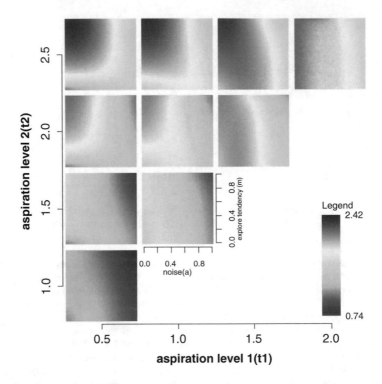

Fig. 2 The effect of aspiration level (t_1 and t_2), exploration tendency (m), and accuracy of judging the fit of options (a) on the objective fit (x_o) of the final choice. For each parameter combination (a total of 41,616), colors indicate the average objective fit of the final choice across 1000 replicate simulations, red indicates choices with a high objective fit, blue indicates choices with a low objective fit. In each subgraph, ranges of 51 values of both m and a are depicted (both varying with step size 0.02, between 0.0 and 1.0). Parameter combinations for which t_1 is equal to or exceeds t_2 have been omitted

Under the current assumptions, the effect of the first aspiration level (t_1, determining whether options are worth exploiting) is more clear; as it increases, choices generally tend to be better. There seem to be some interesting interaction effects for individuals who have high standards for choosing an option (t_2) but relatively low standards for considering one (t_1). If these individuals can accurately estimate the fit of an option (low a), exploiting options does not seem beneficial; this will not improve the subjective fit as this is already close to the objective fit, and the individual has a better chance of finding a good option by exploring a lot. In contrast, individuals who are relatively inaccurate (high a) seem to be better off exploiting options in depth, as this improves their estimation of the objective fit, decreasing their initial inaccuracy and making it more likely to choose a fitting option. In general, it seems to be more beneficial to be more accurate. Surprisingly however, we also observe a small range of conditions, when the exploration tendency (m) and the first threshold (t_1) are low, in which better choices are made by individuals with who are less accurate in estimating the fit of an option.

4 Concluding Remarks

At present, our results are too preliminary to come to any definite conclusions about the workings of our model, and how it may illuminate long-term choice processes. In the coming time, we intend to develop this model further, and aim to investigate the effect of variation in the parameters we have so far kept constant (N, T, and ρ). In later stages of the model, we may consider stronger effects of first experiences (as in Query Theory), and extensions of the model from a more developmentally oriented perspective. For example, it may be interesting to consider if the objective value of an option is not constant, but may change through identity development. In addition, we may consider extensions of the model to include interactions between options, and cognitive biases. Having said that, our first results already suggest that the parameters of our model affect the outcome in relatively complex ways—even for the relatively simple first model presented here. Encouraged by these first results, we are eager to analyze and develop this model further. We aim to connect the model outcomes to verbal theories on career choice and identity development. In this way, we hope to contribute to a deeper, fundamental understanding of career choice processes, which eventually may be used to help young people effectively navigate difficult life decisions.

References

1. Casey, B.J., Jones, R.M., Somerville, L.H.: Braking and accelerating of the adolescent brain. J. Res. Adolesc. **21**, 21–33 (2011). doi:10.1111/j.1532-7795.2010.00712.x
2. ResearchNed.: Monitor beleidsmaatregelen. Nijmegen: ResearchNed (2013)
3. Quinn, J.: Drop-out and completion in higher education in Europe. European Union: Neset. www.nesetweb.edu (2013)
4. Erikson, E.H.: Identity: Youth and Crisis. Norton, New York (1968)
5. Marcia, J.E.: Development and validation of ego-identity status. J. Pers. Soc. Psychol. **3**, 118–133 (1966)
6. Marcia, J.E.: Identity in adolescence. In: Adelson, J. (ed.) Handbook of Adolescent Psychology, pp. 159–187. Wiley, New York (1980)
7. Grotevant, H.D.: Toward a process model of identity formation. J. Adolescent Res. **2**, 203–222 (1987)
8. Luyckx, K., Goossens, L., Soenens, B., Beyers, W.: Unpacking commitment and exploration: preliminary validation of an integrative model of late adolescent identity formation. J. Adolesc. **29**, 361–378 (2006). doi:10.1016/j.adolescence.2005.03.008

9. Germeijs, V., Verschueren, K.: High school students' career decision-making process: development and validation of the study choice task inventory. J. Career Assess. **14**, 449–471 (2006)
10. Lichtwarck-Aschoff, A., Van Geert, P.L.C., Bosma, H.A., Kunnen, E.S.: Time and identity: a framework for research and theory formation. Dev. Rev. **28**, 370–400 (2008). doi:10.1016/j.dr.2008.04.001
11. Oppenheimer, D.M., Kelso, E.: Information processing as a paradigm for decision making. Annu. Rev. Psychol. **66**, 277–294 (2015). doi:10.1146/annurev-psych-010814-015148
12. Busemeyer, J.R., Townsend, J.T.: Decision field theory: a dynamic-cognitive approach to decision making in an uncertain environment. Psychol. Rev. **100**, 432–459 (1993)
13. Johnson, E.J., Haubl, G., Keinan, A.: Aspects of endowment: a query theory of value construction. J. Exp. Psychol. Learn. Mem. Cogn. **33**, 461–474 (2007)

Modelling the Role of Social Media at Street Protests

Annie Waldherr and Nanda Wijermans

Abstract Occupy, the Gezi park movement, the Maidan protests, or the recent solidarity marches for Charlie Hebdo—since the uprisings of the Arab Spring, we could observe many examples of on-site protests on big squares and streets being accompanied by waves of collective action in social media. We present the design stage of an agent-based model that will allow us to explore the following questions: What role does social media play in street protests? How does social media usage influence the dynamics of collective action during street protests? Do social media affect the speed, scale, fluctuation, duration of the protest at large, and in which way? Do they impact specific crowd patterns, e.g., the development of groups within groups? The model builds on and integrates existing models of social media, protests, and crowd behavior to simulate the dynamics of street protests in an urban setting. Our central aim is to compare scenarios with intense, moderate, and no social media usage by the protesters.

Keywords Protest • Social media • Social networks • Crowds • Modelling

1 Introduction

Shortly after the massacre in the Paris editorial office of the Charlie Hebdo magazine, the hashtag #JeSuisCharlie was shared millions of times in online social networks as a symbol of solidarity with the killed journalists. Also on the streets, people soon gathered to protest for freedom of expression which culminated in several millions coming together for funeral marches in Paris and other cities around the world 3 days after the attack. The intriguing question is how these two dynamics

A. Waldherr
Free University of Berlin, Institute for Media and Communication Studies, Berlin, Germany
e-mail: annie.waldherr@fu-berlin.de

N. Wijermans (✉)
Stockholm University, Stockholm Resilience Centre, Stockholm, Sweden
e-mail: nwijermans@gmail.com

© Springer International Publishing AG 2017   445
W. Jager et al. (eds.), *Advances in Social Simulation 2015*, Advances in Intelligent
Systems and Computing 528, DOI 10.1007/978-3-319-47253-9_41

of online social networks and street protests are interconnected. For instance, would the same amount of people have been mobilized to the street without the #JeSuisCharlie hashtag online?

Questions like these have been puzzling social scientists since the uprisings of the Arab spring. There and in many subsequent protests around the world we could observe examples of on-site protests in town squares and streets that were accompanied by collective action in social media (e.g., the Occupy movement, Gezi Park protests in Turkey, or Maidan protests in Ukraine). However, besides much speculation and theoretical discussions about the role and effects of social media on social protests, science has given few answers. Most of them were based on empirical studies, such as surveys of protesters (e.g., [12]). The social simulation community has to date only rarely contributed to these questions. However, particularly agent-based modelling (ABM) approaches are perfectly suitable for investigating the interacting micro–macro-dynamics at work in the phenomenon.

With our modelling project we aim to explore the following questions with an ABM: What role do social media play for street protests? How does social media usage influence dynamics of collective action during street protests? Do social media affect the speed, scale, fluctuation, duration of the protest at large, and in which way? Do they impact specific crowd patterns, e.g., the development of groups within groups? These questions are not only relevant for extraordinary, large-scale protest events (as introduced above), but also for smaller street protests that take place every week in different settings.

This conference contribution presents the design stage of an agent-based protest model that includes the role of social media. The central aim of the model is to compare simulated scenarios of protests without social media to protests with social media. This will enable us to develop informed hypotheses regarding the effects of social media on the dynamic patterns of street protest that can later be tested empirically.

2 Bringing Media, Protest, and Crowd Models Together

Currently, we see two relevant streams of modelling research that we can build on: (1) social media models and (2) protest models.

Social Media Models In recent years, modellers showed growing interest in online social networks and communities such as Twitter, Facebook, or Wikipedia. Existing simulation models on social media communication focus mainly on online behavior such as sharing information in social networks [11, 13]. Other authors focus on online opinion dynamics, such as Iñiguez et al. [4] who model editing wars on Wikipedia, or Sobkowicz and Sobkowicz [9] who study dynamics of polarization in a political discussion forum online. Finally, an emerging strand of research seeks to understand collective emotions online with an ABM approach (e.g., [8, 10]).

Protest Models Research on models of protest typically involves anecdotal references to protests to highlight the importance of looking at social media and their potential for social change and conflict (e.g., [3, 6]). However, there seem to exist only few models of protest. Epstein's [2] model of civil unrest that was used as an illustration of the ABM approach is the best-known model. The aim of this rather abstract model is to simulate rebellion against a central authority (Model I), or violence between two rival groups (ethnic violence) that a central authority seeks to suppress (Model II). Other models of protest appear to be adaptations of the Epstein model, such as Kim and Hanneman's [5] adaption to worker protests.

We find very few models that bring together protest and social media. Casili and Tubaro [1] explore the effect of social media censorship on civil protest by adapting the visual range in the Epstein model to represent the impact of censorship. Makowsky and Rubin [7] use their ABM to show how highly centralized power and widespread connectivity through ICTs interact and generate massive and rapid preference revelations that fuel large-scale social change like the Arab Spring. Hu et al. [3] similarly present an agent-based network model that generally explores the impact of ICT connectivity on collective social action in different cultural contexts.

From the state of research, we infer two main research desiderata. First, there are still too few models that connect social media and protest models to explain the interplay between online and offline behaviors of protests. Second, existing models mainly emphasize large-scale macro-dynamics of rebellion or social change and neglect the specifics of the geographical settings where street protests take place. Thus, the second research desideratum is to connect protest models and crowd models that emphasize movement. This will allow us to model the specific dynamics of street protest in an urban setting.

3 The Conceptual Model

Our model, called SIMPROC, aims to explore the interplay between online and offline behavior and its effects on the crowd in a street protest. In particular, we seek to study the extent to which social media informs individuals' decisions to join, stay at, or leave a protest event and how this leads to different protest dynamics and crowd patterns at large. More specifically, we are interested in size, speed of mobilization, and duration of the protest and specific crowd patterns such as the development of psychological groups. Do these significantly differ in scenarios with intense, moderate, or absent social media usage?

The model is embedded in an urban protest context, i.e., situated as a street protest. This involves protests where the demonstrators walk a certain route and/or gather at a particular location. The environment consists of an on-site (street protest) and off-site (home) area. The on-site area reflects (narrow and wide) streets, (low-high) buildings, and points of interests to meet at (squares, marketplace, and parks). There are three types of agents: citizen agents, law enforcement officer agent (LEOs), and traditional media journalist agents. Each agent can be online or offline which is reflected by a multilayer social network of connections that may affect the agents (social media, traditional media, and local surroundings). Citizen agents can decide to join, stay at, or leave the protest, be online or offline, move to a particular location on the protest site and can communicate about events and their current location. They are affected by others that are physically or socially close and at that time connected to them. LEOs are by default offline, move on-site based on an adaptive scenario that draws on the initial setting of the anticipation of escalation and emerging protest cues (e.g., size, social group formation/asymmetry, and media). Traditional media journalists at first display an adaptive behavior similar to the LEOs. They are by default online and mainly off-site, but some are on-site to report. In correspondence with our aims, the experimental model allows for manipulating social media (on/off) and experimenting with different social structures, such as small vs. large friendship networks, focused vs. dispersed networks, reciprocal friendship networks vs. directed reputation, and different ratios of offline and online relationships (Fig. 1).

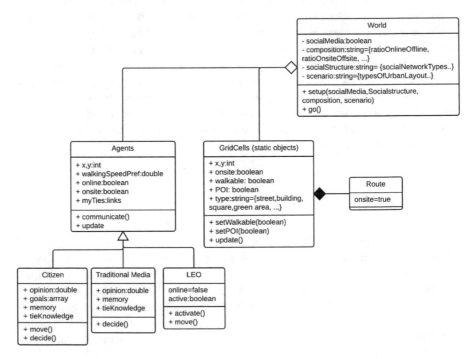

Fig. 1 Class diagram of the protest model SIMPROC

References

1. Casilli, A.A., Tubaro, P.: Social media censorship in times of political unrest—A social simulation experiment with the UK riots. Bull. Sociol. Methodol. **115**(1), 5–20 (2012)
2. Epstein, J.M.: Modeling civil violence: an agent-based computational approach. Proc. Natl. Acad. Sci. **99** (suppl. 3), 7243–7250 (2002)
3. Hu, H.H., Cui, W., Lin, J., Qian, Y.: ICTs, social connectivity, and collective action: a cultural-political perspective. J. Artif. Soc. Soc. Simul. **17**(2), 7 (2014)
4. Iñiguez, G., Török, J., Yasseri, T., Kaski, K., Kertész, J.: Modeling social dynamics in a collaborative environment. EPJ Data Sci. **3**(1), 7 (2014)
5. Kim, J.W., Hanneman, R.: A computational model of worker protest. J. Artif. Soc. Soc. Simul. **14**(3), 1 (2011)
6. Lemos, C., Coelho, H., Lopes, R.J.: Agent-based modeling of social conflict, civil violence and evolution: state-of-the-art-review and further prospects. In Proceedings – EUMAS 2013, pp. 124–138 (2013)
7. Makowsky, M.D., Rubin, J.: An agent-based model of centralized institutions, social network technology, and revolution. PLoS ONE **8**(11), e80380 (2013)
8. Schweitzer, F., Garcia, D.: An agent-based model of collective emotions in online communities. Eur. Phys. J. B **77**(4), 533–545 (2010)
9. Sobkowicz, P., Sobkowicz, A.: Dynamics of hate based internet user networks. Eur. Phys. J. B **73**(4), 633–643 (2010)
10. Tadić, B., Gligorijević, V., Mitrović, M., Šuvakov, M.: Co-evolutionary mechanisms of emotional bursts in online social dynamics and networks. Entropy **15**(12), 5084–5120 (2013)
11. Tubaro, P., Casilli, A.A., Sarabi, Y.: Against the Hypothesis of the End of Privacy: An Agent-Based Modelling Approach to Social Media. Springer, Heidelberg (2014)
12. Tufekci, Z., Wilson, C.: Social media and the decision to participate in political protest: observations from Tahrir Square. J. Commun. **62**(2), 363–379 (2012)
13. Xiong, F., Liu, Y., Jiang Zhang, Z., Zhu, J., Zhang, Y.: An information diffusion model based on retweeting mechanism for online social media. Phys. Lett. A **376**(30–31), 2103–2108 (2012)

AgentBase: Agent Based Modeling in the Browser

Wybo Wiersma

Abstract *AgentBase.org* allows for Agent Based Modeling (ABM) directly in the browser. One can edit, save, and share models without installing any software or even reloading the page. Models use the AgentBase Library, and are written in CoffeeScript, which is instantly interpreted as JavaScript. The AgentBase Library provides a rich set of resources for moving and drawing agents, neighbour detection, and other things expected from ABM toolsets. AgentBase is optimized for making illustrative models. It is opinionated software which values simplicity and clean model code over CPU performance.

Keywords AgentBase • NetLogo • ABM • Agent based modeling • Illustrative models • abm toolset • abm platform • JavaScript

1 Introduction

AgentBase.org is a minimalist Agent Based Modeling (ABM) platform that allows one to quickly build models that run directly in the browser [2]. It follows NetLogo's Agent oriented Programming model and is entirely implemented in CoffeeScript [7, 12, 16]. Its quick feedback between tinkering and model behaviour makes it surprisingly fun to use. And the ability to share models with nothing more than a hyperlink (no software or browser-plugins to install), across all major browsers (IE, Firefox, and Chrome), makes the models very accessible.

After AgentBase.org is described more in depth, its design philosophy will be explained, clarifying how it relates to other ABM toolsets. This will be followed by a brief description of how to build your first model.

W. Wiersma (✉)
Oxford Internet Institute, University of Oxford, Oxford, UK
e-mail: mail@wybowiersma.net

2   AgentBase.org and the AgentBase Library

AgentBase.org provides the web platform, while the AgentBase Library is the toolset for model-building [3]. The toolset is Open Source (GPLv3) and written in CoffeeScript. In layman's terms CoffeeScript is a clearer, less verbose dialect of JavaScript. The AgentBase toolset was derived from AgentScript, a similar Open Source toolset that tries to reimplement NetLogo in JavaScript [5]. The AgentBase Library is more streamlined, and more thoroughly tested than AgentScript, but we will get back to this in the design philosophy section. Besides the CoffeeScript language, that—through JavaScript—allows models to run directly in the browser. HTML canvas is also used to render models in the browser [9]. Models made with the AgentBase library can be exported, and used on any website, not just on AgentBase.org.

The web platform, AgentBase.org, is a site for sharing and editing models. It uses the ACE text editor, which mimics the powerful editors used by programmers, in the browser, and is implemented in JavaScript, so it does not require anything to be installed [1]. The ABM model code held by the editor is instantly (client-side) compiled from CoffeeScript to JavaScript using the *Browser.coffee* library, and then executed, when the user hits run [6]. So changes are immediate. ABM models and any changes to them are stored on GitHub, the site most commonly used for sharing and managing code repositories [8]. They are stored in the so-called Gists, which are mini-repositories. Users of AgentBase.org thus need a GitHub account to save models in their own name. No other accounts (not even an AgentBase.org account) are needed. On the server-side, AgentBase.org is implemented in about 150 lines of NodeJS, and it uses the MongoDB database for keeping track of models' Gists repositories [11, 13].

3   Design Philosophy

There are two types of ABMs that we define here as Illustrative and Predictive. Illustrative ABM models are models that serve to illustrate a sociological or other process to fellow scholars, or other human audiences. A good example is Schelling's segregation model, which illustrated that even slightly racist preferences can lead to complete segregation on a macro scale, and that thus segregation does not imply high levels of racism, as was previously thought. Yet because it was an illustrative model, and kept simple—it used a chessboard type grid, not real city maps—the model could not be used to predict segregation in actual neighbourhoods [14, 15]. Illustrative models are kept intentionally simple so they can be more effectively communicated. While predictive models approach reality much closer in terms of complexity, making them better at predicting outcomes, but too opaque to illustrate individual processes. The test of an illustrative ABM is not predictive power, but whether it elucidates the modelled process to colleagues; whether they think it make sense [10].

AgentBase is optimized for the quick development of illustrative—not predictive—ABM models. And it is exactly because illustrative models benefit from being easily shareable, that AgentBase was made to run in the browser. Similarly, CoffeeScript, a relatively slow scripting language, was chosen for its readability. The beauty and elegance of a language— the so-called syntactic sugar—matters a lot for this, as well as for productivity. CoffeeScript is a lot cleaner than JavaScript, because it leaves out semicolons, and superfluous brackets, and makes object-oriented programming a lot more straightforward. And while it is true that most software engineers are more familiar with JavaScript than with CoffeeScript, illustrative models—and their code—aim to speak to scholars, not to software engineers, and they will find CoffeeScript much easier to read. Other design-choices in terms of function-naming, etc., that make the model code easier to read, but slightly slower, have been favoured as well. AgentBase values readable, pretty code over CPU performance.

Another choice that sets AgentBase apart is that unlike Tortoise and AgentScript, it does not aim to reimplement NetLogo, or parts of it. Though NetLogo is a very rich ABM toolset, many of its functions and some of its design are optimized for the Logo language, and the Desktop environment, and some of it could also be considered bad. Some of its functions are—for example—really badly named for readability, such as 'cp' shortcut for clear patches and 'fd' for forward. Also, CoffeeScript as a language is quite different from Logo (it is more inspired by the C and Java languages than Lisp), and therefore different ways of organizing functionality make more sense. In addition to furthering the illustrative power of models, readable code that can be changed and shared from a webpage as easily as a Wiki document also helps lower thresholds to contribution. This would be expected to help more people get involved, similar to how Wikis did this for text-editing. Which could further models' reusability, and increase their academic impact. Finally, AgentBase also sets itself apart because it is well-tested through good coverage by automated tests.

4 Your First Model

Models for AgentBase only need to implement two methods: setup() and step(). The Template Model on agentbase.org essentially consists of the following [4]:

```
class MyModel extends ABM.Model
  setup: ->
    @patches.create()
```

```
      for agent in @agents.create 25
        agent.shape = u.shapes.names().sample()

    step: ->
      for agent in @agents
        agent.forward()


  new MyModel(div: "model").start()
```

Setup() creates the patches and 25 agents, and gives them each a random shape. Step() then makes the agents move forward. Visit AgentBase.org and tinker with it! (and save it under a different name)

If you want to embed the model in your own webpage, just download and unzip AgentBase. The model can then be directly embedded with:

```
<html>
  <head>
    <script src="agentscript.js"></script>
    <script src="coffee-script.js"></script>
    <script type="text/coffeescript">
    ...your model...
    </script>
  <head>
  <body>
    <div id="model"></div>
  </body>
<html>
```

To conclude, AgentBase is a toolset well-suited for the quick development of models that can easily be shared and extended. Its use of CoffeeScript makes models more readable, without diverging from browser standards. And though AgentBase limits itself to the niche of illustrative models, this is the niche that many, if not most, academic models fall into.

Acknowledgements This research was made possible by an ESRC and Scatcherd PhD Scholarship.

References

1. Ace - Code Editor for the Web (2015). http://ace.c9.io/#nav=about
2. AgentBase (2015). http://agentbase.org/
3. AgentBase Library (2015). http://lib.agentbase.org/
4. AgentBase Template Model (2015). http://agentbase.org/model.html?9d54597f7aafc995d227
5. Agentscript (2015). http://agentscript.org/
6. Browser.coffee (2015). http://coffeescript.org/documentation/docs/browser.html

7. CoffeeScript (2015). http://coffeescript.org/
8. GitHub (2015). https://github.com/
9. HTML Canvas Element (2015). http://en.wikipedia.org/wiki/Canvas_element
10. Miller, J.H., et al.: Complex Adaptive Systems: An Introduction to Computational Models of Social Life: An Introduction to Computational Models of Social Life. Princeton University Press, Princeton (2009)
11. MongoDB (2015). https://www.mongodb.org/
12. NetLogo (2015). https://ccl.northwestern.edu/netlogo/
13. Node.js (2015). https://nodejs.org/
14. Schelling, T.C.: Dynamic models of segregation. J. Math. Sociol. **1**, 143–186 (1971)
15. Schelling, T.C.: Micromotives and Macrobehavior. WW Norton & Company (2006)
16. Tisue, S., et al.: Netlogo: a simple environment for modeling complexity. In: International Conference on Complex Systems, pp. 16–21 (2004)

Index

Printed in the United States
By Bookmasters